HYDROMAGMATIC PROCESSES AND PLATINUM-GROUP ELEMENT DEPOSITS IN LAYERED INTRUSIONS

The role that hydrothermal fluids may have played during the crystallization of layered intrusions and the formation of the ore deposits they contain has long been debated. This book summarizes the evidence for fluid-crystal-liquid (hydromagmatic) interactions and their importance for the understanding of the formation of platinum-group deposits in layered intrusions. It discusses the composition of igneous fluids in mafic magmatic systems, the generation and movement of these fluids in layered intrusions, their impact in altering the mineralogy and composition of the originally precipitated assemblages, and their role in the transport of the platinum-group elements (PGE). Using examples from the Bushveld complex of South Africa and other intrusions, this book provides a comprehensive overview of the hydromagmatic model for the origin of various features of layered intrusions. It is a useful reference for academic researchers and professional geologists working on economic mineral exploration, layered igneous intrusions and hydrothermal metallogenesis.

ALAN BOUDREAU is Professor of Geology at Duke University and an expert on the origins of layered intrusions. He has worked on numerical modelling of crystallization processes such as crystal aging and compaction, and how they give rise to the variety of features observed in these intrusions. He is also interested in the role of igneous fluids in the petrogenesis of platinum-group element (PGE) deposits in layered intrusions, including the understanding of the interaction of igneous fluids with liquid-crystal assemblages to produce the observed features.

HYDROMAGMATIC PROCESSES AND PLATINUM-GROUP ELEMENT DEPOSITS IN LAYERED INTRUSIONS

ALAN BOUDREAU
Duke University

CAMBRIDGE
UNIVERSITY PRESS

CAMBRIDGE
UNIVERSITY PRESS

University Printing House, Cambridge CB2 8BS, United Kingdom

One Liberty Plaza, 20th Floor, New York, NY 10006, USA

477 Williamstown Road, Port Melbourne, VIC 3207, Australia

314–321, 3rd Floor, Plot 3, Splendor Forum, Jasola District Centre, New Delhi – 110025, India

79 Anson Road, #06–04/06, Singapore 079906

Cambridge University Press is part of the University of Cambridge.

It furthers the University's mission by disseminating knowledge in the pursuit of education, learning, and research at the highest international levels of excellence.

www.cambridge.org
Information on this title: www.cambridge.org/9781108416009
DOI: 10.1017/9781108235617

First published 2019

Printed in the United Kingdom by TJ International Ltd. Padstow Cornwall

A catalogue record for this publication is available from the British Library.

Library of Congress Cataloging-in-Publication Data
Names: Boudreau, Alan, 1953- author.
Title: Hydromagmatic processes and platinum-group element deposits in layered intrusions / Alan Boudreau (Duke University, North Carolina).
Description: Cambridge ; New York, NY : Cambridge University Press, 2019. | Includes bibliographical references and index.
Identifiers: LCCN 2018036907 | ISBN 9781108416009 (hardback : alk. paper)
Subjects: LCSH: Platinum ores. | Ore deposits. | Folds (Geology) | Hydrothermal deposits. | Intrusions (Geology)
Classification: LCC QE390.2.P56 B68 2019 | DDC 553.4/22–dc23
LC record available at https://lccn.loc.gov/2018036907

ISBN 978-1-108-41600-9 Hardback

For Kathleen, and the kids

Contents

Preface

The idea for this book originally began as a description of the hydrothermal model for the concentration of platinum-group elements (PGE) in layered intrusions. This is a model that is contrary to the majority opinion of those who work in layered intrusions, and acceptance requires overcoming a certain amount of cognitive dissonance with longstanding interpretations. However, much of the supporting evidence has grown stronger over time and the need to present a full and detailed description of the model is long overdue. Previous work by the author had shown that the evidence for hydrothermal petrogenetic models had grown beyond the ability of single papers and even extended review articles to cover the supporting evidence in a complete manner. Equally important is to answer some common criticisms of the model; it is common for the idea that the PGE were concentrated by magmatic fluids to be discounted in as little as one sentence. Thus, the idea for this book was conceived.

However, it soon became evident that such a book would need to include related evidence on the effect of volatiles on the crystallization of magmas. This includes growing evidence for re-melting of pre-existing minerals and metasomatic reactions involving fluids and the crystal pile. In many cases this also includes the role of crustal fluids and their effects on the isotopic character of the rocks in general and the growing evidence for isotopic disequilibrium. Indeed, one of the ideas expressed in this book is that large layered intrusions, and particularly the Bushveld Complex, are an excellent analogue for subduction zone hydrothermal systems. Both involve fluids derived from dehydrating underlying sediments that manage to preserve their isotopic character as they move through thick sequence of ultramafic rock to reach hotter rock where they may induce melting. These factors and more led to the expanded scope of this book.

It is common for many review books to be a collection of papers whose chapters are authored by perhaps several authors but the overall content of which is largely determined by one or more editors. This book is instead the product of a single author to present a unified view of the subject. In part, this is because the main ideas presented are very much a minority viewpoint. It is by no means meant to be the last word nor even to be correct in some details. However, it is felt that without a strong description of the hypothesis, future advancement will be haphazard at best.

An overview of the hydrothermal systems in layered intrusions and how it differs from well-known hydrothermal systems in porphyry systems is presented in Chapter 1, along with an overview of orthomagmatic and hydromagmatic models for the concentration of the PGE in layered intrusions. Chapter 2 discusses some of the problems of cumulate terminology with examples of how an originally precipitated crystal assemblage can be modified. Because many of the examples used in this book are from the Stillwater, Bushveld and Skaergaard intrusions, a brief overview of these three intrusions is also presented. Those who are not familiar with these intrusions may want to start here. Chapter 3 presents a summary of volatiles in mafic magmas, but also discusses igneous fluids as hydrothermal solutions and mineral solubility considerations. Chapter 4 covers the geochemistry of the PGE in both magmas and hydrothermal fluids, the latter of which is still poorly known in high temperature solutions. Chapter 5 looks at bubble formation and the phenomenon of three phase flow in crystal-silicate liquid-volatile fluid systems. Chapters 6 and 7 summarize the halogen geochemistry evidence for fluid involvement and the evidence from silicate liquid and fluid inclusions, respectively.

The remaining chapters largely look at evidence for the role of fluids in the formation of specific features in layered intrusions. Chapter 8 discusses the formation of discordant features such as pegmatoids and PGE-bearing pipes. Chapter 9 looks at the role of fluids in altering the crystallization behaviour of magmas and volatile fluxing. Chapter 10 considers chromatographic theory and the origin of ore metal offsets. Chapter 11 discusses the role of compaction in forming stratigraphic traps for migrating fluids and the formation of Great-Dyke type deposits at ultramafic-mafic boundaries, and Chapter 12 discusses the potential role of fluids in the formation of chromitites and their associated PGE mineralization. Chapter 13 looks at isotopic evidence and compares examples of isotopic disequilibrium with that observed in mantle samples. Finally the last chapter looks at some common objections of the hydromagmatic interpretation and a response to these criticisms.

A note on the terminology and format of this book. Ideas or terms that may be unfamiliar with students or even some professionals are generally defined with *italic* font. Because a silicate liquid and a supercritical volatile fluid are both fluids in the physical sense, this report will reserve the terms *silicate liquid* or *liquid* to denote a silicate liquid (magma) in general and the terms *melt* or *silicate melt* to denote a silicate liquid produced specifically by the local melting of pre-existing crystals. The terms *volatile fluid*, *fluid* and *vapour* will be used interchangeably to define a volatile-rich fluid phase. If need, the term *transitional fluid* will be used to define those volatile-rich silicate liquids that can evolve into solute-rich volatile fluids without a phase transition. Finally, cumulate terminology (Wager et al., 1960) is not used in rock descriptions. As noted by a number of authors (Hunter, 1987; Higgins; 1991, 1998, 2002; McBirney and Hunter, 1995; McBirney, 2009) layered intrusion have undergone extensive recrystallization and cumulate terminology has interpretive implications that can have a pernicious effect on the understanding of the processes which formed the rocks. This is particularly true for understanding those rock whose compositions may have be influence by volatile fluids. Thus, this report will use respectable but model-neural rock names with modifiers as necessary (e.g., granular harzburgite, melanorite).

This book required more than just the labour of the author. In particular, many thanks go out to reviewers of individual chapters: Steve Prevec, Edmond Mathez, Rais Latypov, James Mungall, James Webster and Jacob Hanley. Errors of fact, omission or conclusion are otherwise those of the author. Editorial assistance and support by Cambridge University Press editors and staff, including Sarah Lambert, Zoë Pruce, Harsha Vardhanan and Emma Kiddle, are much appreciated.

1
Introduction

1.1 Layered Intrusions and Their Volatiles

Layered intrusions are igneous intrusions that exhibit *modal layering* (variation in mineral percentages) and *compositional* or *cryptic layering* that is developed on a millimetre to kilometre scale. This igneous layering is typically best developed during the crystallization of a basaltic magma (broadly defined), but can also be found in intrusions formed from more siliceous parent magmas. Their sizes and shapes can be roughly equant or box-like in some smaller layered intrusions, such as the Skaergaard intrusion of East Greenland, to sill-like intrusions such as the Bushveld Complex in South Africa that is up to 9 km thick and extends over 50,000+ km^2 (See Chapter 2 for a description of the stratigraphy and general geology of the Skaergaard intrusion, the Bushveld Complex and the Stillwater Complex that are the intrusions most used as examples in this book). The extensive record of compositional and mineral changes preserved in the rocks has led them to be studied to understand the processes by which magmas crystallize and differentiate by the process of crystal fractionation. In addition, they contain important economic minerals deposits of chrome, copper, nickel, titanium, vanadium and the platinum-group elements (PGE: Ir, Os, Ru, Rh, Pt, Pd) and the mechanisms that concentrated these elements are still actively debated.

With some notable exceptions, the magmas that formed layered intrusions have been generally considered to have been relatively 'dry' and their volatile contents and the role of degassing during the crystallization of the parent magma has generally been ignored. Parent magma compositions are typically reported without reference to any possible volatile components, although basaltic magmas can have substantial concentrations of H_2O and other volatile species. In part this omission is the result of a lack of evidence of the original volatile content of a parent magma: even in the best circumstances where one has a crystallized chilled margin or contemporaneous sill or dike, the rock has lost much of its volatiles during solidification. In addition, the evidence that the rocks initially formed by the accumulation of crystals on the floor of a magma chamber with an initial liquid fraction that was further decreased by compaction and other processes would suggest that the amount of vapour that could be generated within and move through a crystal-liquid mush would be small, if it could move at all! Yet the evidence for volatile fluids and their role in the crystallization of layered intrusions in general and in the formation of the economic deposits they contain is non-trivial.

The evidence for degassing and vapour interaction in the crystal-liquid mush is increasingly recognized in many intrusions. It is the contention of this work that many of the features commonly attributed to magma replenishments mixing of magmas can instead be the result of fluid interactions with both the magma and in the crystal pile. This is particularly true for the many of the ore deposits that these intrusions contain; much of this book will use the petrogenesis of the PGE deposits they contain as an example of the importance of volatile fluids. This book is an attempt to summarize some of this evidence and hopefully expand the understanding of what is possible during the crystallization of basaltic magmas and the formation of layered intrusions.

1.2 Hydromagmatic Processes as a Special Type of Hydrothermal Activity

Degassing in large layered intrusions, particularly the large, thick sills such as the Bushveld and Stillwater complexes, is qualitatively different from degassing in classical siliceous porphyry systems. In the latter, crystallization is assumed to occur from the top and side margins of a cooling pluton and crystallization fronts move into the magma chamber (Figure 1.1a). Vapour in the form of bubbles may begin to separate at low degrees of crystallinity, and the vapour that does separate moves into colder portions of the pluton or into the country rock that can readily begin to fracture as the rock cools below its solidus.

Some mafic layered intrusions such as the Skaergaard intrusion do have sizable marginal zones that crystallized from the side walls and roof inward in which fluid saturation would be similar to the porphyry system. However, in the Skaergaard and elsewhere, most of the crystalline product accumulates on the floor. In addition, heating of the country rock can lead to dehydration in the metamorphic aureole and be a source of external fluids.

In several respects, degassing and vapour migration in the lower mush zone of a layered intrusion is simply the porphyry system turned upside down. In a large sill-shaped layered intrusion, heat loss and cooling of this lower mush zone is from the bottom (at least initially), and vapour will move upward into hotter rocks and potentially into the main magma body (Figure 1.1b). In this regard, the thermal profile is similar to that in subduction zones, in which fluids generated by heating and dehydration of the descending lithospheric slab move into hotter rocks of the overlying mantle wedge (Figure 1.1c). Indeed, large layered sills are an excellent proxy for understanding the role of fluids in subduction environments.

Fluids in the porphyry system generally move from hotter to cooler rock, and precipitate abundant quartz and sulfides. In contrast, the heating of rising fluids in layered intrusions as they migrate upward toward the main magma body leads to the fluids becoming undersaturated in silica (and other phases). Pyroxene can react with the fluid to form olivine and produce the discordant dunite and troctolite described in Chapter 9 and the migrating vapour can redissolve pre-existing sulfides. For both layered intrusions and subduction zones, ascending fluids can induce incongruent melting as they encounter hotter rocks (e.g., Boudreau, 1988, 1999a; Nicholson and Mathez, 1991).

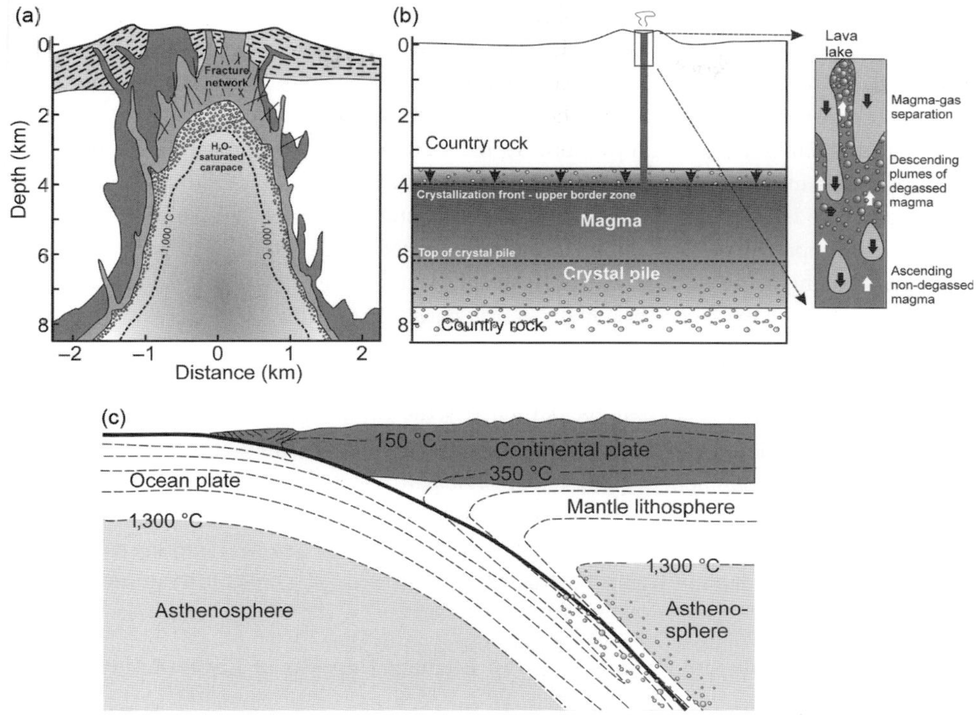

Figure 1.1 Comparison of three types of hydrothermal systems. (a) Simplified model of a volatile fluid-saturated carapace of a typical porphyry hydrothermal system. Redrawn after Burnham (1979), with permission. (b) The two hydrothermal regions in a thick sill that is losing heat from the top and bottom. The upper hydrothermal zone is qualitatively like the fluid-saturated carapace of the porphyry system, whereas the hydrothermal behaviour in the crystal pile at the base of the sill has a number of unique characteristics, including compaction and fluid released by the dehydration of underlying country rock, as discussed in the text. Modified after Boudreau (2016a), with permission. Detail at left shown how shallow level degassing to a persistent lava lake can lead to degassed magma that can moved back down into the magma chamber. Detail after Shinohara (2008), with permission. (c) Subduction zone hydrothermal system, in which fluids released by heating of the subducted lithosphere moves up into the hot overlying mantle wedge. Redrawn after Rüpke et al. (2004), with permission.

Other differences between porphyry and layered intrusion systems include different styles of degassing. Although for both systems degassing can accompany decompression or crystallization, the viscosity difference can affect how this is accomplished. For example, unlike high viscosity porphyry systems which tend to erupt explosively as they approach the surface, deep mafic magma chambers can communicate with shallow surface magma chambers and long lived (decades or more) lava lakes, allowing the magma to at least partially degas under low-pressure conditions (see detail of Figure 1.1b). The loss of S to shallow degassing and remixing with deeper magma chamber is typical of other non-subduction volcanic systems reported by Shinohara (2008), who suggested that the vertical magma mixing can occur between the surface to as deep as 10 km. Li and Boudreau (2017) suggested that low-pressure S loss was responsible for the low S/Cu and S/Se ratios of the Skaergaard intrusion, for example.

The situation is further complicated by the different stress regimes. In the porphyry system, degassing is characterized by tensile fracture on cooling that follows the solidification front (Candela, 1991). In contrast, vapour migration in a thick crystal mush is accompanied by compaction-driven compressive (pure shear) stress. Although of perhaps limited importance in high silica magmas (e.g., Lee and Morton, 2015), there is abundant compositional and textural evidence for compaction in basaltic lava flows and sills (e.g., Philpotts et al., 1996, 1998, 1999; Shirley, 1987) and larger layered intrusions (e.g., Meurer and Boudreau 1996, 1998a,b; Boorman et al., 2004; Tegner et al., 2009). These works demonstrate that compaction is accommodated by crystal rearrangement, compression of three-dimensional linked crystal structures, crystal bending and breaking and recrystallization. The lack of the development of a colonnade-style parting or other evidence that the volume change on cooling is not accommodated in large sills such as the Stillwater and Bushveld complexes suggests that the rocks respond to compaction stresses well into the subsolidus. This different stress regime affects how fluids migrate through a crystal-liquid mush.

1.3 Igneous Fluids and Platinum-Group Element Deposits in Layered Intrusions

Eugene Stumpfl was one of the first to explicitly recognize the evidence for and potential importance of fluids in stratiform platinum-group element (PGE) ore petrogenesis in layered intrusions (e.g., Stumpfl and Tarkian, 1976; Stumpfl and Rucklidge, 1982; Ballhaus and Stumpfl, 1986). These early works noted that common occurrence of the pegmatoidal textures and hydrous minerals such as biotite associated with these deposits. Since this time, there has grown an extensive literature on the evidence regarding the importance of fluids in the petrogenesis of PGE deposits. This includes unusually Cl-rich apatite found in the ores themselves and in the rocks beneath the main PGE zones in the Stillwater and Bushveld Complexes that have been interpreted as indicating Cl enrichment from upwardly migrating fluids (e.g., Boudreau et al., 1986; Boudreau and Kruger, 1990, Willmore et al., 2000, Mathez and Webster, 2005). Pothole structures in the Bushveld Complex have been modelled as fluid escape features (Ballhaus and Stumpfl, 1985; Buntin et al., 1985; Boudreau, 1992), analogous to pockmark structures that can form during degassing in the sedimentary environment. More recently, fluid inclusion evidence has confirmed the Cl-rich nature of the fluids and that they can carry significant amount of the ore metals (e.g., Hanley et al., 2005a).

Thus, there has been a debate between those who view these deposits as having been formed by traditional orthomagmatic mechanisms, typically involving sulfide saturation, and those who suggest that they were formed by hydrothermal transport of the ore elements. These contrasting views have even been given informal identifiers: The orthomagmatic *downers* broadly envision the PGE having been collected by an immiscible sulfide as it settles though a thick column of silicate liquid to settle on the top of the crystal pile on the floor of the magma chamber. In contrast, the hydrothermal *uppers* envision the ore elements having been moved upward though a solidifying crystal-liquid mush to be deposited either at the top of the crystal pile or at some stratigraphic trap within the mush column.

Unfortunately, most existing deposit reviews and book-length description of the PGE deposits in layered intrusions give only a cursory and generally dismissive treatment of the evidence for the fluid transport model and the increasingly detailed hydromagmatic models for how mineralizing fluids move and interact with the silicate liquid + mush system. This book aims to correct this imbalance by concentrating on the evidence and petrogenetic models for a primary role of magmatic fluids in the formation of these deposits.

1.3.1 The Conventional Magmatic Model for the Concentration of the PGE

There has been much written on the conventional magmatic model and only the basics are covered here. By far the most successful treatment is the *R-factor model*, which extends conventional trace element portioning behaviour. The partitioning of a trace element between two phases can be characterized by a distribution coefficient, D, for elements that that follow ideal solution behaviour. For example, for Cu partitioning between a silicate liquid and an immiscible sulfide liquid, the distribution coefficient would be

$$D^{Cu}_{Sulide\ liq/silicate\ liq} = \frac{C^{Cu}_{Sulfide\ liq}}{C^{Cu}_{Silicate\ liq}} \tag{1.1}$$

For compatible elements for which the distribution coefficient and the mass of the precipitating sulfide are both relatively low, one can use the concentration of the silicate liquid prior to the separation of the second phase to find the concentration of the metal in the sulfide liquid:

$$C^{Cu}_{Sulfide\ liq} = D{\cdot}C^{Cu}_{Silicate\ liq} \tag{1.2}$$

However, when the distribution coefficient is very high, the silicate liquid can be rapidly depleted by the equilibration with even a small mass of sulfide as the metals preferentially partition into the sulfide phase. This is particularly true for the PGE, for which the measured distribution coefficient are on the order of 10^4 or higher (e.g., Barnes and Maier, 1999). The result is that the PGE concentration in a sulfide equilibrated with a relatively small mass of silicate liquid will be lower than expected if the sulfide equilibrated with a large (infinite) silicate liquid reservoir. As discussed by Campbell and Naldrett (1979), this depletion effect can be accounted for by the following equation

$$C^{Cu}_{Sulfide\ liq} = \frac{D\ C^{Cu}_{Silicate\ Liq\ init}\ (R+1)}{(R+D)} \tag{1.3}$$

Where

$$R = \frac{Mass\ of\ silicate\ liq}{Mass\ of\ sulfide\ liq} \tag{1.4}$$

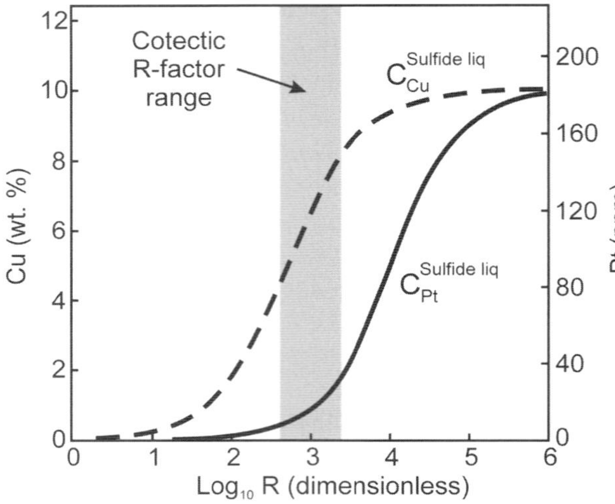

Figure 1.2 The effect of the R-factor on the Cu and Pt concentration in an immiscible sulfide liquid precipitating from a silicate magma. For Cu, $C^{Cu}_{silicate\ liq\ init} = 200$ ppm and $D^{Cu}_{sulfide\ liq/silicate\ liq} = 500$; for Pt, $C^{Pt}_{silicate\ liq\ init} = 0.01$ ppm and $D^{Cu}_{sulfide\ liq/silicate\ liq} = 10,000$. The grey area shows the range of R expected for a multi-phase saturated magma separating an immiscible sulfide liquid in cotectic proportions.

The R-factor is the ratio of the mass of sulfide to the mass of silicate liquid, and $C^{Cu}_{Silicate\ Liq\ init}$ is the initial concentration of Cu in the magma prior to sulfide saturation.

An example is shown for Cu and Pt in Figure 1.2. High concentrations of a metal in the sulfide require both a high distribution coefficient and equilibration with a large volume of magma if the metal is present in very low concentrations in the magma. For example, the J-M Reef of the Stillwater Complex contains about 15 ppm Pd over an erratic but typical thickness of about 1 m. To get the tenor/grade observed, the sulfide would need to have completely removed all the Pd from about a 1 km thick column of magma if the initial magma concentration was 15 ppb Pd.

1.3.2 *The Hydromagmatic Model for the Concentration for the PGE*

As noted in the last section, evidence for a role by high temperature fluids in the transport of the PGE has long been suggested by the unusually coarse grained to pegmatoidal nature of the rocks and abundance of hydrous mineral. To be shown in the following chapters, this has been supplemented with evidence that the hydrous minerals, particularly apatite are unusually enriched in chlorine, an element that is preferentially concentrated in an igneous fluid. Also detailed in later chapters, direct evidence is seen in the presence of fluid inclusions that can have metal ratios similar to those seen in the ore zones. Recent experimentally determined (volatile fluid)/(silicate liquid) partition coefficients for Pt can be as high as 10^3–10^4 and that chloride–carbonate fluids can be effective for moving both the *PPGE* (Rh, Pt, Pd) and the *IPGE* (Ir, Os, Ru).

Bulk rock analysis of the rocks below major PGE zones also shows them to be better source rocks than the parent magma. For example, in the Bushveld complex, the median values for Pt and Pd, respectively, in the rocks beneath the Merensky Reef are 78 and 21 ppb in the Upper Critical zone, 27 and 14 ppb in the Lower Critical zone, and 10 and 12 ppb in the Lower zone (Naldrett et al., 2009). These average values, particularly in the Critical zone rocks, are higher than the parent magma concentrations estimated from marginal chill and contemporary sills of ~30 ppb combined Pt + Pd reported by Davies and Tredoux (1985). Barnes et al. (2009) interpret the rocks below the Merensky Reef of the Bushveld Complex to have been sulphide-saturated based on moderate enrichments in the PGE, and have suggested that the loss of S to a high temperature degassing event as one possible explanation. Adding the metals from the ore zones back into 1,000 m of average footwall would increase the footwall concentrations from 8% to 28%, with the higher values for the PPGE (Figure 1.3). The preferential loss of the PPGE from the footwall also can explain the flatter mantle-normalized patterns of the footwall rocks compared with either the reefs or the parent liquid.

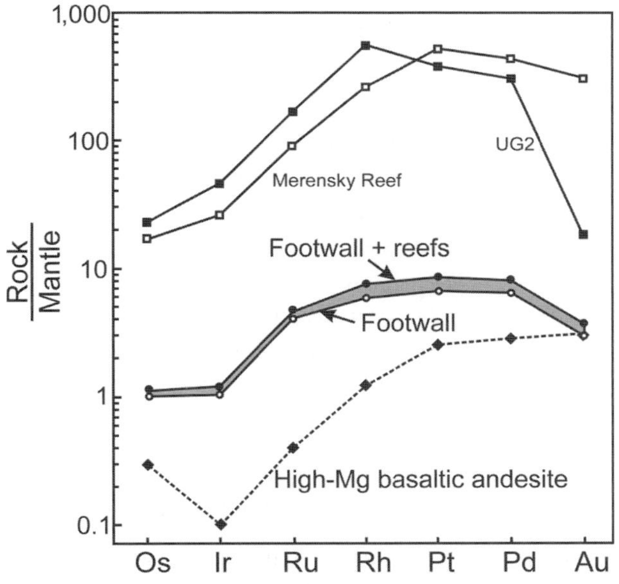

Figure 1.3 Mantle-normalized noble metal concentrations illustrating the change in the average footwall assuming the footwall was the source of the Bushveld reef metals. ‘Footwall’ = thickness-normalized average concentrations of the Lower and Critical Zones, excluding mineralize zones. ‘Footwall + reefs’ = average footwall + 2 × (UG2 + Merensky Reef) added to 1,000 m of footwall rock. ‘High–Mg basaltic andesite’ (dash line) = average composition of the high-Mg basaltic andesite from the Bushveld marginal sill-dike suite, shown for reference. From Kanitpanyacharoen and Boudreau (2013), reproduced with permission.

A first-order model by which sulfur and the PGE are concentrated by a vapour refining process in a growing and solidifying crystal pile is shown schematically in Figure 1.4. At some point during the solidification of the crystal pile, a vapour is exsolved from the crystallizing interstitial liquid. The separation and loss of this vapour removes sulfur and, if present, can result in partial resorption of any pre-existing sulfide and removal of some of the metals they contain. This vapour will migrate upward where it will encounter hotter, vapour undersaturated silicate liquid. If the fluid is composed predominately of a low solubility component like CO_2, then the redissolution of vapour will quickly drive the small amount of interstitial liquid to become vapour-saturated, leading to a rapid upward migration of the vapour-saturation front in the crystal pile. Eventually the vapour saturation front may reach the top of the crystal pile. Owing to its larger mass, the magma above the pile can readily absorb the relatively small amount of vapour escaping from the crystal pile without becoming vapour saturated. However, the increase in S concentration as the fluid redissolves in the magma can drive a thin layer of magma just above the top of the pile to sulfide saturation. The precipitation of sulfide can then lead to modest enrichments of the PGE being incorporated into the growing pile.

As the pile grows, sulfur and other ore elements are continuously recycled, moving from the interior of the crystal pile to the top to produce a thin, ephemeral zone of sulfide-saturated magma. This vapour refining front can be trapped to form an ore zone by one of two general mechanisms. First, as the pile grows, the mass of fluid migration out of the pile can be high enough to temporarily inhibit crystallization of the overlying magma, or even induce melting in the mush at the floor of the magma chamber. Vapour continuously

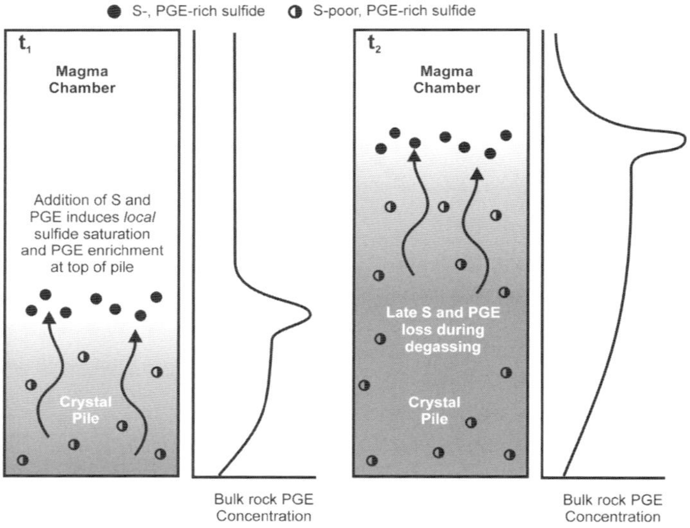

Figure 1.4 Cartoon of the ore element transport and sulfur recycling in a growing crystal pile at the two time steps t_1 and t_2. See text for discussion. From Kanitpanyacharoen and Boudreau (2013), reproduced with permission.

entering the bottom of the magma chamber precipitates PGE-sulfides at the top of the pile until the underlying mush zone has more or less fully degassed, at which time crystallization of the main magma body can continue. Second, the vapour migration front may encounter stratigraphic discontinuities present within the crystal pile. These stratigraphic boundaries may be defined by compositional or modal changes that react with upward migrating fluids to induce sulfide precipitation. Coupled with the compositional changes are potential changes in the permeability-porosity that inhibits fluid flow. Examples of both mechanisms are presented in the later chapters.

2

Layered Intrusions

An Overview

2.1 Cumulate Terminology and Its Problems

Layered intrusions preserve a record of how mafic magmas crystallize and differentiate; the problem is how to interpret this record. Most works on layered intrusions make extensive use of cumulus terminally to describe the rocks (Wager et al., 1960). In brief, this terminology grew out of assumed similarities in the formation of igneous layering as for clastic sedimentary bedding. Thus, *cumulus* minerals are the equivalent of clastic sediment, and accumulated by settling to the floor of the magma chamber to form a sequence of *cumulate* rocks. *Intercumulus* minerals are those that grew from the interstitial liquid between the cumulus minerals as the liquid + crystal assemblage cooled and crystallized. In many ways, it is a quick and efficient way to describe rock textures. In contrast, the use of traditional textural descriptions can drive one back to one's petrology textbooks, as seen in this summary of rock textures of the Ultramafic series of the Stillwater Complex from Jackson (1961):

> The textures of the rocks of the Ultramafic (series) show every gradation between automorphic-poikilitic and xenomorphic, and this textural variation is largely dependent on the amount of secondary enlargement that has occurred. Rocks with little enlargement have interposition fabrics with euhedral settled crystals and contain relatively large amounts of predominantly poikilitic interstitial material; rocks with much enlargement have mosaic fabrics and contain little or no interstitial material.

However, this book will avoid the use of cumulus terminology in the descriptions of the rocks. It is the opinion of the author that using cumulus terminology, which traditionally has interpretive implications, has pernicious effects on the understanding of the processes that formed the rocks. For example, the commonly used terms *orthocumulate*, *mesocumulate* and *adcumulate* are used to describe rocks that have modal abundances of interstitial or oikocrystic minerals assumed to have crystallized from a mush with decreasing amounts of trapped liquid, respectively. However, as discussed more fully in Chapter 11, this simple two-component model of the rocks as being composed of cumulus minerals and a trapped liquid component fails when the rocks are undergoing concurrent compaction and crystallization. In this case, earlier crystallizing interstitial phases can react with and record higher amounts of trapped liquid than do later phases, because liquid can be lost before the later minerals have a chance to crystallize.

The situation can be further complicated if minerals have different nucleation rates. The cotectic crystallization of plagioclase and orthopyroxene can form either a 'plagioclase-orthopyroxene cumulate' if both mineral phases have similar nucleation rates and crystallize to form a granular-textured rock, or an 'plagioclase cumulate with poikilitic orthopyroxene' if the orthopyroxene had only a few nuclei from which to grow and instead formed large oikocrystic minerals. For the latter, one might note that the modal proportions of plagioclase and pyroxene are still in expected cotectic proportions and with the aid of the petrographic microscope one might determine that the orthopyroxene is compositionally unzoned. In this case, the proper cumulus terminology for what is a poikilitic norite would be a *heteradcumulate*, but this would require some interpretation of how the rock formed.

McCallum et al. (1980) has an even more extreme example involving the recrystallization that accompanied the inversion of pigeonite to orthopyroxene on cooling from the Stillwater Complex. In this case, recrystallization caused initially isolated pigeonite grains to become crystallographically aligned, such that multiple pigeonite grains coalesced to form large grains with a poikilitic habit, although the granular nature of the original texture is evident from the random orientations of clinopyroxene exsolution lamallae. This textural change led Segerstrom and Carlson (1982) to map the inverted pigeonite unit as an anorthosite cumulate (plagioclase as the only 'cumulus' mineral), whereas the modes and preserved original texture are consistent with the cotectic crystallization of plagioclase + pigeonite + clinopyroxene, or a pigeonite gabbro.

Other examples include the possibility olivine in certain cases is the result of either incongruent melting owing to the fluxing effects of H_2O (e.g., Boudreau, 1999a) or the subsolidus replacement of pyroxene with olivine (e.g., Meurer et al., 1997). For the former, the rocks should be called *restites* (residual solid assemblages) and not cumulates at all, and the latter *metasomatic replacement*, yet the literature is still filled with descriptions of these rock as being composed of 'cumulus' olivine. As will be apply demonstrated, these textural ambiguities are particularly common in the PGE zones. Thus, following the suggestion of McBirney and Hunter (1995) and McBirney (2009), this report will instead use respectable but model-neural rock names with modifiers as necessary (e.g., granular harzburgite or mela-norite).

2.2 Processes of Accumulation of Crystals on the Floor of a Magma Chamber

Several mechanisms have been proposed to explain the formation of thick sequences of crystals that appear to have formed from the bottom up. These can be represented by four endmember processes illustrated in Figure 2.1. These can be summarized as follows:

Crystal settling. Conventional cumulus terminology is based, in part, on the assumption that layered intrusions formed by settling of minerals from a crystallizing magma in a manner analogous to the formation of clastic sedimentary rocks (e.g., Wager et al., 1960). In this view, heat is lost out from the roof and side walls of an intrusions, leading to nucleation and initial growth of crystals that then fall to the floor of the magma chamber. Hydromechanical sorting of mineral grains of different size and density can result in modal

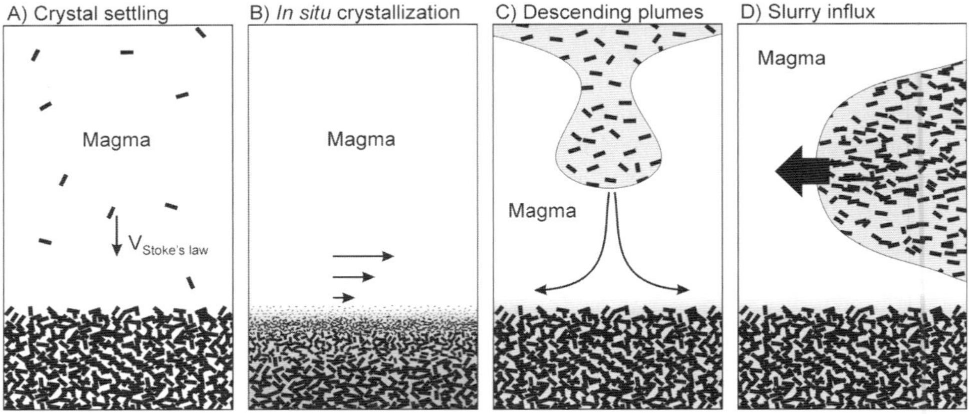

Figure 2.1 Crystal accumulation mechanisms. (a) Crystal settling. (b) *In situ* crystallization. (c) Descending crystal-phyric plumes. (d) Injection of crystal-rich slurries. See text for discussion.

and size-graded igneous layering. Variations on this process can include minerals carried into the chamber by crystal-phyric magmas. The crystals are assumed to settle according to *Stoke's Law* for the terminal settling velocity, *V*, for roughly equant crystals that can be approximated as a sphere:

$$V = \frac{gd^2 \, \Delta\rho}{18\mu} \tag{2.1}$$

where g is the acceleration of gravity (cms^{-2}), d is the grain diameter (cm), $\Delta\rho$ is the density difference of the crystal and the liquid (gcm^{-3}), and μ is the viscosity of the liquid ($gm^{-1}s$). For a 1.0 and a 10 mm crystal with a density difference of 0.3 gcm^{-3} settling in a silicate magma with a viscosity of 100 poise, the settling velocity will be 0.0016 and 0.16 cms^{-1}, respectively.

While there is no doubt that the mechanism is entirely appropriate in any number of instances, problems with simple crystal settling mechanisms arose almost from the start. For example, Jackson (1961) noted that co-existing olivine and chromite in the Peridotite Zone of the Stillwater Complex are not hydraulically equivalent. That is, although chromite is the denser phase, the olivine is so much larger that it would have the higher settling velocity, yet both appear together. However, this earlier work did not consider later modification events such as crystal aging that can alter crystal size, as noted below.

More critically, plagioclase has a neutral to positive buoyancy in most basaltic liquids and thus could actually 'float' (e.g., Campbell, 1978; Campbell et al., 1978; McBirney and Noyes, 1979; Scoates, 2000). Indeed, this propensity of plagioclase to not settle has been used explain features such as the thick anorthosite units of the Stillwater Complex where it has been suggest that they represent plagioclase that did not settle out from the magma during formation of the Ultramafic Zone but instead accumulated at a density-neutral zone in the magma chamber (McCallum et al., 1980). For many intrusions, however, it appears that the most abundant 'settled' mineral is one that should have floated (McBirney and Noyes, 1979).

There is also the apparent problem that modal layering can occur along the roof and vertical walls of a number of intrusions, suggesting that gradients other than gravity can lead to layer formation. Also, it is observed that crystals do not settle on the tops of xenolithic blocks or even on footwall highs. This has been informally termed the *Snoopy's doghouse effect* by Norman Page of the USGS; snow is observed on the ground around the doghouse (the xenolith), but not on the roof.

In situ crystallization. Owing to these problems with crystal settling, a number of authors have suggested that the crystals instead nucleated and grew where they are now found by *in situ* (in place) crystallization (e.g., Campbell, 1978; McBirney and Noyes, 1979). This interpretation makes use of the fact that, for most minerals, an increase in pressure lowers the liquids temperature at which the mineral will saturate. In these models, the magma in the chamber again loses heat from the roof and sidewalls. Convection of this cooler magma down to the floor of the chamber then favours crystallization both by the higher pressure and the presence of pre-existing crystals on which to nucleate. Some of the evidence that has been used to support this is the observation that in rocks with a poikilitic textures, crystals enclosed in the larger minerals are smaller than are those that are outside the oikocryst. This has been interpreted that the smaller grains had their growth arrested once they became entrapped in the larger oikocryst whereas those outside could continue to grow (e.g., McBirney and Noyes, 1979).

This model assumes that thermal convection is robust enough to carry nominally supersaturated liquids to the floor, which some authors have disputed (e.g., Marsh, 2013). It also has its own version of the plagioclase problem. In this case, the lower density of plagioclase relative to the silicate liquid means that crystallization is favoured by lower pressure and should preferentially crystallize under the roof. The question again arises, why do we get plagioclase on the floor?

Descending plumes of crystal-phyric magma. This mechanism again assumes that the magma loses its heat and begins to crystallize along the roof and sidewalls of the magma chamber. In this case, it is not the crystals that settle out from the cooling fronts, but instead it is the partially crystallized, cooler, and denser crystal + liquid mixture that descend as crystal-bearing plumes to the floor of the magma chamber (e.g., Jaupart and Brandeis, 1986). Separation of crystals and liquid then occurs over longer time scales by compaction or advection of interstitial liquid from the mush. It has been suggested that this mechanism can occur in thick flood basalt lavas; cooling from the roof down produces an unstable crystallization front that brakes free and descends to the bottom of the still-liquid part of the lava flow (Philpotts et al., 1996).

This mechanism has a number of similarities with the formation of sedimentary evaporates. In this case, evaporation leads to the nucleation and growth of salts either on the surface (as a salty crust) or in the near-surface layer. This denser layer then becomes unstable and descends as a plume to the bottom, where the salts eventually separate from the water. In both the sedimentary and the igneous systems, it is the higher bulk density of the crystallizing liquid + crystal mixture relative to the underlying liquid that causes the crystallization boundary to become gravitationally unstable and form descending plumes and not the individual crystals.

This model does have a number of features to recommend it. First, it eliminates the plagioclase buoyancy problem, as it is the crystal + liquid mixture that has a relatively high bulk density, in part by being cooler than the underlying liquid. Second, the descending plumes will flow to the lowest levels, consistent with the observation that the more primitive rocks tend to occur in footwall depressions. This solves the Snoopy's doghouse effect. Third, descending plumes can effectively mix and homogenize the magma, even if the magma is not otherwise undergoing convection. Forth, it suggests a simple but fundamental difference between how mafic and more siliceous magmas crystallize. For the latter, the magmas undergo crystallization from the margins inward; the combination of higher viscosity and the smaller density contrast between partially crystallized and the underlying rhyolitic magma would both tend to inhibit the formation of descending plumes.

Injection of crystal-rich slurries. Mechanically similar to the descending plumes mechanism is the injection of crystal-rich slurries into the magma chamber. Marsh (2013) argued that crystal-free magmas would be unable to differentiate owing to crystals becoming trapped in advancing crystallization fronts before they can grow large enough to escape. He suggested that magma differentiated intrusion requires the influx of crystal-rich slurries, which are injected away from the crystallization fronts at the cooler margins and thus the crystals can both continue to grow and separate from the liquid and cause compositional evolution of the magma. This model has been criticized by many, particularly in regards as a fractionation mechanism (e.g., Latypov et al., 2015). The injection of crystal-phyric magmas is clearly justified in many instances. The case for the injection of an orthopyroxene-rich magma in the Basement Sill, Antarctica, is an excellent example (e.g., Bédard et al., 2007).

There are some fundamental consequences for these four models. For conventional crystal settling, the entire magma chamber is fractionating and the liquid initially trapped in the mush zone is the same composition as the host magma. For *in situ* crystallization and the plume model, crystallization in a narrow solidification zone implies that the interstitial liquid will be considerably more evolved than is that of the main body of magma in the chamber. The injection of crystal-rich slurries can be of markedly different composition from the fractionated magma in the chamber, but is otherwise similar to the descending plume model. For all but the crystal settling model, if the liquid did not move out of the mush and mix back into the main magma body the intrusion would not fractionate. It is the mixing of this small volume of evolved liquid from the crystallization zone (in the case of *in situ* crystallization) or from the descended plume back into the main magma body that leads to the compositional evolution of the system. This is qualitatively different than is conventional crystal settling and the Raleigh equations that describe whole chamber crystal fractionation. Langmuir (1989) has modelled how fractional crystallization differs from standard Raleigh fractionation during *in situ* crystallization. The general idea is that liquid in the solidification zone undergoes some degree of crystallization before it is returned and mixed back into the main magma body. For both incompatible but especially compatible elements, the change in concentration in the magma is less as compared to Raleigh fractionation for any given amount of solidification. Langmuir has suggested that this can explain the observed compositional trends observed in the Kiglapait intrusion.

2.3 Processes Occurring within the Crystal-Liquid Mush

A fundamental problem in the interpretation of the mineral compositions and textures is unravelling changes that occur as the rocks slowly cool over what may be many thousands of years. In sedimentary rocks the term *diagenesis* include all processes that modify the original sediments to produce a sedimentary rock, including cementation, compaction and the generation of oil and gas. No such general term exists for modifications of the originally precipitated assemblage in layered intrusions, although the term *igneous diagenesis* may be used in an analogous manner as for sedimentary term. Indeed, a number of *hard ground* interpretations suggest that the minerals, textures and igneous layering formed in a very sharp interface between the magma and fully solidified rock such the rocks preserve a direct record of magma chamber processes. Others take the view that the rocks can evolve over the long cooling history and that layered intrusion have many similarities with the formation of metamorphic rocks in that the texture and even modal layering can develop over time. In this view using the final rock to infer magmatic processes is as difficult as using a metamorphic pelitic schist to understand sedimentary processes. A description of some of these processes are described below.

2.3.1 Crystal Aging

Crystal aging, also known as *crystal coarsening* and *Ostwald ripening*, is the process by which favoured mineral grains grow larger at the expense of less favoured grains owing to variations in grain size or the local environment a crystal finds itself. Classical interpretations of the aging process considered a fixed *surface energy* term. Because the contribution of surface free energy to the overall free energy of a crystal is always positive, a given mass of large crystals will have a lower free energy than does an equal mass of small crystals because of the larger surface area of the latter. (The effect of surface energy on bubble nucleation is discussed in Chapter 5.) This results in smaller crystals having a higher solubility (or solubility product for crystals which are composed of more than one liquid component) than do larger crystals. Thus crystal systems evolve, in time, to fewer but larger crystals. This grain size-dependence of the equilibrium solution concentration for grains that can be approximated by a sphere is expressed by the Gibbs-Thomson equation:

$$C(r) = C^{\infty}\frac{2\sigma}{\rho RTr} = \frac{k}{r} \quad \text{(Gibbs-Thomson equation)} \tag{2.2}$$

Where $C(r)$ is the solution concentration in equilibrium with a crystal of radius r, C^{∞} is the equilibrium for a nominally large (infinite) radius grain, σ = the surface energy, ρ is the mineral density, R the gas constant, T is temperature and k is the collection of all the constants that leads to the simple constant temperature expression on the right-hand side of Equation 2.1.

Because smaller crystals require higher solution concentrations to remain stable, material will tend to migrate to regions of larger grains. This leads to a positive feedback effect in which the smaller crystals get smaller and cause local solution concertation to further increase which then causes the larger grain to grow yet faster. An example of the aging growth of Sn grains in a Sn-Pb liquid is illustrated in Figure 2.2. The process of crystal aging not only reduces the number of mineral grains, it also tends to produce a characteristic crystal size distribution in both igneous and metamorphic rocks (e.g., Cashman and Ferry, 1988; Cashman, 1993; Waters and Boudreau, 1996; Higgins, 1998, 2002, 2006, 2011).

The aging phenomenon has important implications in the crystallization behaviour of layered intrusions. After grains have nucleated, crystal growth by solidification of the liquid is limited to the rate at which the heat of crystallization is lost; crystallites may remain very small and unstable for a long time in slowly cooled intrusions. However, although the crystal mass will increase only slowly with time, the crystal assemblages can continue to coarsen by the aging process because it is largely neutral with respect to the

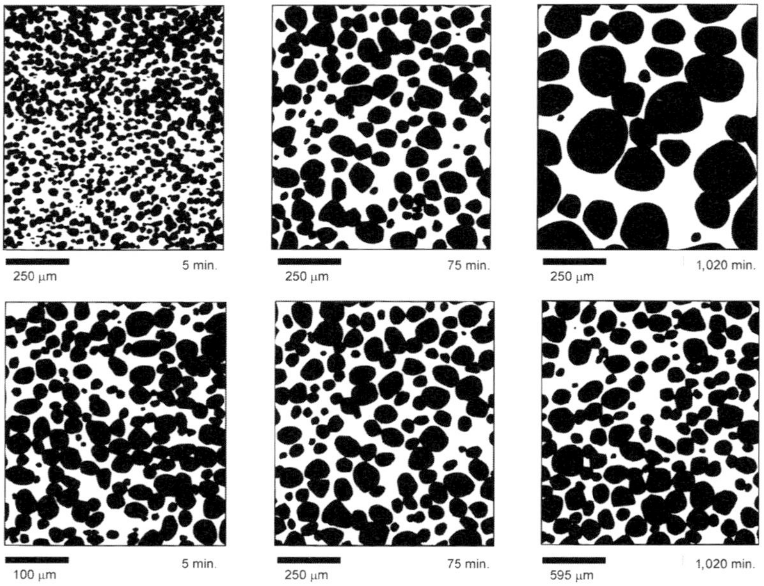

Figure 2.2 Example of the crystal aging growth of Sn-rich crystals in a Sn-Pb liquid. Top row shows the stem at three different times, all at the same scale. The bottom row changes the scale at the different times to illustrate the characteristic size distribution remains self-similar with time. Redrawn after Hardy and Voorhees (1988), with permission.

overall heat budget; growth of favoured grains is compensated by the dissolution of a thermally equivalent amount of unflavoured, smaller grains. Thus, the aging process can be a potentially significant cause of crystal growth in slowly cooled intrusions. Because a growing crystallization front will have gradient in grain size, numerical modelling has shown that aging can lead to a variety of consequences in the development of igneous layering, including the development of fine-scale layering, size-graded layers, 'inch-scale' doublet layering and sharply defined modal layering (e.g., Boudreau, 1994, 1995, 2011; Boudreau and McBirney, 1997).

2.3.2 Compaction

Beginning with the work of Irvine (1980), an increasing number of studies have begun to illustrate the importance of compaction in igneous rocks in general and in layered intrusions in particular. For example, thick mafic lava flows and sills may range from largely homogenous and undifferentiated to those that show some significant differentiation. Philpotts et al. (1996) demonstrated that compositional trends in the Holyoke flood basalt are correlated with compaction-driven deformation of an original crystal framework defined mainly by plagioclase chains. Shirley (1987), has suggested that compositional trends in the Palisades sill can be explained by a compaction-driven migration process. These small igneous bodies can be described as a mixture of crystals plus some residual trapped (and now crystallized) interstitial liquid that was not lost to compaction.

More complete loss of interstitial liquid can lead to the formation of rocks with a very small trapped liquid component that is typical of much larger layered intrusions such as the Stillwater Complex. Evidence for this has been presented by Meurer and Boudreau (1996, 1998a,b), who have demonstrated correlations between the development of foliation and incompatible element concentrations in the Stillwater Complex and attributed them to concurrent crystallization and compaction-driven interstitial liquid migration.

Meurer and Boudreau (1996) have modelled the effect of sharp stratigraphic density changes in the solid assemblage, such as occur at ultramafic-mafic contacts in layered intrusions, on the compaction-driven distribution of interstitial liquid. In this case, interstitial liquid tends to collect as porosity waves that propagate from the interface. This model can explain the high abundance of incompatible elements and relatively evolved rock compositions that occur just below the ultramafic-mafic boundary that is found in a number of intrusions including the Great Dyke of Zimbabwe, and the Jimberlana and the Munni Munni Complex of Australia. The role of compaction in the formation of Great Dyke type PGE deposits is explored in some detail in Chapter 11.

2.3.3 Crystallization of Interstitial Liquid and Subsolidus Re-Equilibration

As summarized by Barnes (1986), the interpretation of the chemical evolution of layered intrusions involves inversion of mineral composition trends to infer compositions of parent magmas. Three principal factors must be taken into account when interpreting mineral compositions in cumulate rocks: (1) changes in the composition of the parent magma due to fractional crystallization, contamination, magma mixing etc.; (2) superimposed modification of mineral compositions as a result of crystallization of trapped interstitial liquid; (3) redistribution of elements due to sub-solidus re-equilibration between minerals.

The second of these effects, the re-equilibration of liquid and crystals on cooling, was initially described by several authors (Henderson, 1970; Raedeke and McCallum, 1980; Grant and Molling, 1981; Cawthorn, 1982; Longhi, 1982). They considered the crystal mush to be a two-component mixture of (1) initially precipitated crystals and (2) interstitial silicate liquid, all assumed to be initially in equilibrium with one another at near-liquidus temperatures. The initial proportion of *trapped liquid* in the rock is referred to as the *initial porosity* (Irvine, 1982). The trapped liquid then crystallizes to produce overgrowths on the original crystals (normal zoning), plus the precipitation of additional interstitial phases such as apatite containing components excluded from the originally precipitated minerals. If diffusion can homogenize the core-rim mineral compositions, the final composition of the minerals will be the average of the original composition and that of the overgrowth: Ferromagnesian phases will become more iron-rich and more depleted in compatible elements such as Ni and Cr, and plagioclase will become more sodic. The higher the initial porosity, the greater will be the difference the initially precipitated mineral composition and that after complete solidification. This difference is referred to as the *trapped liquid shift effect.*

The magnitude of the trapped liquid shift can be calculated knowing the initial cumulus mineral composition and proportion of trapped liquid by making the *closed system* assumption that the bulk composition remains constant and equal to the original proportions of solid and liquid. This situation can be complicated if compaction is concurrent with crystallization of the interstitial liquid. In this case, the original two-component assumption is not valid, as some liquid is lost over the course of crystallization. As noted by Meurer and Boudreau (1996) and Meurer and Meurer (2006), this can cause estimates of the original trapped liquid to vary depending on how incompatible certain elements behave. Indeed, it can be even more complicated as compaction can result in porosity to go up and down over time as 'porosity waves' move through the mush. This is discussed in more detail in Chapter 11. The same thing can happen if there is advective loss of an evolved low density liquid out of the mush late in the solidification of the interstitial liquid.

In addition to minerals re-equilibrating with the solidifying interstitial liquid, the minerals can also re-equilibrate with each other. A common example or this is observed in Fe-Mg exchange between co-existing olivine and chromite such that olivine becomes more Mg-rich and chromite more Fe-rich on cooling (Irvine, 1965):

$$\underset{\textit{chromite}}{MgCr_2O_4} + \underset{\textit{olivine}}{Fe_2SiO_4} \rightarrow \underset{\textit{chromite}}{FeCr_2O_4} + \underset{\textit{olivine}}{Mg_2SiO_4} \qquad (2.3)$$

The effect of mineral-mineral re-equilibration is strongly a function of mineral modes. Olivine in a dunite containing a small amounts of chromite might show little change whereas the minor chromite can be significantly re-equilibrated to more Fe-rich compositions, for example.

2.3.4 Metasomatic Changes

Metasomatic changes are those that involve substantial change in bulk composition and mineralogy of a rock and are among the more controversial mechanisms proposed to operate in layered intrusions. The agent causing the metasomatism can be a silicate liquid or a volatile fluid, and examples of the latter make up a substantial part of this book. The results of this process are most clearly displayed when the metasomatic replacement produces an isolated, discordant body of rock such as a podiform or a pipe-like feature with contrasting mineral assemblage from the host rock (Chapter 8). More uncertainty is present when the replacement is concordant with igneous layering.

Early examples of the include the concept of infiltration metasomatism proposed by Irvine (1980) to explain the upward offset of discontinuities in mineral compositional trends relative to contacts between cyclic units in the Muskox intrusion. He suggested that the offsets of Mg/(Mg + Fe) discontinuities in olivine and chromite the minerals and interstitial liquid that was displaced upwards across a mineral compositional discontinuity as a result of compaction. More recent examples include replacement of troctolite to make anorthosite in the Stillwater Complex (Boudreau and McCallum, 1992b), the formation of troctolite after gabbro in the Rum Layered suite (Holness et al., 2007), and metasomatic models for chromitite formation (e.g., Mathez and Kinzler, 2017) in general and the Merensky Reef in particular (Mathez, 1995).

2.4 Layered Intrusions and Their PGE Deposits: Examples

While this book will use examples from a number of intrusions, three intrusion in particular form the backbone of this work. They are the Bushveld Complex of South Africa, the Stillwater Complex of Montana, and the Skaergaard intrusion of East Greenland. Each of these intrusions and their more important ore deposits are described below. Two other intrusions, the Great Dyke of Zimbabwe and the Munni Munni intrusion of Australia, are described in Chapter 11.

2.4.1 The Bushveld Complex

The Bushveld Complex of the Republic of South Africa is, for historical reasons, composed of two major units, the Rustenburg Layered Series and the Bushveld Granite (Figure 2.3). While temporally related to the same mantle thermal anomaly, the Bushveld Granite is the result of a separate crustal melting event and is not the product of fractional crystallization of the mantle magma that crystallized the Rustenburg Layered Series. Most authors use the Bushveld Complex moniker as synonymous with the Rustenburg Layered Series and this convention is used here. It is the world's largest continental layered intrusion, with an aerial extent of about 66,000 km^2 and an age of 2.06 Ga (Walraven, 1988). It is also economically a sizeable resource; it contains about half of the world's know PGE reserves. It has been described as a lopolithic intrusion with a maximum

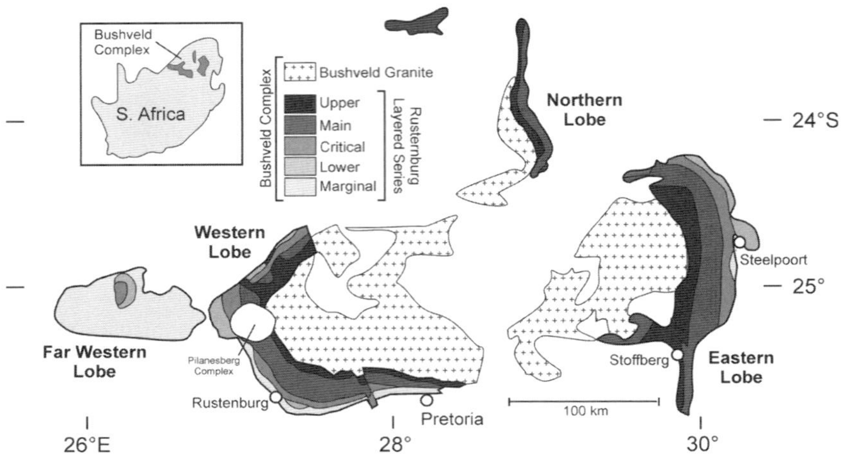

Figure 2.3 Generalized geology map of the Bushveld Complex, Republic of South Africa. Redrawn after Roelofse and Ashwall (2012), with permission.

thickness of about 7–9 km, although with as aspect ratio of >40:1 it may be considered a variably thick sill.

Regionally, the Bushveld Complex is subdivided into the eastern, western, far western, southern and northern lobes. The stratigraphy of the Bushveld Complex, defined largely from the eastern and western limbs, has been subdivided into a basal Marginal Zone, overlain sequentially by the Lower Zone, Critical Zone, Main Zone and Upper Zone (Figure 2.4). The intrusion of the Bushveld magma produced a metamorphic aureole and local melting in the underlying Transvaal Supergroup sedimentary rocks that locally exceed 3 km in thickness (e.g., Harris et al., 2003). The Marginal Zone comprises sills of locally quench-textured pyroxenite and micropyroxenite whose compositions have been interpreted as parental to the main Bushveld magmas (e.g., Sharpe, 1985). The Lower Zone is approximately 800–1,000 metres thick and consists primarily of layers of pyroxenite, dunite, and harzburgite with some chromite with small amounts of interstitial plagioclase, biotite, and clinopyroxene. The boundary between the Lower Zone and the Critical Zone has been defined by what is, to the casual observer, a rather modest increase from about 2% to about 6% in the modal abundance of interstitial plagioclase (Cameron, 1978), although others define the boundary by a decrease in olivine abundance (Teigler and Eales, 1996).

The Critical Zone is subdivided into the Lower Critical and Upper Critical Zones, with the boundary between the two occurring where plagioclase changes from an interstitial mineral to an abundant, euhedral phase. Thirteen major chromitite seams have been identified in the Critical Zone, namely the Lower Group (LG1–LG7), the Middle Group (MG1–MG4), and the Upper Group (UG1 and the PGE-rich UG2) chromitites, although in detail each group can be composed of several chromitite seams interlayered with silicate rock.

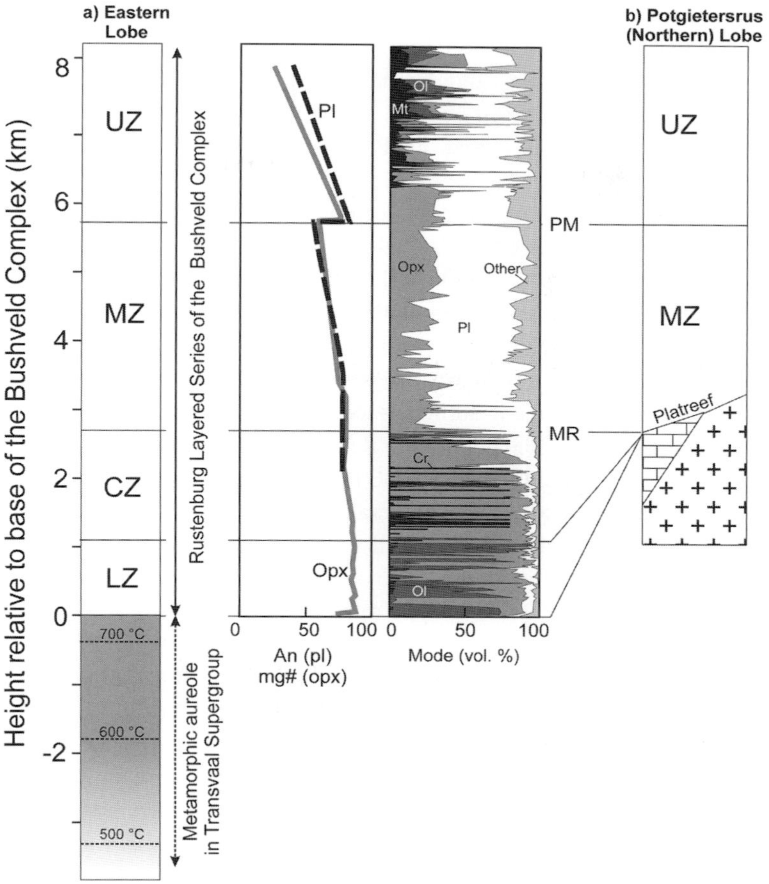

Figure 2.4 The Rustenburg Layered Series of the Bushveld Complex. (a) Stratigraphic sections from the eastern lobe and its underlying metamorphic aureole showing the main stratigraphic units, general pyroxene and plagioclase compositional trends and modal variations. Maximum thermal isograds in the aureole in the footwall rocks are labelled. Abbreviations: LZ, CZ, MZ and UZ = Lower, Critical, Main and Upper Zones of the Bushveld Complex, respectively; MR = Merensky Reef; PM = Pyroxenite marker near top of the Main Zone; Ol = olivine; Cr = chromite; Opx = orthopyroxene; Pl = plagioclase; Mt = magnetite. (b) Shorter section of the Rustenburg Layered Series seen in the northern lobe, in which rocks below the Main Zone are absent. Lateral correlation of the Merensky Reef (MR) and the Platreef are inferred but not demonstrated. Modified after Maier and Barnes (1998), Harris et al. (2003) and White (1994) with permission.

The boundary between the Critical Zone and the Main Zone is usually taken to be the top of the Giant Mottled Anorthosite layer, an anorthosite with large oikocrysts of pyroxene. The Main Zone consists of a thick succession of norites and gabbronorites devoid of olivine or chromium spinel. The base of the Upper Zone is generally taken as occurring at the Pyroxenite Marker horizon (Kruger, 1990), and is also marked by the appearance of euhedral magnetite (Eales and Cawthorn, 1996). This Zone is characterized by generally well-defined modal layering, with particularly well-developed magnetite and anorthosite layers.

As noted above, the Bushveld Complex has some impressive PGE mineralized Zones. These include the Merensky Reef, PGE mineralization associate with the UG2 chromitite, the Platreef, and a number of discordant PGE-bearing replacement pipes that occur in the Critical Zone.

The Merensky Reef. In its most common occurrence, in what is generally termed *normal reef*, the Merensky Reef consists of a distinctly coarse-grained to pegmatoidal pyroxenite 20–25 cm thick with a basal 1–3 cm thick chromitite seam (Figure 2.5). A less persistent chromite seam locally may be present along the top of the pegmatoid. This reef package is sandwiched between an underlying anorthositic to noritic rock with an undulating contact and an overlying non-pegmatoidal pyroxenite (Figure 2.6a). Mineralization in normal reef consist of a few volume percent sulfide (most commonly pyrrhotite, pentlandite and chalcopyrite) and associated PGE minerals. Recalculating the average ore grades of the Merensky Reef to 100% sulfide average ~500 ppm combined PGE. Local variations in thickness, ore grade, and the location of mineralization are common (e.g., Cawthorn et al., 2002), notably around and within footwall disturbances known as potholes, discussed in more detail in Chapter 8.

The UG2 Chromitite. The UG2 is a platiniferous massive 0.4–2.5 m thick chromitite layer that occurs from 20 to 400 m below the Merensky Reef (Figure 2.6). The PGE and base metals are mainly concentrated along the bottom, mid-level and top of the chromitite as shown in Figure 2.6, but otherwise base metal sulfides are less abundant and considerably smaller in size than is found in normal Merensky Reef.

The Platreef. In the northern limb of the Bushveld Complex, Main Zone rocks onlap onto country rocks, the later composed of Archean granites and sedimentary rocks of the Transvaal Sequence. The Platreef, a 10–300-m thick package of PGE-rich pyroxenite and gabbro, is developed in the basal part of the Main Zone along the contact with the country rock. Base metal sulfides and PGE concentrations are variable; the mineralized zone can be up to 300 m thick, and grade generally increases as one moves into the Bushveld Complex and away from the country rock. As summarized by McDonald et al. (2005), earlier workers considered the Platreef as an equivalent to the Merensky Reef, while more recent work has highlighted the differences.

Mineralized discordant pipes. Occurring sporadically throughout the Bushveld Complex are a number of unusually large discordant pipe-like bodies. In the eastern lobe, three of these, the Onverwacht, Mooihoek and Driekop pipes, cut Critical Zone rocks and contained cores of unusually high-grade PGE-sulfide mineralization (up to 2,000 g/ton Pt), prior to being mined out in the first half of the 1900's. The pipes can be in excess of 200 m in diameter, with the mineralized, iron-rich olivine cores (Fo $<$ 50 mole %) up to about 25 m in diameter (e.g., Stumpfl and Rucklidge, 1982). Although small discordant and podiform mineralized bodies can be found in a number of intrusions, the large size and degree of PGE enrichment observed in these pipes appears to be unique to the Bushveld Complex and are discussed in more detail in Chapter 8.

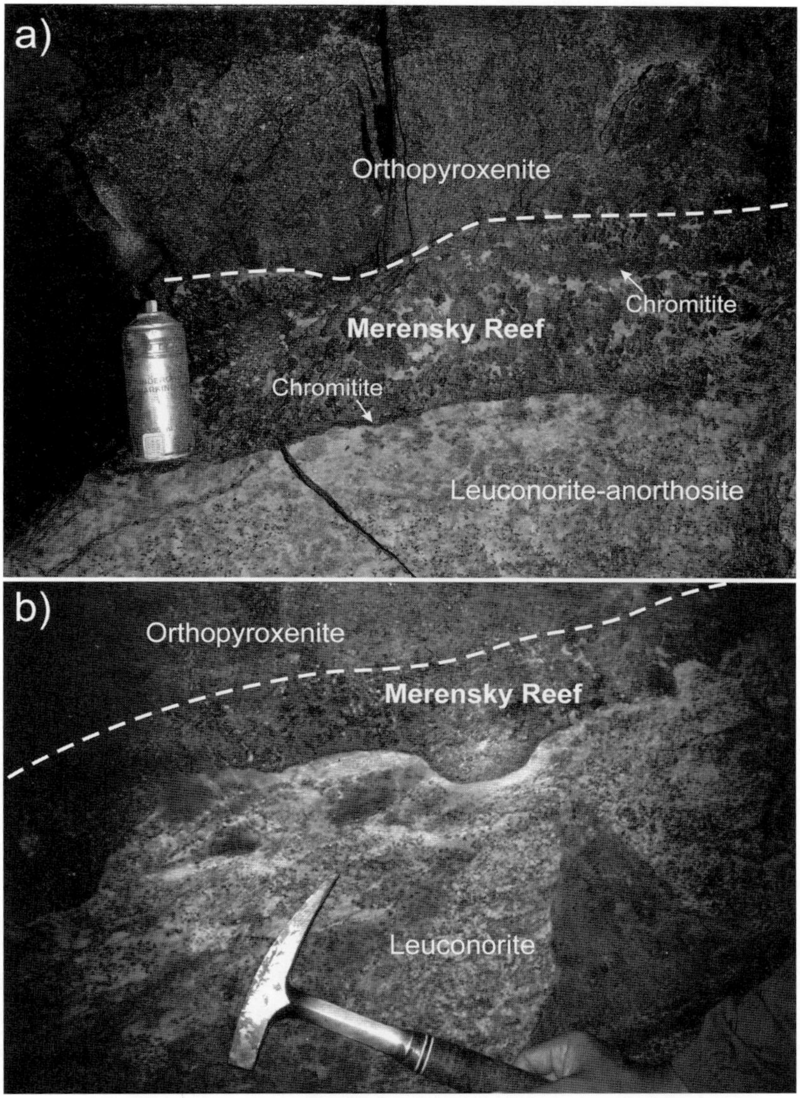

Figure 2.5 The Merensky Reef, Bushveld Complex. (a) Example of normal Merensky Reef, Thembelani Mine. Note coarser grain size compared of the reef rocks with both the underlying plagioclase-rich rocks and especially the overlying orthopyroxenite. Photo curtesy of Rais Latypov, University of the Witwatersrand. (b) Example of dimpling or 'teats' where Merensky Reef protrudes into underlying leuconorite-anorthosite, Western Platinum Mine. Photo by author. A black-and-white version of this figure appears in some formats. For the colour version, please refer to the plate section.

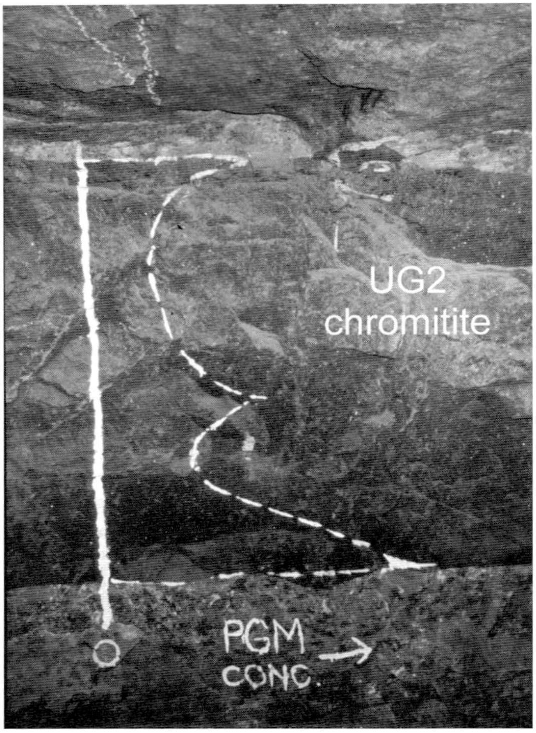

Figure 2.6 Photograph of the UG2 underground at Lonmin mine, east of Rustenburg. The chromitite of this often-photograph underground exposure is about 1 m thick. Painted on the chromitite is the relative PGE concentration as a function of height. Photo by author.

2.4.2 The Stillwater Complex

The 2709–2012Ma (e.g., DePaolo and Wasserburg, 1979; Wall and Scoates, 2016; Wall et al., 2018) Stillwater Complex covers an area of 180 km^2 on the northern edge of the Beartooth Mountains in Montana (Figure 2.7). The magma was emplaced into Archaean meta-sedimentary rocks. Like the Bushveld Complex, a wide metamorphic aureole is developed in the footwall rocks that likely exceeds 2 km thick but its true thickness is uncertain owing to extensive faulting and lack of marker horizons. The upper portion of the intrusion was unroofed, eroded, and eventually covered by pre-Middle Cambrian sedimentary rocks.

The complex has been subdivided into five Series: the Basal, Ultramafic, Lower Banded, Middle Banded and the Upper Banded Series (McCallum et al., 1980) (Figure 2.8). The base of the Ultramafic Series (US) is placed at the first significant appearance of olivine and its top at the first appearance of abundant plagioclase feldspar. The Ultramafic series is divided into the Peridotite Zone (PZ) and the Bronzitite Zone

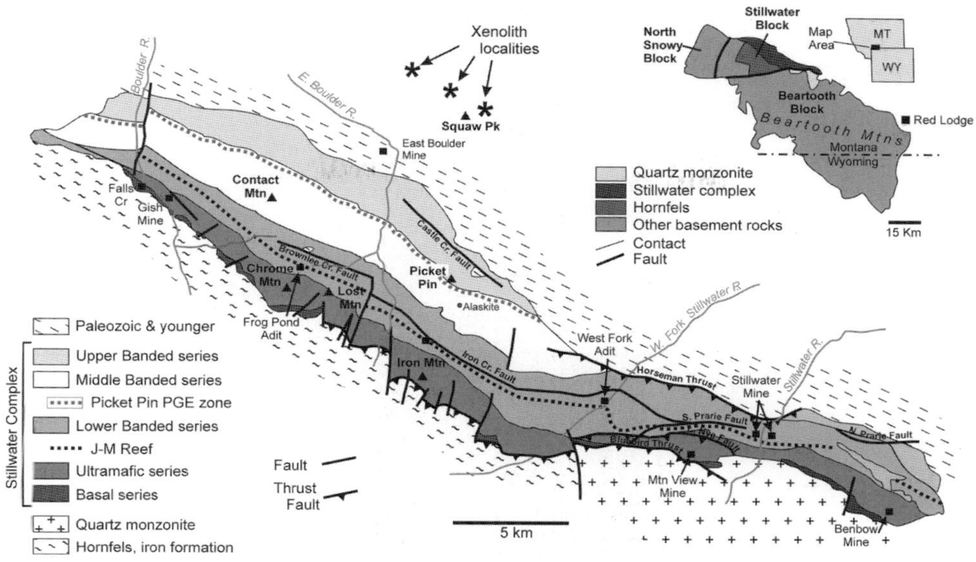

Figure 2.7 Geology of the Stillwater Complex. Inset shows the relationship of the Stillwater Complex to the three Archean blocks that make up the Beartooth Mountains. After Zientek (1993), with permission.

(BZ). The PZ consists of up to 21 cyclic sequences of dunite (olivine ± chromite), harzburgite (olivine + orthopyroxene ± chromite) and bronzitite (orthopyroxene), which may be only partly complete. A number of chromitites occur in the PZ, and are labelled A–K sequentially upward and chromitite zones G and H being the thickest. The BZ is, for the most part, a relatively uniform thick orthopyroxenite (but see Figure 11.6).

The Banded Series comprises all rocks above the Ultramafic series, the base of which is defined at the level plagioclase becomes a major modal mineral. Rock types include troctolite, norite, gabbronorite, olivine gabbro and anorthosite. The boundaries of the Middle Banded Series are defined by the two thick (up to 600 m) anorthosites that form the bottom and top of this zone. Olivine-Bearing Zone I (OB-I) is the first stratigraphic reappearance of olivine in the Lower Banded Series and this unit contains the main PGE-sulfide ore zone, the *J-M Reef*. The lower part of OB-I contains abundant troctolite and subordinate dunite, which typically have a heterolithic, coarse-grained to pegmatoidal texture (Figure 2.9). These are interlayered with minor anorthosites, norites and gabbronorites, one to several metres in thickness. In contrast, the upper portion of OB-I comprises medium-grained anorthosites and subordinate anorthositic troctolites (McCallum et al., 1980).

The J-M Reef. The J-M Reef is a broadly stratabound zone of 0.5–3% disseminated PGE-rich sulfides that averages 18 ppm Pt + Pd with a Pt:Pd ratio of 1:3 (Todd et al., 1982; Barnes and Naldrett, 1985). Typically, the reef package is a zone of about 1–3m

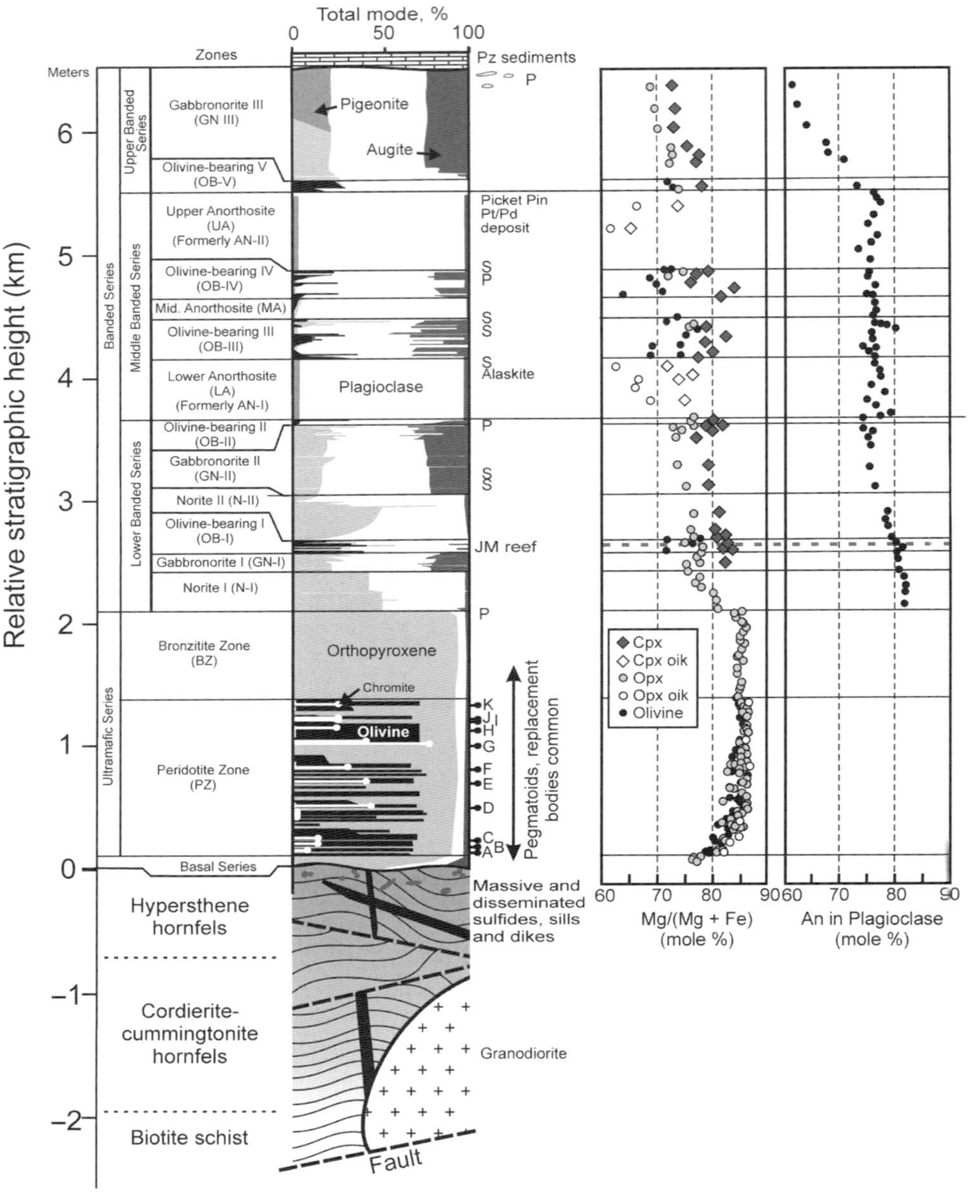

Figure 2.8 Major mineral modes in the Stillwater Complex and mineral compositional trends as a function of stratigraphic height. Modal olivine in poikilitic and granular harzburgite from the Ultramafic series is in part schematic, based on average modes. Also shown are the locations of the A through K chromitites (labelled A–K), zones where pegmatoids and sulfides are common (labelled 'P' and 'S', respectively), the alaskite body described by Czamanske et al. (1991), and the locations of the J-M Reef and the Picket Pin PGE Zone (labelled). Modal data after McCallum et al. (1980) and Raedeke and McCallum (1984). Modified after McCallum et al. (1985) and Raedeke and McCallum (1985), with permission.

Figure 2.9 Example of the texturally and modally mixed nature of the olivine-bearing rocks of Olivine-Bearing Zone I, host zone of the J-M Reef of the Stillwater Complex. Photo by author. A black-and-white version of this figure appears in some formats. For the colour version, please refer to the plate section.

thickness which is PGE-enriched and comprises dunite, troctolite, anorthosite and norite of pegmatoidal texture. Although traceable along the ~48km strike of the complex at about 400m above the base of the Banded Series, mineralization is laterally discontinuous. This variable grade is illustrated in the grade-thickness map shown in Figure 2.10. The discontinuous nature of the mineralization requires definition drilling at ~15 m centres all along the reef to avoid excessive mining of barren or poorly mineralized rock.

Locally, the reef package appears to cut down into the underlying sequences almost to the base of the Banded Series (Zientek et al., 2002; Boudreau et al., 2019) (Figure 2.11). Similar to the Merensky Reef, pothole structures are locally developed. Pegmatoids, some of which host small bodies of massive sulfide, are also locally present (Figure 2.12). It is also common for mineralization to swell to form what mine geologist term *ballrooms*, regions in which disseminated mineralized rock may extend up to 15m into underlying rock and it can extend across lithologic layers (Figure 2.13).

Figure 2.10 Long section of *grade x thickness* along the J-M Reef illustrating the non-uniform nature of mineralization along the mineralized surface. Note especially the lack of mineralization in the region of the Dow Meadow Depression (Figure 2.11). A black-and-white version of this figure appears in some formats. For the colour version, please refer to the plate section.

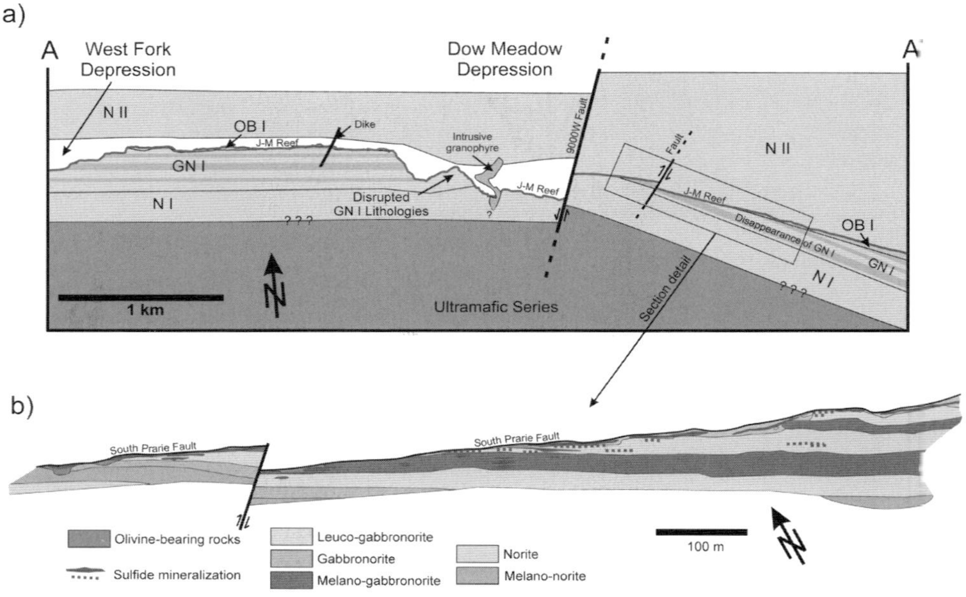

Figure 2.11 Regional downcutting of the J-M Reef into the underlying footwall rocks at the 'Dow Meadow Depression' illustrating stratigraphic variability of the J-M Reef. (a) Regional section constructed along the line of section A–A' shown in Figure 2.12. The reef cuts down or replaces original footwall lithologies. From Boudreau et al. (in review), with permission. (b) Detail of the reef development into the underlying GN-I and N-I Zones showing the presence of olivine-bearing rocks along the unconformity and the irregular distribution of sulfide mineralization that defines the J-M Reef. Redrawn after Zientek et al. (2002), with permission from the Canadian Institute of Mining, Metallurgy and Petroleum. A black-and-white version of this figure appears in some formats. For the colour version, please refer to the plate section.

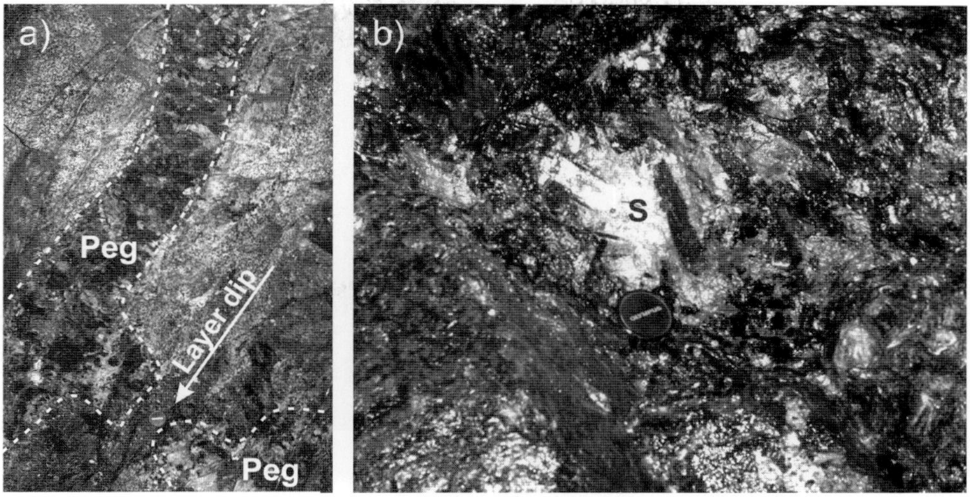

Figure 2.12 Pegmatoids associated with the J-M Reef, Stillwater Complex. (a) Crudely layer-parallel to discordant/crosscutting pegmatoids developed in gabbronorite from Olivine-Bearing Zone I. Arrow shows local dip of layering and also points to lens cap for scale. (b) Local core of massive sulfide in pegmatoid from the same region as in (a). Cast of surrounding pegmatoidal minerals are preserved in the massive sulfide. From Boudreau (2016), with the permission of the Mineralogical Society of Great Britain & Ireland. A black-and-white version of this figure appears in some formats. For the colour version, please refer to the plate section.

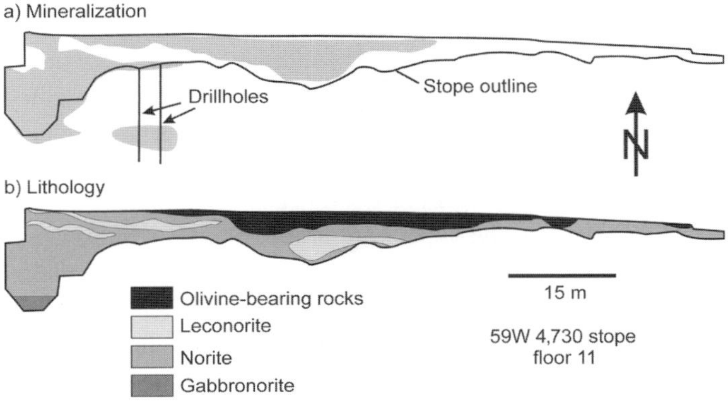

Figure 2.13 Example of 'ballroom' mineralization in the J-M Reef. (a) Mineralization exposed in the 59W 4,730 stope, floor 11, shown in grey. Owing to the near-vertical dip, the plan view shows a part of the stratigraphic section. (b) Lithology of the host rocks. Note that mineralization developed in the ballroom is largely indifferent to host rock type. Redrawn after Zientek et al. (2002), with permission from the Canadian Institute of Mining, Metallurgy and Petroleum.

Figure 2.14 The Picket Pin PGE Zone of the Stillwater Complex. (a) Discordant, pipe-like body of disseminated sulfide located approximately 150 In below the top of the Upper Anorthosite of the Middle Banded Series. Stratigraphic 'up' is towards top and slightly into the plane of the photograph. The pipe has separated along a joint which runs diagonally through its long axis; this allows one to see that the sulfide-rich pod has a roughly circular section in plan view. From Boudreau and McCallum (1986), with permission. (b) Photo of the typical weathered podiform occurrence of the disseminated PGE-sulfide mineralization of the Picket Pin Pt-Pd deposit that occurs in the upper part of the Upper Anorthosite Zone at the top of the Middle Banded Series. Note hammer (at arrow) for scale. Photo by author. A black-and-white version of this figure appears in some formats. For the colour version, please refer to the plate section.

The most extensive magmatic model for the petrogenesis of the J-M Reef and associated rocks involves the influx an 'A' magma with anorthositic affinities that mixed with a fractionated 'U' magma that was largely responsible for the Ultramafic series and rocks of the Lower Banded Series up to Olivine-Bearing Zone I (e.g., Irvine et al., 1983). Hydromagmatic models are discussed in Chapter 9.

The Picket Pin Deposit. The Picket Pin Pt/Pd deposit is a zone of disseminated, PGE-bearing sulfide which occurs in the upper 150 m of the ~600 m thick Upper Anorthosite unit at the top of the Middle Banded Series. Although PGE-bearing sulfides locally occur up to the contact with the overlying troctolite, sulfides most commonly occur in a poorly defined zone, the top of which occurs 5–10 m below the top of the anorthosite at a distinct textural and mineralogical transition. Below this transition, the anorthosite is characterized by relatively coarse plagioclase and up to 20% oikocrystic pyroxene (augite and inverted pigeonite). Sulfides occur as podiform and lenticular accumulations which are conformable but discontinuous along strike, similar to what is observed in the J-M Reef. The podiform nature of the mineralization is broadly analogous to the laterally discontinuous nature of mineralization seen in the J-M Reef. Below this zone, discordant sulfide-bearing pods and pipes occur to a depth of 150 m in the anorthosite (Figure 2.14).

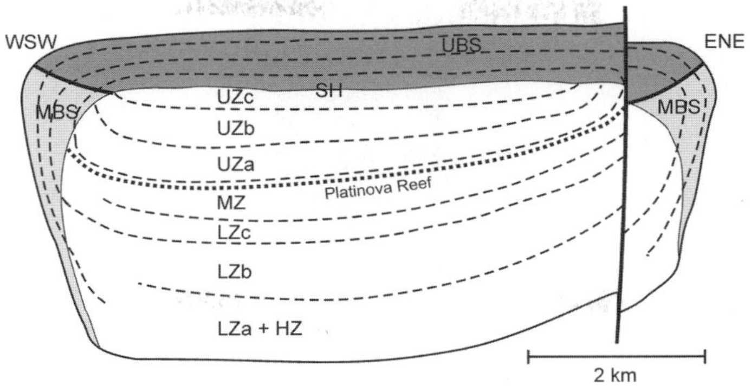

Figure 2.15 Cross-section through Skaergaard intrusion (no vertical exaggeration). Stratigraphic abbreviations: UBS = Upper Border Series; MBS = Marginal Border Series; HZ = Hidden Zone; LZa, b and c = Lower Zones a, b and c; MZ = Middle Zone; UZa, b and c = Upper Zones a, b and c; SH = Sandwich Horizon. The Platinova Pd-Au reef is shown as the dashed line near the top of the Middle Zone. Modified from Nielsen (2004), with permission.

2.4.3 Skaergaard Intrusion

The Skaergaard intrusion is located in East Greenland, and is part of the North Atlantic Large Igneous Province. Although magmatism of the East Greenland rifted margin spanned tens of millions of years from 61 to 13 Ma, the vast majority erupted as an enormous flood basalt succession at ~55 Ma coinciding with continental rupture between Greenland and Eurasia (e.g., Pedersen et al., 1997). The Skaergaard intrusion was emplaced partly in the lower part of these flood basalts at a relatively shallow emplacement depth of ~2 km (e.g., Nielsen, 2004).

The intrusion is traditionally divided into three series (Figure 2.15): the Layered Series (LS; ~2,500 m thick), the Marginal Border Series (MBS; variable widths) and the Upper Border Series (UBS; ~1,000 m thick). The Layered Series accumulated up from the floor, the Upper Border Series down from the roof and the MBS inward from the sidewalls of magma chamber. The Layered Series and the Upper Border Series meet at ~700 m below the roof of the intrusion in the Sandwich Horizon (SH). The crystallization of the rocks occurred within a closed system, from a single, homogenous batch of magma (e.g., Wager et al., 1957; Wager, 1960; McBirney, 1996).

Platinova Reef. The Platinova Reef occurs in the uppermost 100 m of the Middle Zone and is characterized by a Pd-Au-Cu-dominant, Ni- Pt-poor disseminated sulfide assemblage (Figure 2.16). The stratabound reef is associated with a group of three ~15–20 m thick prominent layering units known as the Triple Group, which occurs at the uppermost 100 m of the Middle Zone in the layered series (Wager and Deer, 1939), and the mineralization appears to be concordant with the modal layering (Andersen, 2006).

Models for the formation of the Platinova Reef range from rather complex orthomagmatic models to those that envision a role for fluids. For example, Nielsen et al. (2015)

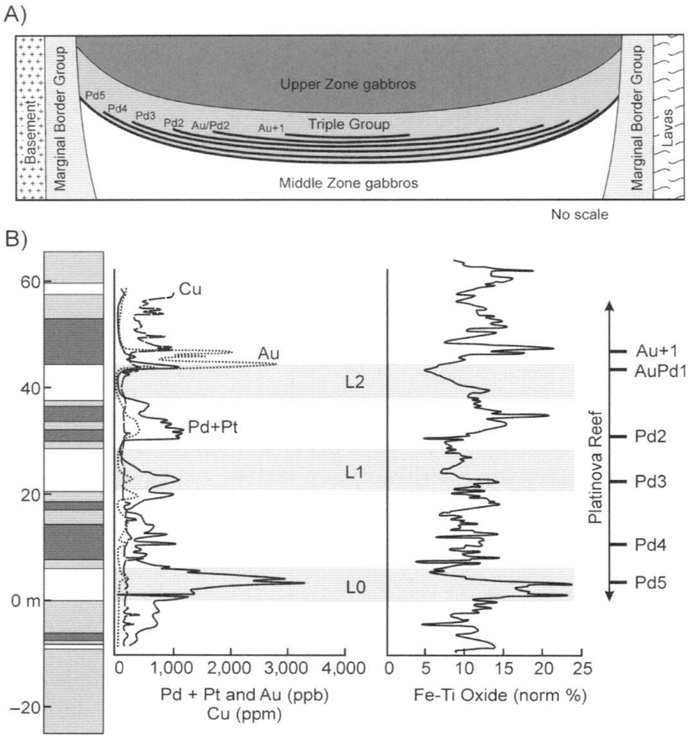

Figure 2.16 The Platinova Reef, Skaergaard intrusion. (a) Cartoon of the intrusion-scale structure of the mineralization. Gold is concentrated at the uppermost layer and along the upturned margins of the other layers, forming a stack of gold-rimmed plates of decreasing size. (b) Detail of the stratigraphy and mineralization of the Platinova Reef from the central portion of the Skaergaard intrusion. Shading in the column on the left is a function of increasing normative mafic mineral abundance, ranging from leucogabbro (white) to melanogabbro (dark grey). Metal concentrations are shown for Pd + Pt (solid line) Au (dotted line) and Cu (dashed line). The normative Fe-Ti oxide % is shown in a separate column. Various ore metal peaks are labelled Au+1, Au and Pd1 to Pd5. The three prominent leucogabbro layers are labelled L0, L1 and L2. Note the concentration of Au and Cu towards the top, and the general decrease in Pd going up section. Redrawn after Nielsen et al. (2015), with permission.

proposed a complex model involving the gravitational separation of immiscible silicate liquids, one Fe-rich the other Si-rich, as well as segregation of immiscible sulfide liquid preferentially redissolving in the layers enriched in the Fe-rich silicate liquid. The hydromagmatic models envision either metal precipitation from upward moving fluid at different redox fronts represented by the variable oxide distribution (Andersen, 2006) and those that have suggested that the stratigraphic metal distribution was the result of the fluid-driven chromatographic separation of metals (Boudreau and Meurer, 1999).

The Skaergaard intrusion is unusual in a number of respects. One is the relatively S-poor nature of the intrusion; even the Platinova Reef has S-concentrations lower than expected for a sulfide-saturated magma (Wager et al., 1957; Andersen, 2006; Keays and Tegner, 2016). Related to this is the distinctly Cu-rich sulfide assemblage, being composed mainly of chalcopyrite ($CuFeS_2$), bornite (Cu_5FeS_4) and locally native Cu that are rarely observed as significant components of the disseminated sulfide assemblages in other layered intrusions (e.g., Aird et al., 2017). Keays and Tegner (2016) estimate the initial Skaergaard magma contained 240 ppm Cu, 89 ppm S, and 90 ppb Se with an S/Se ratio of 937–1,180 being much lower than average chrondritic and presumed mantle values of 2,560 $\pm$ 150 (Dreibus et al., 1995). Although other estimates based on Skaergaard chilled margins, Skaergaard-associated dikes and mass balance considerations can vary widely (see summary in Keays and Tegner, 2016), the S concentration overall is anomalously low compared with mid-ocean ridge, ocean island and flood basalts (800–1,500; e.g., Wallace and Anderson, 1998; Gerlach and Graber, 1985; Self et al., 2008; Keays and Lightfoot, 2010).

There have been a number of proposed explanations for the unusually low Skaergaard S-concentrations. Andersen (2006) suggested extensive late magmatic S losses within and below the Platinova Reef, and Li and Boudreau (2017) showed that sulfide-oxide mode variations, particularly in the Lower Zone, were consistent with a $f(O_2)$-neutral sulfur loss by the following reaction:

$$\underset{\textit{chalcopyrite}}{3.75CuFeS_2} + \underset{\textit{vapour}}{4.33H_2O} \rightarrow \underset{\textit{magnetite}}{Fe_3O_4} + \underset{\textit{bornite}}{0.75Cu_5FeS_4} + \underbrace{4.33H_2S + 0.16SO_2}_{\textit{fluid}} \tag{2.4}$$

For this reaction, all metals are conserved and only S is lost. They used the Kilauea system as a model to suggest that the Skaergaard magma chamber lost significant S to low pressure degassing to high level, low P magma chambers and long-lived lava lakes. In contrast, Keays and Tegner (2016) suggested that the high Cu/S ratio reflected anomalous low S-concentrations in the mantle source and the influence of volatiles was minor.

3

Magmatic Volatiles and Fluids

3.1 Introduction

The presence of hydrous minerals, graphite, high temperature carbonates and even fluid inclusions are evidence of the contribution of volatiles and volatile fluids to the crystallization history of layered intrusions. The problem is that volatile fluids are fugitive, and except for the rare fluid inclusion they may leave little evidence of their fleeting presence. The role of igneous fluids can be further complicated by the influx of country fluids or the assimilation of volatile-bearing country rocks. Nonetheless, magmatic volatiles play an essential role in magmatic systems from initiating melting and controlling melt migration in the mantle (e.g., Keller and Katz, 2016; Keller et al., 2017) to determining the style of eruption at the surface. They can also play a significant role in the physical properties of magmas, from their effect on viscosity to defining liquidus and solidus relationships and in defining the *liquid line of descent*, the order in which minerals crystallize from a magma. Finally, igneous hydrothermal fluids can transport significant concentrations of economically important ore elements.

3.2 Volatile Components of Magmas

The major volatile species of interest in basaltic magmas include H_2O, CO_2, S and the two halogens, Cl and F. Depending on the initial magma composition, fluids that exsolve from a magma can range from H_2O- to CO_2-rich, a hydrosaline (Cl-rich) liquid or brine, to an immiscible sulfide liquid. As noted previously, the terms *fluid* or *vapour* is used here to denote any volatile-rich fluid phase (ranging from a low density vapour to a high density saline fluid or aqueous fluid), *liquid* for a silicate-rich liquid or magma and *immiscible sulfide liquid* for an S-rich liquid that separates from the silicate liquid. The term *solubility* refers to the concentration of a volatile component at which the silicate liquid is saturated in one or more fluid or sulfide phases

3.2.1 H_2O and CO_2

As summarized by Wallace et al. (2015) concentrations of H_2O range from <0.4 wt. in normal mid-ocean ridge basalts (*N-MORB*) to as high as 1.5 wt.% in enriched MORB (*E-MORB*) magmas and up to 2.5 wt.% in rare mid-ocean andesites and dacites. In general

H_2O concentrations correlate positively with other non-volatile incompatible elements such as K_2O. Ocean Island 'hot spot' basalts generally overlap with MORB but extend to higher values similar to that seen in E-MORB. In contrast, CO_2 concentrations in all variety of MORB are much lower than H_2O concentrations, ranging from 0.005 to 0.04 wt.%. Although the liquid concentrations of CO_2 are low, quenched MORB glasses can have vapour bubbles containing >90% CO_2, implying the magma was saturated in CO_2. Both H_2O and CO_2 are higher in arc magmas and boninites owing to the role of fluids in the generation of these magmas in the mantle source regions.

The general solubility behaviour of H_2O and CO_2 are illustrated in Figure 3.1 for a Bushveld parent magma. The solubility of H_2O is strongly a function of pressure otherwise is similar for liquid compositions ranging from basalt to rhyolites and over a wide range in temperature. A similar pressure dependence also occurs for CO_2, but its solubility is 1–2 orders of magnitude times less soluble than is H_2O. The solubility of H_2O is decreased by the presence of CO_2 for mixed H_2O-CO_2 systems, and the amount of CO_2 in the fluid is high relative to the concentrations of H_2O and CO_2 in the basaltic liquid. An example is shown in Figure 3.1b; at point 'A', a basaltic liquid containing 1.9 wt.% H_2O and 0.13 wt.% CO_2 will have a relative mole fraction, $X_{CO_2}^{liquid} = \frac{moles_{CO_2}^{liquid}}{moles_{CO_2}^{liquid} + moles_{H_2O}^{liquid}} = 0.027$ will be saturated in a fluid at 2 kbar pressure in which the mole fraction of CO_2 in the fluid, $X_{CO_2}^{fluid} = 0.6$.

Owing to its low molecular weight and high solubility, H_2O can comprise a high mole fraction of the silicate liquid components even at low wt.% concentrations, and thus can have a strong effect on the physical properties of the magma, affecting the liquidus and solidus, the density, and the viscosity (the latter by disruption of the silica polymer network of the magma). In subduction environments H_2O is the driving mechanism for mantle melting (i.e., water acts as a flux in lowering the solidus of the mantle assemblage). High H_2O concentrations will increase element diffusivity in a magma, leading to increased crystal size. While many reviews of H_2O concentrate on its solubility and dissolution

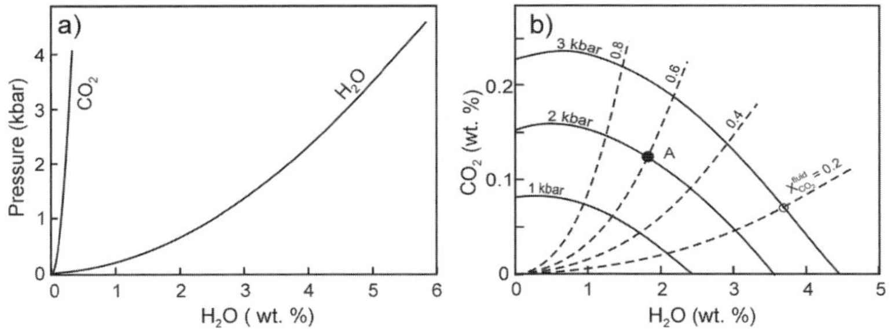

Figure 3.1 (a) Solubility of pure H_2O and CO_2 in a Bushveld parent magma at 1,410 °C (dry liquidus at 2 kbar). (b) Solubility of mixed H_2O-CO_2 fluid (solid lines) in a Bushveld parent magma at 1,410 °C and at 1–3 kbar pressure. Dashed lines are the mole fractions of CO_2 in the vapour as discussed in the text. Both plots calculated using the program Pele (Boudreau, 1999b) for the B1 Bushveld parent liquid composition of Maier et al. (2000).

mechanisms, it is perhaps its effect on mineral stability that is of potentially the most consequential during the crystallization of layered intrusions. As described in more detail in Chapter 9, gain or loss of H_2O can cause substantial change the crystallization sequence during both crystallizing of a magma or during flux melting in the crystal-liquid mush. In contrast, owing to its lower solubility, the effect of CO_2 on silicate liquid properties is much less, particularly at shallow crustal pressures.

3.2.2 Sulfur

The solubility of sulfur in a silicate liquid is complicated by the fact that sulfur can exists in two common oxidation state, S^{+6} and S^{-2}, the former preferred in more oxidized magmas. Further, a magma can become saturated in a sulfide liquid, sulfide minerals or sulfate minerals (e.g., anhydrite) as well as an S-bearing volatile fluid phase composed of H_2S, SO_2 and S_2, again generally a function of the oxidation state. In general, however, basaltic liquids tend to saturate in S-bearing fluids and/or immiscible sulfide liquid. Indeed it is the possibility of sulfide saturation form a magma or transport of S by hydrothermal fluids that is the heart of the debate over the origin of PGE deposits in layered intrusions.

Sulfur solubility between silicate liquids and various S-bearing phase has been the subject of numerous works and reviews (c.f., Behrens and Webster, 2011, for a good collection of review articles). Sulfur concentration in MORB generally range from 0.08–0.14 wt.%, and show a positively correlation with bulk total Fe concentration. In contrast, some ocean island basalts can contain up to 0.3 wt.% S without significant Fe enrichment. This is attributed to a higher f_{O_2} and its effect on destabilizing an immiscible sulfide liquid and subsequent higher S concentration in the silicate liquid required to reach saturation in an immiscible sulfide liquid (Wallace et al., 2015).

For basaltic systems, much of the early work was concerned with the understanding the controls on the concentration of S in the silicate liquid at the point the liquid becomes saturated in, typically, an immiscible sulfide liquid. This saturation concentration is termed the *sulfur concentration at sulfide saturation*, or the *SCSS*. Early experimental work such as that by Haughton et al. (1974) showed that the SCSS was strong positive function of the total iron content of the silicate liquid. More recent work has included more complete silicate liquid composition control. An example is the model for the SCSS presented by Liu at al. (2007):

$$\ln S_{SCSS}^{liquid} = 11.35 - \frac{4,454.6}{T} - 0.0319\frac{P}{T} + 0.71\ln(MFM) - 1.98(MFM)X_{H_2O}^{liquid} + 0.219\ln X_{H_2O}^{liquid} + 0.362\ln X_{FeO}^{liquid} \qquad (3.1)$$

Where $\ln S_{SCSS}^{liquid}$ is the sulfur concentration in the silicate liquid at sulfide saturation, T is in °C, P is pressure in bars, $X_{H_2O}^{liquid}$ and X_{FeO}^{liquid} are the mole fraction of H_2O and FeO in the silicate liquid respectively, and MFM is a modified FM parameter of Ryerson and Watson (1987) that accounts for all the other silicate liquid effects of the SCSS:

$$MFM = \frac{Na + K + 2(Ca + Mg + Fe^{2+})}{Si(Al + Fe^{3+})} \qquad (3.2)$$

Where the element symbol represent the mole fractions of the anion. It is claimed to have a precision of about 10% for $\log f(O_2) = FMQ \pm 2$ and for pressure range from 1 atm to 9.0 GPa (90 kbar). Note the positive temperature and negative pressure dependence of the SCSS; a magma saturated at depth will not be sulfide saturated as it moves to the surface if cooling is negligible. Similar equations include those of Li and Ripley (2005, 2009). Smythe et al. (2016) have expanded this to include the effect of the composition of the immiscible Fe-Ni-Cu-S sulfide phase.

Partitioning of S between vapour and a silicate liquid is non-trivial owing to the role of liquid composition and changing $f(O_2)$on sulfur speciation (S^{2-} and S^{6+}, and the gas species H_2S and SO_2) as well as the difficulty of undertaking partitioning studies at temperatures appropriate to basaltic systems. As summarized by Webster and Botcharnikov (2011), sulfur tends to strongly partition into an exsolving fluid at crustal pressure pertinent to most layered intrusions, with experimental values of $D_S^{fluid/liquid} = 40 - 300$. Furthermore, D_S increases with decreasing T, the molar $\frac{S^{6+}}{(S^{6+}+S^{2-})}$ ratio of the silicate liquid (which itself increases with increasing f_{O_2}), and the CaO content of the silicate liquid. In basaltic systems, D_S increases markedly in H_2O-poor silicate liquids (Figure 3.2), and Teague et al. (2008) also note that S has a strong affinity for a CO_2-rich fluids in andesitic liquids. For example, in a study on S partitioning between H_2O-S±Cl fluids and basaltic liquids undergoing decompression degassing, Fiege et al. (2014) found that the MgO, H_2O and Cl content of the initial liquid had no significant effect on D_S, but that a positive

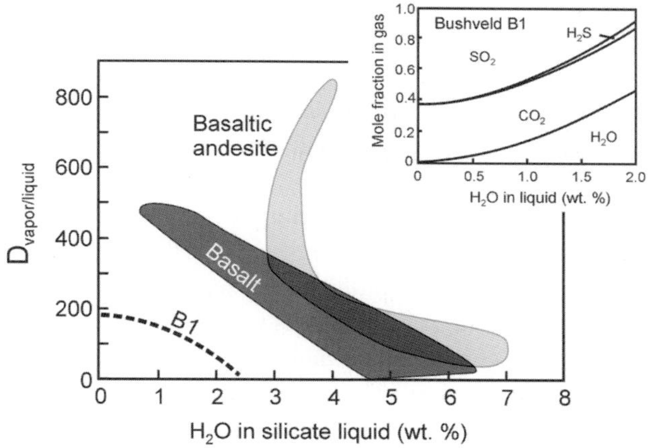

Figure 3.2 Variation of the partition coefficient for S between a volatile fluid and a silicate liquid as a function of the H_2O content of basalt and basaltic andesite at 1,050 °C, 200–300 MPa, and $\log f_{O_2} = FMQ - 0.3$ *to* $FMQ + 0.4$. Redrawn after Webster and Botcharnikov (2011), with permission. The labelled dashed line shows a calculated D and the inset Figure shows the mole fraction of gas species in the vapour in equilibrium with the Bushveld B1 parent magma, also as a function of changing H_2O concentration in the silicate liquid. For the inset Figure, the initial liquid has a fixed concentration of CO_2 = 0.40 wt.% and S = 0.15 wt.%, calculated at 1,410 °C (dry liquidus), 2 kbar pressure, and $\log f(O_2) = FMQ,$ *calculated* with the program PELE. (Boudreau, 1999b)

correlation was observed between D_S and T in the range of 1,150–1,250 °C for oxidizing ($D_S = 52 - 76$ at QFM+4) and intermediate ($D_S = 94 - 209$ at QFM+1.5) redox conditions. This is consistent with the summary of Webster and Botcharnikov (2011) that the presence of Cl does not markedly influence S partitioning.

3.2.3 Halogens

The halogens include the elements F, Cl, Br and I, with Cl and F being the compositionally most abundant in basaltic magmas. F and Cl can reach into the wt.% values, whereas Br and I are generally <10's ppm. In general, however, Cl in uncontaminated and undegassed MORB is generally low and lower than F concentrations (10–50 ppm and 100–600 ppm, respectively). Cl and F concentrations in ocean island basalt overlap and modestly extend the range seen in MORB. Considerably higher concentrations occur in arc-related magmas and in boninites, both of which involve the introduction of fluids into the mantle source regions.

Halogen compositional variations have been used to infer the degassing and migration of vapour in magmas in general and in layered intrusions in particular. Although halogen partitioning in vapour–liquid systems is complex as summarized below, these studies are based on the observation that chlorine is broadly vapourphile and will preferentially fractionate into a vapour phase if present. In contrast, although a volatile element, fluorine is a magmaphile element and prefers to remain with the liquid during vapour loss (c.f summary of Webster and Piccoli, 2015). Thus, variations in the Cl/F ratio of the magma or its proxies (particularly apatite) is a potentially good indicator of degassing in igneous systems.

In H_2O-poor experiments where HCl is the main volatile species, Cl is moderately solubility in silicate liquids, being intermediate between that of H_2O and the relatively insoluble CO_2. Basalts have a much higher solubility than do more siliceous liquids at upper crustal pressures (Figure 3.3). Chlorine partitioning between a silicate liquid and a volatile phase generally favours the vapour, especially in the more silica-rich liquids, but not always. Experimentally determined $D_{Cl}^{fluid/liquid}$ partition coefficients ranging from <1–6 in basalt and andesite to values of ~100 in granitic systems (e.g., Shinohara et al., 1989).

Solubility relationships become more complex in H-O-Cl-silicate liquid systems. For example, the experiments of Stelling et al. (2008) on the solubility of C-O-H fluids in basaltic liquids found that the molar partition coefficient of Cl between one or more volatile-rich fluids (hydrous fluid, saline fluid or both) and a basaltic liquid increases from 0.04 to 14 with increasing concentration of Cl in the liquid, indicating that at low bulk Cl contents (up to about 1 wt.% Cl in the melt) chlorine will preferentially partition into the basaltic liquid whereas at higher Cl concentrations it will enrich the fluid phase(s). This implies that basaltic magmas that reach vapour saturation early (in a low solubility CO_2-rich fluid, for example), Cl can behave as an incompatible element with a bulk distribution coefficient, $\bar{D}_{Cl}^{fluid/liquid} \ll 1.0$, and Cl will become enriched in the residual silicate liquid.

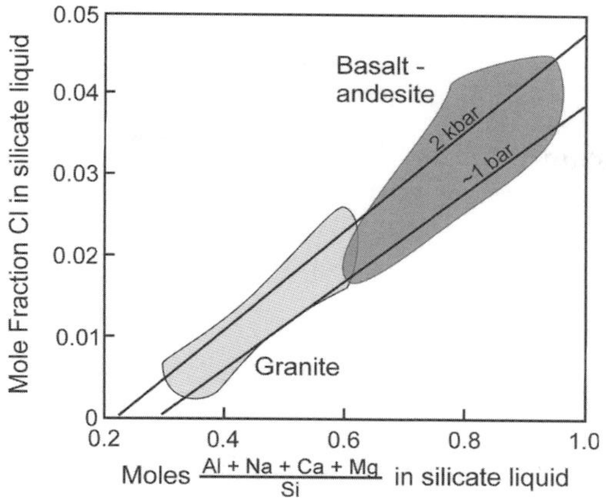

Figure 3.3 Concentration of Cl in H_2O-poor silicate liquids at vapour saturation at ~1 and 2,000 bars, illustrating the strong compositional and weak pressure dependence on Cl solubility. Shaded regions show the experimental data for basalts and andesites (dark grey) and granitic silicate liquids (light grey). Redrawn after Webster et al. (1999), with permission.

In contrast, later fluids will evolve from more H_2O-rich silicate liquids will be considerably more Cl-enriched.

Experimentally it is found that the solubility of H_2O can depends strongly on the silicate liquid Cl content at high but not at low Cl concentrations; similarly, the solubility of Cl depends strongly on melt H_2O at high but not at low H_2O concentrations. Because of this, the solubility curves as a function of H_2O and Cl concentrations in the liquid are notably box-like (Figure 3.4). The volatile phase in equilibrium for a basaltic liquid containing less than about 5 wt.% dissolved H_2O is a brine (the nearly vertical solubility solid line shown in Figure 3.4); at higher H_2O concentrations the volatile phase is H_2O-rich with dissolved Cl (the nearly horizontal solid line of Figure 3.4). As discussed by Webster et al. (2015), the composition of mixed H_2O-HCl-NaCl vapour exsolving from a silicate liquids is complicated by the fact that the Cl anion readily bonds with other cations such as Na to form stable complexes such as NaCl at high temperature.

Furthermore, solvus relationship in the H_2O-NaCl system that can extend to supersolidus temperatures in igneous rocks: in the pure H_2O-NaCl system at 2 kbar the critical point is at ~920°C and 40 wt.% NaCl. As temperature increases (or pressure decreases) the solvus opens between progressively more H_2O- and Cl-rich fluids (brines) (Bodnar et al., 1985; Chou, 1987). This is a consequence of NaCl having both a higher melting temperature and a much lower vapour pressure than does H_2O. At pressures above the solvus, the fluid shows complete miscibility between H_2O-rich to NaCl-rich (brine), as shown by the dashed line of the inset Figure of Figure 3.4. In this case, the fluid saturation boundary as a function of the Cl and H_2O concentration of the magma is a smoothly varying curve as shown by the dashed line in the main graph of Figure 3.4. In this case, the Cl concentration

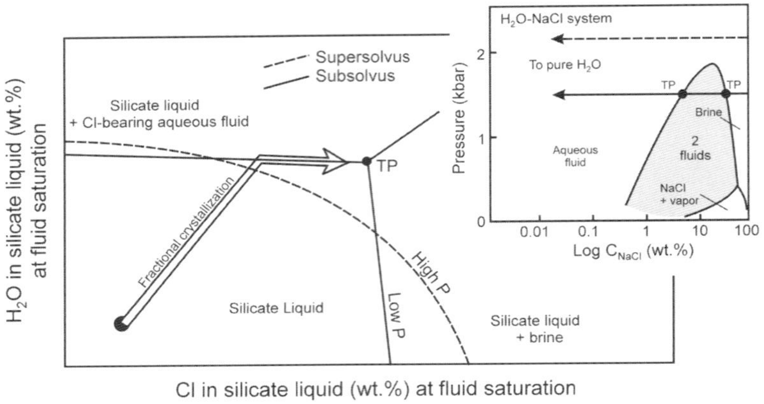

Figure 3.4 Schematic vapour saturation curves as a function of Cl and H_2O concentrations at pressures above (dashed lines) and below (solid lines) the solvus of the H_2O-NaCl systems shown in the inset (solvus region = '2 fluids' region, shaded grey; note log scale on inset graph). TP = triple point for the presence of silicate liquid, an aqueous fluid and a NaCl-rich brine. The arrows illustrate how a magma may evolve to produce a Cl-bearing, H_2O-rich vapour that becomes more Cl-rich by fractional crystallization of H_2O- and Cl-free minerals. Redrawn after Webster et al. (2015), with permission.

in the fluid follows relatively simple portioning behaviour between vapour and liquid (see below). At lower pressures, however, the fluid can enter the solvus region. The box-like saturation surface in Figure 3.4 the result of the exsolved fluids being either being brines present on one side of the solvus region or much more dilute Cl-bearing aqueous fluids on the other side.

3.3 Mineral Solubility Considerations

Volatile fluids are more than mixtures of gas species - they can contain abundant dissolved solids as well, the amount of which can be a strong function of the chlorine concentration. An example is illustrated by the solubility of SiO_2, which increases rapidly as a function of T and P in the SiO_2-H_2O system (Figure 3.5a). It is the cooling of hydrothermal fluids that leads to the extensive precipitation of quartz in vein assemblages. In a layered intrusion, the upward migration of fluid in the mush column into hotter rock will result in a fluid becoming increasingly undersaturated in silica and it will react with the mineral assemblage to produce a relatively silica-poor mineral assemblage seen in a number of discordant bodies described in Chapter 8. Furthermore, at low pressures, the presence of modest amounts of NaCl can enhance the solubility of SiO_2, but decreases the solubility relative to the SiO_2-H_2O system at higher NaCl concentrations and at pressures above about 2 kb (Figure 3.5b). In SiO_2-H_2O-CO_2 systems at 800 °C and 10 kb, silica solubility is even lower, such that SiO_2 will partition very strongly into a concentrated salt solution in preference to a coexisting immiscible CO_2-rich fluid (Newton and Manning, 2000). In this regard, the observation of that most replacement bodies involved replacement by

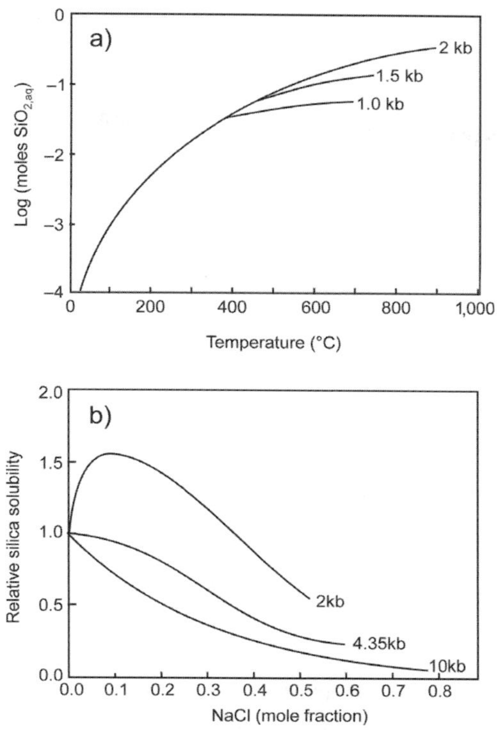

Figure 3.5 Silica solubility curves. (a) Solubility of quartz in the H_2O-SiO_2 system as a function of temperature and pressure. Redrawn after Manning (1994), with permission. (b) Silica solubility as a function of fluid NaCl concentration relative to that in SiO_2-H_2O system at the same pressure and temperature conditions. NaCl enhances solubility at 2 kbar in modest NaCl concentration solutions, whereas silica solubility decreases with increasing pressure for equivalent NaCl concentrations. Redrawn after Newton and Manning (2000), with permission.

silica-poor minerals (e.g., olivine ± chromite replacing pyroxene or intermediate plagioclase replaced by a more anorthitic plagioclase) described in Chapter 8 and 12 is consistent with this. Fluid reactions at relatively low upper crustal pressures with modest concentrations of NaCl would be most favourable to this replacement.

Given its potentially high concentration in the fluid phase, Cl is an important complexing agent and can enhance the solubility of certain elements considerably. Yardely (2005) notes that metal concentrations in saline solutions increase with both temperature and chlorine concentration. For example, fluid inclusion evidence suggest that fluids can carry in excess of 10 wt.% Fe at 700 °C and presumably even higher amounts at higher temperatures (Figure 3.6). These Cl-rich fluids can become enriched in Fe, Mg and Ca by leaching these elements from the host rock, extensive evidence for which it presented in Chapter 8.

Holland (1972) noted that, for many metals, their concentration in a fluid in equilibrium with a silicate liquid is an exponential function of the Cl concentrations. As a first-order approximation, Holland considered the partitioning of chlorine and a divalent metal cation

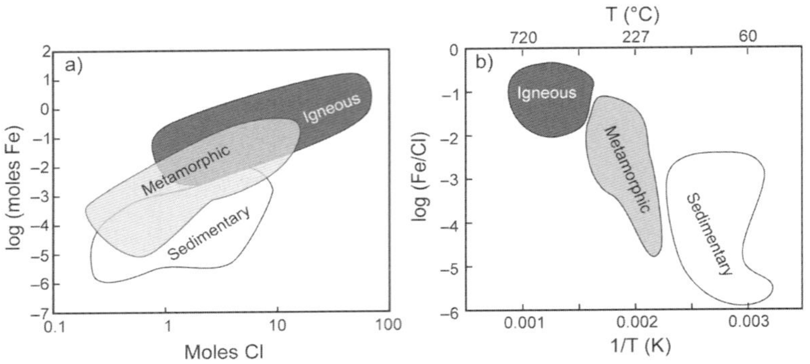

Figure 3.6 Plots of Fe–Cl relationships in fluid inclusion brines from sedimentary (white field), metamorphic (light grey) and igneous rocks (dark grey). Redrawn after Yardley (2005), with permission.

such as Mg^{+2} between a silicate liquid and a Cl-bearing volatile fluid. In the case where chlorine partitions into the fluid phase with NaCl as a dominant aqueous chloride complex, NaCl will control the fluid-liquid exchange of minor species involving trace metals. A plausible exchange reaction for Mg between fluid and silicate liquid could be written as follows:

$$\underset{liquid}{Mg^{+2}} + \underset{fluid}{2NaCl} = \underset{fluid}{MgCl_2} + \underset{liquid}{2Na^+} \tag{3.3}$$

For this exchange reaction, the equilibrium constant can be expressed as:

$$K = \frac{a^{fluid}_{MgCl_1} \cdot \left(a^{liquid}_{Na^+}\right)^2}{a^{liquid}_{Mg^{+2}} \cdot \left(a^{fluid}_{NaCl}\right)^2} \tag{3.4}$$

For the case in which the Na content of liquid is constant, the concentration of Mg in the fluid (as the neutral species $MgCl_2$) will be proportional to the square of the Cl concentration in the fluid (as NaCl), if all other activities remain constant. A doubling of the chlorine content of the fluid means that the fluid will be able to contain four times more Mg. Boudreau and McCallum (1992) suggested that a similar solubility relationship might explain why high grade PGE deposits are associated with Cl-rich intrusions described in Chapter 6.

The enhanced solubility of Mg and Ca in Cl-bearing solutions along with the presence of CO_2 leads to the possibility for the precipitation of high temperature carbonate minerals (e.g., Aird and Boudreau, 2013). A phase diagram for the calcite-H_2O-NaCl system is shown in Figure 3.7. A calcite-saturated fluid starting at point 'A' would begin with a relatively low NaCl content. On cooling and precipitating calcite the fluid would eventually intersect the two-fluid field at the point F_1. The fluid would then split into two fluids, one a carbonatitic liquid (F_1) and the other a NaCl-rich aqueous solution (F_2). The former would become slightly more carbonate-rich on further cooling while the aqueous solution

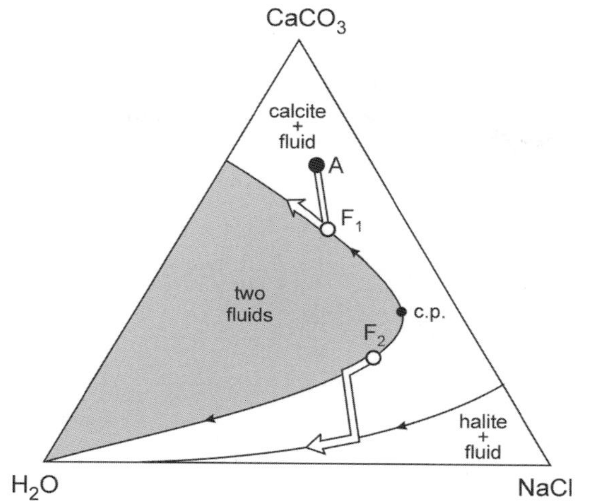

Figure 3.7 The calcite-H_2O-NaCl system, showing a possible crystallization path for a Bushveld and Stillwater carbonatitic liquid/fluid. Two fluid region shown shaded. See text for details. Modified after Newton and Manning (2002), with permission.

would become less saline and less carbonatitic. These two fluids would continue to precipitate calcite until the carbonate fluid is used up. Once the carbonatitic fluid is exhausted, the remaining aqueous solution continues to precipitate calcite until the fluid reaches halite saturation, at which point the fluid precipitates both calcite and halite and continues to become richer in H_2O.

Azimov and Bushmin (2007) have summarized the solubility behaviour of minerals in an aqueous fluid as a function pressure, temperature and the HCl-NaOH (acid-alkalinity or pH) conditions. They have suggested that mineral solubility can be classified into one of six major types (Figure 3.8):

Type 1. This includes quartz and alkali-rich minerals such as K feldspars. These have generally high solubility and are largely independent of the HCl-NaOH composition of the fluid. The solubility generally increasing with temperature.

Type 2. This type is characteristic of Al-rich minerals (muscovite and the aluminosilicate polymorphs of Al_2SiO_5) and Fe^{3+} minerals such as hematite. These generally have low solubility in acidic to pure water and slightly increases in more alkaline solutions. The solubility is modest at low pressure (~1 kbar) but generally increasing with temperature at higher pressures. This mineral solubility type is perhaps least important in layered intrusions as these minerals are not common minerals in mafic rocks.

Type 3. This is characteristic of minerals rich in the amphoteric elements (e.g., Al and Fe; elements able to react both as an acid and a base). Common mafic igneous minerals include annite (Fe-biotite), fayalite, ferrosilite, and Fe- and Al-rich spinel (e.g., magnetite). As expected from their amphoteric natures, they are least soluble in pure H_2O. Solubility

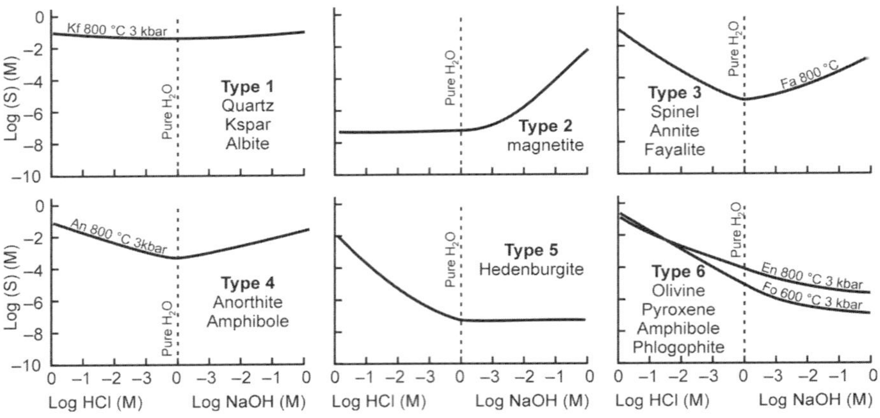

Figure 3.8 Six types of mineral solubility behaviour as a function of the HCl-NaOH (acidic - basic) composition of the aqueous fluid. Heavy lines labelled with a mineral endmember, temperature and pressure are examples, whereas those that are not labelled are generalized. Abbreviations Kf = K feldspar; Fa = fayalite; An = anorthite, En = enstatite; Fo = forsterite. See text for discussion. Redrawn after Azimov and Bushmin (2007), with permission.

need not be a strong function of temperature, and can increase or decrease depending on the mineral; for fayalite, solubility decreases with increasing pressure but is only a weak function of temperature.

Type 4. This type is characteristic of Al- or Fe-rich silicates of Ca and Mg. The overall solubility range is a more modest expression of the Type 3. The common igneous minerals would include anorthite and hornblende. At low pressures (1 kbar) the solubility of anorthite in pure H_2O slightly decreases with increasing temperature, and it is about an order of magnitude more soluble in acid as compared with alkali solutions.

Type 5. This group is again Al- and Fe-rich with high Ca and Mg contents, and includes hedenburgite and some amphiboles.

Type 6. This group includes Al- and Fe-poor silicate minerals such as forsterite, enstatite, diopside, Mg-biotite (phlogophite), Na-bearing magnesian amphiboles (e.g., pargasite). This minerals show a generally monotonic decrease in solubility with increasing alkalinity of the solution.

3.4 Volatile-Rich Silicate Liquids, Solute-Rich Volatile Fluids and Transitional Fluids

As illustrated by the high Fe concentration observed in brine fluid inclusions (Figure 3.6), at high temperatures and pressures a Cl- and solute-rich fluid can begin to compositionally approach that of an equilibrium volatile-rich silicate liquid. For solid assemblages that have re-equilibrated on cooling with either one, it can be difficult to establish the nature of the metasomatic agent absent fluid or melt inclusion evidence.

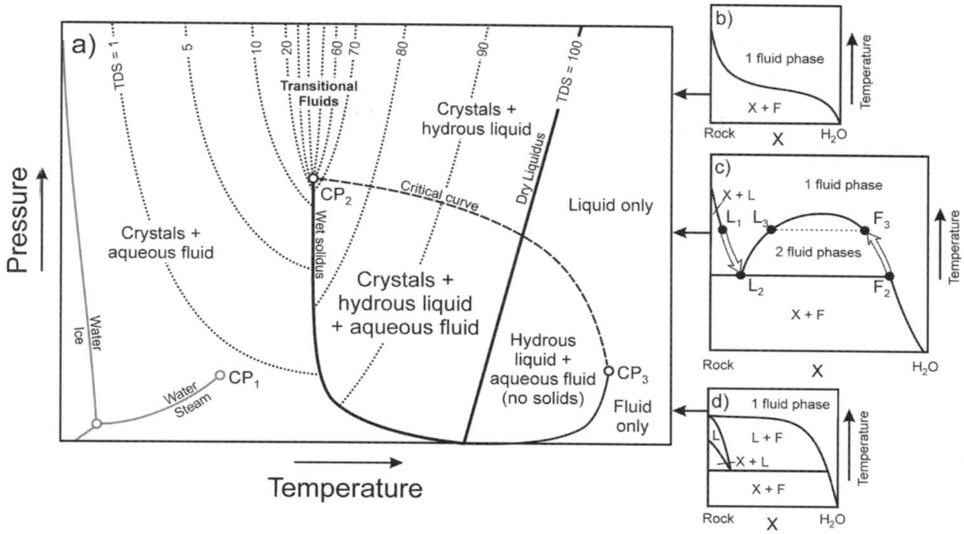

Figure 3.9 Fluid behaviour in igneous systems. (a) A generalized P-T diagram illustrating the phase relationships in a rock-H_2O binary system. Also shown as the grey lines on the left-hand side are the phase boundaries for the one component H_2O system (ice polymorphs are ignored). (b)–(d) Simple isobaric $T\text{-}X_{H_2O}$ sections illustrating phase relations at different pressures. Abbreviations: TDS = total dissolved solids in the fluid phase (broadly define from silicate liquid to aqueous fluid), L = hydrous silicate liquid, X = crystals, F = aqueous fluid. See text for additional discussion. Redrawn after Hermann et al. (2006).

An example of the complications is illustrated in Figure 3.9, which shows a generalized P-T phase diagram representing a simple binary silicate – H_2O system (modified after Hermann, 2006). Also shown as the grey phase boundaries on the left of Figure 3.9a are the phase boundaries for the one component H_2O system. The system as shown has three critical points that mark the merging of two phases into one. The first critical point (labelled CP_1 on the diagram) is the critical point of water along the equilibrium boundary between water and steam, also known as the *steam curve*. As temperature increases along the steam curve, the molar volume of water increases, whereas conversely the increase in the pressure along the steam curve leads to the molar volume of steam to decrease. The molar volume of both phases become equal at the critical point CP_1 and only one *supercritical fluid* is present above this point.

For rock-fluid systems for which H_2O is the only volatile component, the amount of total dissolved solids in the volatile fluid phase increases with both temperature and pressure (doted lines of Figure 3.9a). For a silicate liquid, the amount of dissolved H_2O will increase with increasing pressure. Along the wet solidus (labelled 'wet solidus' in Figure 3.9a), one has a fluid in equilibrium with a hydrous silicate liquid, and both are equilibrium with the solid assemblage. As one moves to higher pressures along the wet solidus, the amount of H_2O dissolved in the silicate liquid increases and amount of solids dissolved in the fluid both increase. At the second critical point (CP_2 of Figure 3.9a) the amount of solids dissolved in the hydrous silicate liquid and the volatile fluid are equal and one has a single volatile-rich fluid phase. Note that at pressures above CP_2, the idea of an igneous solidus in the conventional sense does not exists. Instead, there is a smooth transition from crystals in equilibrium with a H_2O-rich silicate liquid (hydrous magma) at high temperature to a more solute-rich volatile fluid at lower temperature. Hermann et al. (2006) suggested that one use the terms *aqueous fluid* for an H_2O-rich fluid phase with $\leq$30 wt.% dissolved solids, *hydrous melt* (or preferred here *hydrous liquid*) for a fluid phase with $\geq$65 wt.% dissolved solids, and *transitional fluid* for those in between.

At lower pressures and at very high temperatures well above the dry liquids curve (labelled 'dry liquidus' in Figure 3.9a), the system can consist entirely of vapour a situation not likely to be commonly encountered except in the immediate aftermath of a meteorite strike. The boundary between all vapour and vapour + silicate liquid field follows a similar physical behaviour as does pure H_2O along the stream curve to form yet another critical point at CP_3. The labelled critical curve between CP_3 and CP_2 defines regions in which there is either a single solute-rich, volatile fluid and a region where two fluids phases (a hydrous liquid and an aqueous fluid) are possible. Both regions may have crystals at temperatures below the dry liquids.

For systems for which H_2O is the main volatile component, CP_2 is estimated to occur only at lower crustal-upper mantle pressures. Cl-rich silicate liquids become fluid-saturated at lower concentrations than do H_2O-rich liquids. Thus it is unlikely that true critical point behaviour between silicate liquid and volatile fluids at shallow crustal pressures. However, the presence of a strong complexing agent such as Cl can markedly increase the solubility of silicate rock components as noted above. Given the Cl-rich nature of the Bushveld and Stillwater systems discussed in Chapter 6, it is not inconceivable that the initial fluid that separates might be considered to be a solute-rich transitional fluid. These solute-rich fluids would be particularly effective in producing the metasomatic features described in Chapters 8 and 12.

3.5 Application to Layered Intrusions

From the summary above, it is clear that knowledge of the volatile behaviour in igneous systems is increasingly well-understood. However, there are several complications that limit these models to understanding the role of volatiles in the crystallization of layered intrusions. These include the uncertainty of the volatile concentration of the parent magmas, the role of country fluids and the complex fluid evolution as it migrates through a crystal-liquid mush. Although much is uncertain, some generalities can be made

3.5.1 Volatile Concentrations in Parent Liquids

Lacking melt inclusions that could have trapped samples of the parent magma prior to degassing, one must look for suitable proxies to estimate volatile concentrations in parent magmas. In some cases, this is rather straightforward. For example, one might assume the Skaergaard parent magma had a volatile concentration similar to the Iceland hot spot magmas. This becomes less tenable when one goes back into the Proterozoic and Archean, and even more so if the magmas underwent extensive contamination prior to emplacement.

For example, Mungall (2016) made the following inferences in regard to the volatile content of the Bushveld parent magma. He assumed that the initial mantle melt was a komatiitic magma derived from a depleted mantle source similar to the MORB source. Assuming typical MORB ratios H_2O/Ce = 170, CO_2/Nb = 530, K/Cl = 13 and assuming that a typical Al-undepleted komatiite contained 0.33 ppm Nb, 0.09% K2O and 0.7 ppm Ce, Mungall estimated the komatiite mantle melt contained approximately 175 ppm CO_2, 119 ppm H_2O and 57 ppm Cl. Complicating this, however, is that this initial mantle melt is estimated by some to have assimilated approximately 40% by weight of lower crustal rock. Using estimated lower crustal concentrations of 0.5% CO_2, 1.5% H_2O and 200 ppm Cl, this assimilation would produce a Bushveld parent magma containing 0.6% H_2O, 0.2% CO_2 and 114 ppm Cl. Mungall notes that the average Bushveld B1 marginal sill composition contains about 400 ppm Cl, which he suggested was an upper limit. Owing to the high concentration of CO_2, this magma would be vapour saturated with a CO_2-rich vapour, resulting in about half of the original CO_2 lost to degassing during emplacement of the parent magma at the crustal level of the Bushveld Complex but with only a minimal effect on the H_2O or Cl budget of the magma. His models also assumes that significant amount of Cl is the result of assimilation, but the Cl isotopes of the Bushveld show typical mantle, not crustal, values (e.g., Willmore et al., 2002).

For the Stillwater and Bushveld Complexes, analysis of sill-dike rocks and other proposed parent liquids (McCallum, 1996) can be described as light REE-enriched, Ti-poor, siliceous high magnesium basalts (SHMB), of which boninites would be the modern day equivalent (e.g., Sun et al., 1989). Boninites crystallize abundant pyroxene, have late plagioclase saturation, and can have olivine and chromite as phenocrysts (e.g., Crawford et al., 1989), suggesting their volatile concentrations would be similar as well. Boudreau et al. (1997) suggested that the second stage mantle hydration melting event postulated for boninite formation would also add Cl at the source region and could explain the unusually high Cl of the Stillwater magma.

3.5.2 A Country Fluid Component?

For many intrusions, the country rocks are relatively dry crustal granitoid or gneissic rocks in which fluid generation by heating is assumed to be minimal. (Fluids that may invade during any later, post-solidification metamorphic events are ignored). However, for large sills that intruded unmetamorphosed or only weakly altered sedimentary sequences, fluid generation during heating of the country rock could be substantial. Typical crustal rocks can lose 3–6 wt.% H_2O to dehydration during metamorphism (Connolly, 2010). The plot shown in Figure 3.10 illustrates the amount of H_2O that remains as a typical pelite (shale) protolith undergoes heating. For shallow intrusions, most of the H_2O in a pelite is lost during the early stages of metamorphism, up to ~400 °C. However, even a small amount of

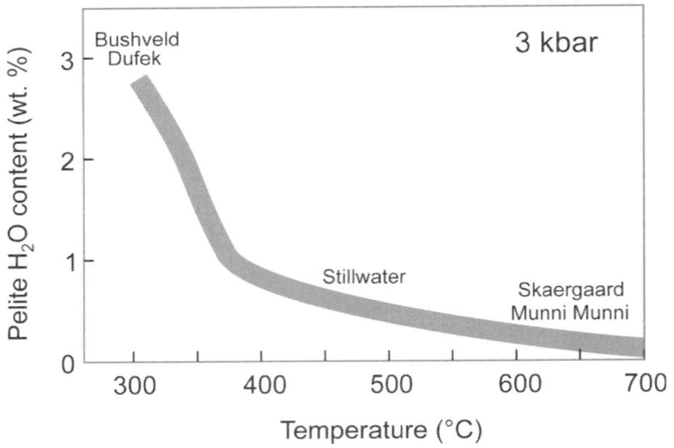

Figure 3.10 Amount of H_2O remaining in a metapelite as a function of temperature at 3 kbar pressure, along with the approximate degree of metamorphism in the country rocks of selected intrusions (labeled). Generalized after Connolly (2010), with permission.

country fluid infiltrating into the base of a sill may prime the pump by making the initial generation and movement of bubbles and fluids from the crystal mush easier.

For those intrusions that intruded previously unaltered crustal rocks, country fluids may play a much more significant role. Boudreau (2016c) have suggested that the hydrothermal systems defined by large sills and their underlying country rock is thus an ideal proxy for understanding fluid movement and interaction in subduction zones. In the latter, it is the movement of fluids from dehydrating a descending lithospheric slab into hotter ultramafic mantle rocks is the preferred mechanism driving subduction zone melting and arc magmatism. In the subduction case, the slab fluids can largely retain their geochemical and isotopic characteristics as they migrate into the hotter parts of the mantle wedge to induce hydration melting (e.g., Hawkesworth et al., 1993; Kimura and Yoshida, 2006).

The Bushveld Complex is perhaps the best example of this. The Bushveld magma intruded the Proterozoic Pretoria Group, Transvaal sequence rocks composed dominantly of shale, quartzite, evaporites and lava filling the Transvaal paleobasin (to 2.5 Ga in age) (Eriksson et al., 1993, 1995). These rocks underwent very little deformation or thermal metamorphism prior to the emplacement of the Bushveld magmas. Once intruded, however, the heating of the country rock formed a metamorphic aureole that extends at least 3.5 km beneath the base of the complex. Heating of the country rock was intense enough to produce localized melting near the base of the complex (Harris et al., 2003). In Chapter 13 it is suggested that some of the isotopic variation and disequilibrium seen in the Bushveld Complex may be due to the influx of country fluids

This situation of a thick magma intruding into an unaltered sedimentary sequence is rare. For example, the Stillwater Complex is similar in size and mineralogy to the Bushveld Complex and has a metamorphic aureole several km wide at its base in which the rocks closest to the complex have undergone extensive dehydration to a pyroxene-cordierite hornfels. However, the country rocks farthest from the contact with the Stillwater Complex are composed, in part, of a biotite schist, interpreted by Labotka and Kath (2001) to have formed during an earlier low-pressure regional metamorphic event (525–550 °C and 2 kbar) that affected a volcaniclastic shale, greywacke and breccia, quartzite and banded iron formation protolith. Comparing the H_2O contents of pelite as a function of temperature (Figure 3.10), a biotite schist would have but a fraction of the H_2O content of a lower temperature shale or slate intruded by the Bushveld magma.

The only other large sill that is potentially similar is the nominally 8–9 km thick, 182 Ga age Dufek intrusion of Antarctica that was emplaced into Paleozoic sedimentary sequence. However, much of the intrusion itself covered by ice, the lower sections and the underlying country rock in particular are not well exposed (e.g., Ferris et al., 1998). Furthermore, magnetic constraints suggest that the Dufek massif is a separate intrusion from the igneous rocks exposed in the Forrestal Range, the latter generally previously interpreted to be stratigraphically above the Dufek massif (Gee et al., 2016). This implies a much smaller thermal imprint on the country rock.

4

Geochemistry of the Platinum-Group Elements

4.1 Introduction

The six group VIII and group VIIIa transition elements platinum (Pt), palladium (Pd), rhodium (Rh), iridium (Ir), ruthenium (Ru) and osmium (Os) together make up the platinum-group elements (*PGE*). Based on their geochemical behaviour, the PGE can be subdivided into the iridium group PGE (*IPGE*) which consists of Ir, Os and Ru and the palladium group PGE (*PPGE*) which consists of Pt, Pd and Rh (Barnes et al., 1985): the IPGE tend to be more associated with chromitite, and the PPGE with sulfide-dominated deposits. The two groups are also distinguished by the IPGE having higher melting temperatures and are generally less soluble (e.g., Cabri, 1981).

Along with Re and Au, the PGE make up the group of highly siderophile (iron-loving) elements (*HSE*) in the Goldschmidt classification as they have a strong affinity for forming metallic bonds and preferentially partition into a metal alloy, if present. The PGE are also moderate to highly chalcophile (copper- or sulfur-loving) and can form covalent bonds with sulfur as do other chalcophile elements such as Cu. Owing to their weak preference for ionic bonding common in silicate minerals, the PGE generally behave as incompatible elements until saturation in a metal alloy or in an immiscible sulfide phase (c.f., Keays, 1995). For example, the PGE are strongly fractionated from chondritic abundances during core formation, and their low mantle concentrations provide insights into planetary accretion and differentiation processes. In crustal magma systems in general and in layered intrusions in particular, their abundance is most strongly associated with a sulfide phase. However, their siderophile nature is seen in their tendency to form various PGE-Fe metal alloy in system that have lost sulfur (e.g., Merensky Reef potholes; Kinloch, 1982; Kinloch and Peyerl, 1990).

Estimates of total HSE in primitive upper mantle range from 23 to 32 ppb (e.g. Palme and O'Neill, 2003). Because of their importance in planetary and ore deposit studies, there have been a number of studies on the solubility and the distribution coefficients for the HSE between sulfide melt and silicate liquid (e.g., Peach et al., 1990, 1994; Fleet et al., 1993; Ballhaus and Ulmer, 1995; Makovicky and Karup-Moller, 2000), between mineral and silicate liquid (e.g., Brenan et al., 2003, 2005; Righter et al., 2004) and between metal and silicate liquid (e.g., Righter et al., 2008; Van Orman et al., 2008). These studies have resulted in a good understanding of the solubility of PGE in silicate liquids and their partitioning between silicate liquid, metal and sulfide phases, although there is still some debate on the origin of micronuggets and the true solubility of the PGE in silicate liquid (e.g., Blaine et al., 2011). Only recently have studies expanded to study PGE solubility outside of dry, volatile-free systems and as a result there are still relatively few partitioning studies on fluid-melt partitioning of the PGE.

4.2 Solubility of the PGE in Silicate Liquids

Blaine et al. (2005) have summarized the experimental work on platinum solubility in wet and dry *haplobasaltic* (Fe-free) silicate liquids. They note that increasing the H_2O concentrations of the silicate liquid up to saturation concentrations affected platinum solubility only indirectly, as increasing H_2O content at fixed $f(H_2)$ increases the oxygen fugacity of the experimental runs. They otherwise note that Pt solubility is roughly constant at upper crustal pressures and that Pt solubility is related to oxygen fugacity (in bars) at 1,523 K (1,250 °C) by the equation:

$$C_{Pt}\ (ppb) = 1,389 \times f(O_2) + 7,531 \times f(O_2)^{1/2} \tag{4.1}$$

For the quartz-fayalite-magnetite buffer (QFM) at 1,250 °C, $f(O_2) = 10^{-7.86}$ and C_{Pt} = 0.88 ppb. In a follow-up study, Blaine et al. (2011) found that Cl also has no effect on Pt solubility up to 2.75 wt.% Cl, the saturation concentration in the silicate liquid. They also suggested that there was no evidence for Pt-Cl complexing in the silicate liquid.

For Ir, Borisov and Plame (1995) looked at the solubility of Ir-Pt alloy at 1300 and 1480 °C at atmospheric pressure and found the following solubility relationship:

$$\log\left({}^{C_{Ir}}/{}_{a_{Ir}}\right)(ppb) = 0.23 \times \log f(O_2) + 2.46 \tag{4.2}$$

Where a_{Ir} is the activity of Ir in the alloy. For a pure metal, the solubility at 1300 °C and for *f(O2)* = QFM ($10^{-7.38}$) would be 0.76 ppb.

The general result of all of these studies is that the PGE dissolve in Fe- and S-free silicate liquids as oxides (e.g., Laurenz et al., 2013) and increases with increasing $f(O_2)$. Basalts typically have an oxygen fugacity ranging from QFM-1 (mid-ocean ridge basalts) to QFM+2 (ocean island basalts) and have Pd contents of from 1–20 ppb (e.g. Peach et al., 1990; Bezos et al., 2005). Naldrett et al. (2008) note that the abundances of PPGE in natural basalts are above the saturation level determined by the experiments in haplobasaltic liquids. They also noted that natural basalts do not contain any visible PPGE alloys, and concluded that silicate magmas may carry higher PGE concentrations than what would be predicted from Pt and Pd solubility in these simple systems.

More recent studies have looked at more natural compositions to assess the role of Fe and S. For example, Laurenz et al., (2010) examined Pd solubility in picritic silicate liquids and found that at equivalent run condition as for the iron free systems the presence of iron enhances Pd considerably, particularly in the $f(O_2)$ range of natural basalts. They conclude that primitive mantle melts are Pd-undersaturated. Similarly, Laurenz et al. (2013) found that the presence of reduced sulfur in the silicate liquid (as S^{2-}) enhances the solubility of Ru by over an order of magnitude. They note, however, that primitive mantle melts tend to fall below the Ru content of the primitive upper mantle (Palme and O'Neill, 2003; Becker et al., 2006) and should be saturated with a refractory Ru-bearing phase. This is consistent with the common presence of IPGE inclusion in chromite and olivine described in Chapter 12. Thus, for the refractory IPGE, the assumed metal-undersaturation in mantle melts does not seem to be generally valid.

4.3 Aqueous Geochemistry of the PGE

Much of the early experimental work on the solubility of the PGE in hydrothermal fluids was done at relatively low temperatures. For example, Pan and Wood (1994) in an experimental Pt and Pd sulfides and Au metal in aqueous bisulfide solutions at temperatures from 200–350 °C and at saturated vapour pressure found the solubility of Au to be an order of magnitude higher that for the two PGE, with individual solubility ranging from 4–800 ppb for Pt, 1–400 ppb for Pd and 2–300 ppm for Au. They suggested that for these conditions, the dissolution of the Pt and Pd sulfide formed aqueous bisulfide complexes similar to that commonly proposed for Au; for Pt the reaction would be:

$$\underset{sulfide}{PtS} + \underset{\text{--- }fluid\text{ ---}}{HS^- + H^+} = \underset{fluid}{Pt(HS)_2^0} \tag{4.3}$$

In summarizing this and other work on the solubility of the PGE at these low temperatures, Pan and Wood noted that the formation of chloride complexes requires strongly oxidizing, acidic and saline conditions (e.g., Gammons et al., 1992), whereas bisulfide complexing is dominant under mildly reducing and near neutral pH conditions. At higher temperatures, however, Cl complexes such as $PdCl_3^-$ and especially $PdCl_4^{2-}$ become increasingly important such that Pd can be transported at ppm levels in chloride solutions at temperatures >600 °C (e.g., Tagirov et al., 2013). While broadly agreeing with this, Bazarkina et al. (2014) found that that Cl complexing is only significant in sulfide-free oxidizing conditions and suggested that, complexing with other S-bearing ligands other than HS- are required to transport Pd for T <~500 °C.

The temperature ranges for all of these studies is still quite low compared with degassing in a crystallizing basaltic magma where fluid temperatures >1,000 °C can be expected. Based mainly on theoretical considerations, Sassani and Shock (1998) suggested substantial solubility (ppm level) for the PGE at 1,000 °C. Otherwise, the experimental evidence for the transport of the PGE at magmatic temperatures is very limited. Fleet and Wu (1993, 1994) looked at the mobility of Pt and Pd in vacuum sealed tubes containing various ligands. They found that these two elements were most mobile in the presence of sulfide + NaCl but less so in Cl-free systems. They also not that the mobility can be strongly affected by the presence of components such as arsenic that tend to form refractory precipitates such as $PtAs_2$.

More direct investigation of PGE solubility in high temperature hydrothermal fluid included that of Ballhaus et al. (1994), who investigated partitioning of Pt and other metals between sulfide, metal and a Cl-bearing fluid trapped in synthetic fluid inclusions hosted in quartz at 600 and 900 °C. They found highly variable Pt concentration in the fluid (up to the percent level), likely due to an uncertain contamination mechanism from the Pt capsule (!). Hanley et al. (2005c) investigated the partitioning of Pt and Au between brines and a peraluminous liquids from 600–800 °C and found concentrations in the brine for both element ranges from $1–10^3$ ppm and that fluid-liquid partition coefficients, also for both elements, are on the order of $10^2–10^3$. Simon and Pettke (2009) experimentally determined Pt solubility in felsic liquid-aqueous vapour-brine system at 800 °C, 1.0 and 1.4 kbar pressure. Pt solubility ranged from 0.28–0.38 ppm in the silicate liquid, 0.37–0.91 ppm for the aqueous vapour and 0.3–16 ppm in the brine. The partition coefficient values for vapour/liquid and brine/liquid ranged from 1.0–2.9 and 6.8–67, respectively.

Blaine et al. (2011), in a study mainly focused on the effect of Cl on Pt solubility in basaltic liquids at 1,250 °C and 2 kbar pressure, estimated Pt concentrations in the Cl-bearing fluid ranged from a few tens to a few hundred ppm, with estimated fluid/liquid partition coefficients of $10^3–10^4$. Compared with the results of Simon and Pettke, these suggest the composition of the silicate liquid has a significant control of Pt partitioning. In his unpublished dissertation results, Blaine (2010) also reports a lower but still significant partition coefficient of 1,100 for Ir partitioning into Cl-, CO_2-rich fluids. The high values for vapour/silicate liquid partition coefficients are on the order of sulfide liquid/silicate liquid partition coefficients, and imply that the fluid can be as efficient at sequestering the PGE as is a sulfide phase. Furthermore, the relatively high PGE concentrations of the fluid phase are consistent with the relatively high concentrations found in Cl-rich fluid inclusions (Chapter 7) and in PGE-rich, sulfide-bearing discordant pegmatoids (Chapter 8) found in layered intrusions. It is clear, however, that more work needs to be done to fully understand the behaviour of the PGE as a function of temperature, pressure and fluid composition.

5

Generation and Movement of Bubbles and Volatile Fluids in a Crystal-Liquid Mush

5.1 Bubble Nucleation

If one ignores for the moment the possible influx of country fluids into a growing crystal + liquid mush at the floor of a crystallizing basaltic magma chamber, then the initial appearance of a volatile fluid phase must start with the nucleation and growth of bubbles from the interstitial liquid.

Bubble nucleation, growth and migration in igneous systems is a well-studied process and its applications to explosive volcanic eruptions (e.g., Toramaru, 1995; Proussevitch and Sahagian, 1998; Davydov, 2012). Fluid movement is also well understood in compacting or deforming rocks in which the fluid is a continuous wetting phases, such as ground water systems and metamorphic devolatilization (e.g., Connolly and Podladchikov, 1998; Connolly 2010). However, the formation and movement in bubbles in crystal-rich systems, with and without compaction, has only recently begun to receive significant attention (Mungall, 2015; Boudreau 2016a; Huber and Parmigiani, 2018). Even in classical hydrothermal porphyry systems, the transition from a magma nucleating isolated bubbles in a crystal + liquid mush to a continuous fluid phase are still subjects of innovative study (e.g., Candela, 1991; Parmigiani et al., 2011).

In magmatic systems, there is a distinction between vapour generated by decompression (*first boiling*) and vapour that occurs due to the increase in volatile components in the magma during crystallization (*second boiling*). The physics of bubble generation are otherwise much the same in both instances. One can further define *homogeneous nucleation* in which bubbles nucleate as isolated bubbles in a silicate liquid and *heterogeneous nucleation* in which bubbles nucleation on a pre-existing surface. Bubble formation can also be considered to occur in a system that has a fixed volume or a fixed pressure on the system, in which the relevant thermodynamic variables are the Helmholtz free energy, F, or the Gibbs free energy, G, respectively. An example where the former would apply would be bubble formation in a fluid inclusion trapped in a mineral grain. In contrast, a bubble forming in a low viscosity magma is an example of the latter.

The total free energy change, ΔG, to form a spherical bubble at constant external pressure can be expressed by two critical parameters. The first is the free energy of the surface, which is a function of the surface area ($= 4\pi r^2$, where r is the bubble radius) and the surface tension between vapour and liquid, σ_{vl}. The second parameter is the free energy difference per unit volume between the volatile component in the vapour and as dissolved in the liquid, ΔG_{volume}, which varies as the volume of the bubble (= $4/3\pi r^3$). Because both are a function of bubble radius, one has

$$\Delta G = \Delta G(r) = 4\pi r^2 \sigma_{vl} + \frac{4}{3}\pi r^3 \Delta G_{volume} \quad (5.1)$$

The ΔG_{volume} term can be positive (if the liquid is undersaturated with respect to a vapour, in which case the bubble will dissolve), negative (in which case the liquid will want to exsolved a vapour) or zero (the liquid is in equilibrium with a vapour). However, because the surface energy term is always positive, a liquid nominally in equilibrium with a vapour (e.g., water and stream at 100 °C and 1 bar pressure) will not nucleate bubbles because the $\Delta G(r)$ term will be positive for all values of r. The surface tension term of Equation 5.1 dominates at small values of the radius and initially causes $\Delta G(r)$ of the bubble to increase as the bubble gets larger. As the bubble grows, however, the ΔG_{volume} term increasingly predominates and the total free energy of the bubble begins to decrease. The bubble radius at which the $\Delta G(r)$ reaches its maximum value is the critical radius, r^*, and occurs where the first derivative of the total free energy with respect to radius, $\left(\frac{\partial \Delta G}{\partial r}\right)_{P_{vapor}}$, is zero (Figure 5.1):

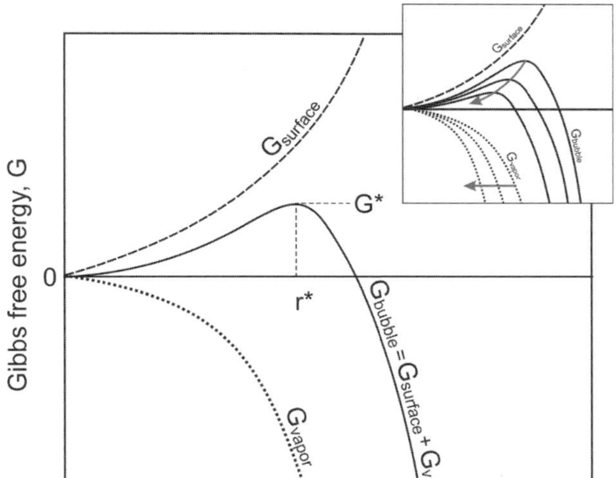

Figure 5.1 Total free energy of a growing bubble as a function of bubble radius, r. Total free energy, G_{bubble} (solid curve), is the linear function of the total free energy of the surface $G_{surface}$ (dashed curve), and the total free energy of the vapour, G_{vapour} (dotted curve). The critical radius, r^*, occurs at the bubble radius that produces the maximum value of the total bubble free energy, G^*. Inset Figure illustrates how G_{vapour} and G_{bubble} change with increasing supersaturation of the volatile component in the liquid (grey arrows), leading to a lower critical radius and easier bubble nucleation.

$$r^* = -\frac{2\sigma_{vl}}{\Delta G_{volume}} \quad (5.2)$$

For bubbles smaller than the critical radius, continued growth will require an increase in the bubble free energy and they will tend to redissolve. For nuclei larger than the critical radius, continued growth lowers the bulk free energy and they will continue to grow. Note also as the liquid becomes increasingly oversaturated, the ΔG_{volume} term decreases more rapidly as a function of r, the critical radius decreases, bubble nucleation becomes easier, and the nucleation rate increases as shown below.

Equations 5.1 and 5.2 can be recast in terms of a *supersaturation pressure*, ΔP, defined as the pressure difference between the internal pressure in the bubble and the actual pressure on the system. For a vapour-saturated system, the internal bubble pressure will be higher than that on the system owing to the surface tension squeezing the bubble. In this case, the ΔG_{volume} term can be expressed as

$$\Delta G_{volume} = \Delta G_{\Delta P} - \Delta P \quad (5.3)$$

and for which,

$$\Delta G_{\Delta P} = \frac{1}{V_{vapor}} \left[\mu_{vapor}(P_{vapor}) - \mu_{liquid}(P_{liquid})\right] \quad (5.4)$$

where $\mu_{vapour}(P_{vapour})$ and $\mu_{liquid}(P_{liquid})$ are the chemical potentials of the gas and liquid defined at P_{vapour} and P_{liquid}, respectively, and v_{vapour} is the molar volume of the vapour. The critical pressure and critical radius are related by the requirement that the free energy curve is at a maximum at the critical values:

$$\left(\frac{\partial \Delta G}{\partial r}\right)_{P_{vapor}} = \left(\frac{\partial \Delta G}{\partial P_g}\right)_r = 0.0 \quad (5.5)$$

The critical pressure, ΔP^*, and the critical radius are related by:

$$\Delta P^* = \frac{2\sigma_{vl}}{r^*} \quad (5.6)$$

A simple way to look at this is that the pressure inside a bubble produces a force over a cross-sectional area equal to $\Delta P \pi r^2$ whereas the surface tension around the circumference is $2\pi r\sigma$. For a bubble at the critical pressure and radius, these two are equal and solving for ΔP results in Equation 5.6. The critical pressure required for homogeneous nucleation is dependent on the liquid composition and ranges from >120 MPa in rhyolitic liquids to less than 35 MPa for dacitic liquids (e.g., Mourtada-Bonnefoi and Laporte, 1999; Mangan et al., 2004). This critical supersaturation pressure can occur by lowering the confining pressure on the magma (first boiling), or by increasing the saturation pressure by increasing the volatile concentrations in the liquid by crystallization (second boiling). Reported values for surface tension do not vary much as a function of silicate liquid composition ranging from basalts to rhyolite, but can increase modestly with temperature. In general, surface tension in magmas are about the same as for air bubbles formed in corn syrup, all ranging between 0.04–0.4 N/m (Murase and McBirney, 1973; Khitarov et al., 1979; Bagdassarov et al., 2000; Mangan and Sisson, 2005; Belien et al., 2010).

Random, microscopic variations in solution concentrations can lead to the formation of bubble nuclei. From classic nucleation theory, bubble nucleation is assumed random and to obey statistical laws. The homogeneous nucleation rate, J, is the number of stable bubble nuclei produced per unit time. It is a function of a constant, J_O, and an activation energy, A, required to form a spherical surface, which for a constant pressure system is a function of the Gibbs free energy:

$$J = {J_0}^A = J_0 e^{\left(-\frac{\Delta G}{kT}\right)} \tag{5.7}$$

Where k is the Boltzmann constant and T is temperature. Through a number of simplifications, the exponent can be related to the critical overpressure, ΔP^* (e.g., Blander and Katz, 1975):

$$J = J_0 e^{\left(\frac{-16\pi\sigma_{vl}^3}{3kT(\Delta P^*)^2}\right)} \tag{5.8}$$

Equation 5.8 also implies that the heterogeneous bubble nucleation rate will increase as the system undergoes decompression. Comparing the exponential term in Equations 5.7 and 5.8, one can define the critical supersaturation pressure:

$$\Delta P^* = \sqrt{\frac{16\pi\sigma_{vl}^3}{3\Delta G^*}} \tag{5.9}$$

For bubbles nucleating on an existing surface (heterogeneous nucleation), the activation energy needed to form a spherical surface is reduced by the wetting factor, $\phi(\theta)$:

$$\Delta P^* = \sqrt{\frac{16\pi\sigma^3\phi(\theta)}{3\Delta G^*}} \tag{5.10}$$

where $\phi(\theta)$ is a function of the wetting angle, θ, between the silicate liquid, the crystal surface and the bubble. For a smooth flat surface, the wetting factor is defined as:

$$\phi(\theta) = 0.25(2 - \cos\theta)(1 + \cos\theta)^2 \tag{5.11}$$

(e.g., Lubetkin, 2003). Without going into details, Equation 5.11 can be derived by consideration of the effect of three interfacial (surface tension) energies involved in heterogeneous nucleation. Besides the vapour–liquid surface tension term, σ_{vl}, there are in addition terms for vapour-solid σ_{vs} and liquid–solid, σ_{ls}, interfacial energies. The wetting angle can also be found to be a function of these three terms:

$$\cos\theta = \frac{\sigma_{vs} - \sigma_{ls}}{\sigma_{vl}} \tag{5.12}$$

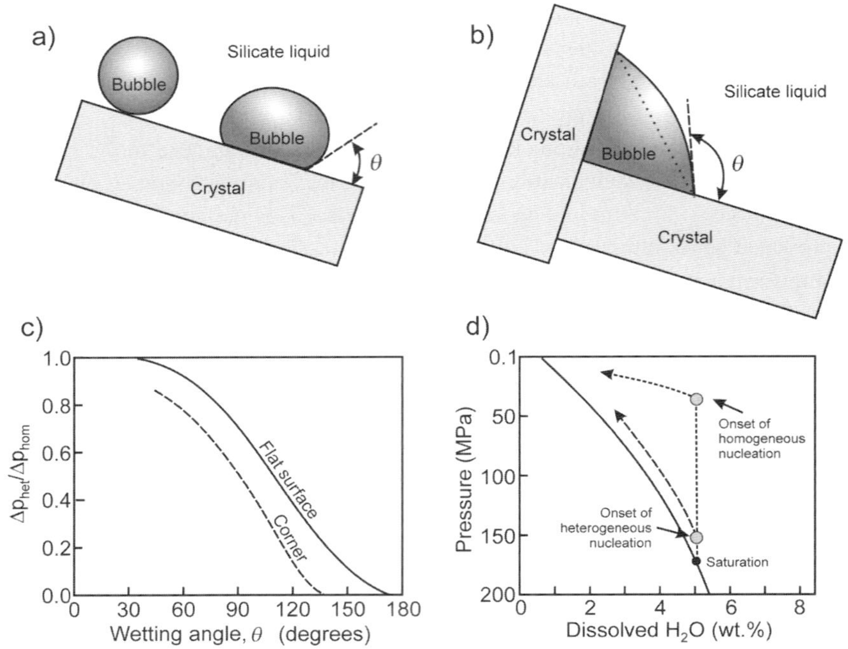

Figure 5.2 Examples of wetting angels on heterogeneous bubble nuceation behaviour. (a) Bubbles on a flat surface illustrating a perfectly sphical bubble with a wetting angel, θ, of zero (left) and one with an acute wetting angle between the bubble (dashed line) and the crystal (right). (b) Bubble nucleation at a corner with an obtuse wetting angel. Because of the symmetry of the bubble, a flat interface between the bubble and the silicate liquid is formed when $\theta = 135°$ and no activation energy is needed for the formation of a bubble. (c) Critical supersaturation pressure needed for a constant ΔG as a function of the wetting angle, θ, normalized to the supersaturation required for homogeneous nucleation. Plots are shown for a flat surface (solid line) and for nucleation in a corner (dashed line). Redrawn after Hurwitz and Navon (1969), with permission. (d) Degassing paths for homogeneous and heterogeneous nucleation of bubbles, illustrating the greater degree of supersaturation required for homogenous nucleation. Generalized after Mangan et al. (2004), with permission.

Wetting angles are a function of vapour composition, the (silicate) liquid composition and mineral type. When $\theta = 0.0$, the bubble is perfectly spherical and tangent to the crystal, the wetting factor is zero, and one has the same conditions as for homogeneous nucleation. Otherwise, the wetting angel can be acute to obtuse: the larger the angle the greater the tendency of the bubble to wet the crystal (Figure 5.2a). In general, the saturation pressure for heterogeneous nucleation relative to that for homogenous nucleation, $\Delta P_{het}/\Delta P_{hom}$, varies as ϕ^2.

For nucleation in a corner, bubble symmetry implies that the maximum wetting angel, $\theta_{max} = (180° - 45°) = 135°$ and for which no activation energy is required for the formation of a bubble (Figure 5.2b). Overall, the saturation pressure is lower for nucleation in a corner than on a flat surface (Figure 5.2c). This explains why bubbles are continuously released from certain spots on the inside of a glass of beer. Similarly, boiling chips (irregular, porous pieces of an inert material) are used in industrial and chemical processes to prevent superheating of liquids and unwanted flashing of liquid to vapour (e.g., water flashing to steam).

The above considerations are consistent with bubble nucleation experiments that show that homogeneous nucleation requires significant decompression to initiate nucleation, but then rapidly returns to the solubility curve. In contrast, heterogeneous nucleation requires smaller supersaturations and tend to follow the vapour solubility curves (Gardner et al., 1999; Mangan et al., 2004) (Figure 5.2d). A cooling and crystallizing crystal + liquid mush is thus a favourable environment for bubble nucleation.

5.2 Bubble Growth

Once a stable bubble nuclei has formed it can continue to grow by two mechanisms: First, the bubble can grow by diffusive growth of volatile components from the liquid to the bubble, much as a crystal will grow as the silicate liquid solidifies. As for crystals, growth by this mechanism is controlled by concentration and diffusion of volatiles in the liquid: a liquid saturated in CO_2 will produce smaller bubbles than one saturated in H_2O owing to the lower dissolved concentration of CO_2, all other factors being the same. Second, and unlike crystals, bubbles are not constrained to a fixed molar volume; they can grow by gas expansion and viscous deformation of the silicate liquid. Bubbles can grow faster in basaltic liquids relative to rhyolitic liquids owing to the generally higher diffusion rates and lower viscosity of basalts. Because of this, basaltic magmas tend to produce fewer, larger bubbles for the same volatile concentration in the liquid than do rhyolitic liquids under similar physical conditions. Surface tension is also a factor, but as noted above, this is does not change significantly with liquid composition

For isobaric bubble growth from a supersaturated liquid (e.g., no decompression of the system), volatiles are initially quickly transferred from the liquid to the bubble while the bubbles are small. Growth is otherwise viscosity limited. Although volatiles are diffusing into the bubble, the bubble still has to expend energy to expand into the surrounding liquid, but this becomes easier as the bubble becomes larger and surface tension effects become less important in defining the internal pressure of the bubble. (One can experience this while blowing up a balloon; it initially requires one to blow very hard to get the balloon to start to expand, but then becomes easier as the balloon fills with air). The net consequence is that bubble growth will tend to follow the viscosity limited, exponential growth rate curve in Figure 5.3, and the growth rate with increase over time.

Rather quickly in the bubble growth history, however, the internal bubble pressure drops closer to the ambient pressure on the system and growth is diffusion limited by the rate that diffusion can supply distal liquid volatile components to the bubble. It also takes progressively more volatiles to cause a unit increase in the bubble radius. At this stage, bubble growth tends to follow the parabolic growth curve of Figure 5.3, and growth slows with time. A growing bubble also must compete for volatile components with other

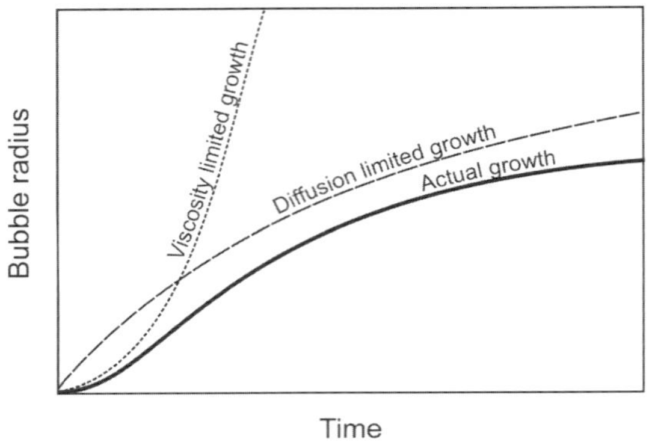

Figure 5.3 Schematic diagram illustrating growth history of a bubble following an early exponential viscosity-limited growth curved, followed by a parabolic diffusion-limited growth curve. The actual growth curve is also affected by competition between neighbouring bubbles. Simplified from Navon and Lyakhovski (1998), with permission.

bubbles, further slowing the growth of an individual bubble, and thus a real world bubble would tend to grow like the solid line of Figure 5.3. In general, the growth of a bubble grows as the square root of time, t, as follows:

$$r(t) = \varepsilon\sqrt{\frac{2\left(C_{\infty}^{liq} - C_{sat}^{liq}\right)Dt}{C_{gas}}} \tag{5.13}$$

Where C_{∞}^{liq} and C_{sat}^{liq} are the volatile concentrations in the silicate liquid away from the bubble at its saturation value next to the bubble, D is the mass diffusion coefficient for the volatile component, and C_{gas} is the concentration of the volatile component in the gas bubble, and ε corrects for the presence of a surface (Boudreau, 2012).

After the period of heterogeneous nucleation and initial growth in a pocket of liquid, a non-wetting bubble growing in a crystal mush will eventually interact with surrounding minerals grains. There are four mechanisms by which the bubble can expand into the surrounding crystal matrix (Boudreau, 2012) (Figure 5.4). While Boudreau discussed these in relation for degassing in sedimentary systems, they are applicable to the igneous situation.

1. Fluidization. In this case, the surrounding solid + silicate liquid matrix can act as a fluid under the stress exerted by a growing bubble and the matrix is simply pushed away from the bubble by viscous deformation. In this regard, growth and movement of a bubble is much the same as growth in a crystal-free magma. The formation of spherical vugs and pipe vesicles in crystal-phyric lava flows would be an example of this mechanism. It also may occur during degassing in the upper parts of a crystal mush system where the crystals are only loosely aggregated.

2. Displacing pore liquid. At the other extreme, the crystal matrix can be rigid and a growing bubble can expand in to the pore space by pushing the pore fluid through

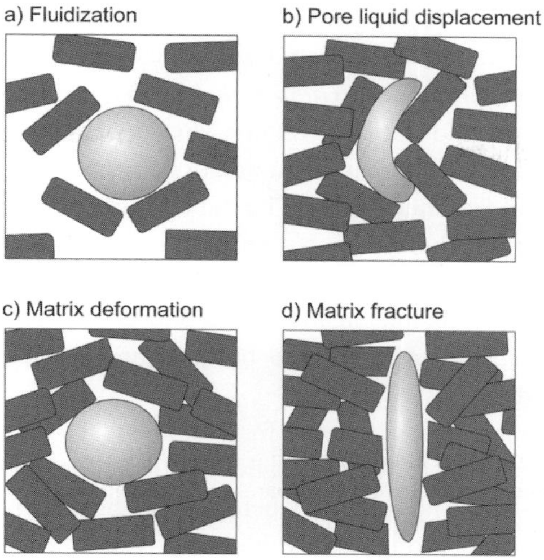

Figure 5.4 Examples of non-wetting bubble growth mechanisms in three-phase systems. (a) Fluidization: crystal-liquid matrix behaves as a pure liquid. (b) Pore liquid displacement: pore liquid is displaced by a deforming bubble in an otherwise rigid crystal framework. (c) Matrix deformation: bubble deforms the crystal-liquid matrix. (d) Matrix fracture: Crystal-liquid matrix behaves as a brittle solid and fractures to accommodate bubble growth. See text for additional details.

connected pores without matrix deformation. In the sedimentary literature, this has been termed *invasion percolation* (Clennell et al., 2000) or *capillary invasion* (Jain and Juanes, 2009). This is the case discussed by Mungall (2015) in regards to degassing in an igneous crystal mush. Gas can penetrate into the adjacent pores if the internal gas pressure, P_{int}, can overcome the surface tension, σ, and the pressure in the pore fluid, P_{fluid}, i.e., penetration will occur if

$$P_{cap} = P_{int}\text{-}P_{fluid} \geq \beta\sigma\cos(\theta) \qquad \text{(Young-Laplace equation)} \tag{5.14}$$

where P_{cap} is the excess capillary pressure, θ is the contact angle between the fluid and the gas bubble as discussed above and β is a numerical constant that depends of pore geometry. Generally, $\theta \cong 0.0$ and $\cos(\theta) \cong 1.0$. For a cylindrical pore, $\beta = 2/r_{throat}$ where r_{throat} is the pore throat radius; for a slit-like pore (such as might occur between angular mineral grains), $\beta = 1/r_{throat}$. Boudreau (2012) suggested that the magnitude of β in real mush systems should be similar to these simple geometries. Jain and Juanes (2009) found that β in Equation 5.14 was proportional to the inverse of the radius of the grains, i.e., $\beta = \sim 1/r_{grain}$ implying that the pressure required to penetrate to another pore is inversely related to the grain size:

$$P_{cap} \propto 1/r_{grain} \tag{5.15}$$

Bubble migration by capillary invasion in a coarse-grained plutonic rock would be easier that in a fine-grained dike. The experiments of Boudreau (2016a) suggested that the capillary pressure needed to be no more than about 5–10% high than the hydrostatic pressure to cause bubbles to move for grain sizes typical of layered intrusions.

3. Deformation of the bulk matrix (liquid plus solids). An intermediate mechanism between fluid-like matrix and a ridged matrix are systems for which there is some elastic/plastic deformation of the matrix that can occur to create space (e.g., Wheeler, 1988, 1990; Wheeler and Gardner, 1989). The bubbles are ~spherical gas-filled cavities, which are forced open by the gas pressure working against the capillary pressure (between grains) and the total load on the bubble. The solid + silicate liquid is considered as a single deformable medium as opposed to separate phases, and it is considered a viable mechanism when the bubbles are greater than about 10 grain diameters in size (Boudreau, 2012).

4. Matrix fracture. In this instance, space for a bubble is made by tensile fracture of the matrix, similar to the mechanism for magmatic dike injections (e.g., Rivalta and Dahm, 2006). In this case, the solid + liquid matrix behaves as a brittle solid on short time scales. Fracturing under tensile stress is observed both in muddy sediments (e.g., Jones, 1994) and in crystal-liquid igneous assemblages (e.g., Philpotts et al., 1996). As summarized by Boudreau (2012), fracture of the crystal-liquid matrix will occur if the total gas pressure in a growing bubble, P_{bubble}, exceeds the sum of the minimum compressive stress, assumed horizontal, σ_H, and the tensile strength, σ_{TS}, of the mush:

$$P_{bubble} > \sigma_H + \sigma_{TS} \tag{5.16}$$

or,

$$P_{fracture} > T \tag{5.17}$$

where $P_{fracture}$ is the critical excess gas pressure required for fracture, $P_{fracture} = P_{bubble} - \sigma_H$. A simplification assumes that the tensile fracture is governed by *linear elastic fracture mechanics* (LEFM) (Broek, 1982; Johnson et al., 2002; Jain and Juanes, 2009). This theory assumes that a solid is not internally perfect, but contains random flaws where fracture is initiated. For a disc-shaped fracture in LEFM, the tensile strength is given by

$$T = cK(\pi\alpha)^{-\frac{1}{2}} \tag{5.18}$$

where K is the tensile fracture toughness of the medium, α is the half-length of the crack (~half the bubble length), and c is a numerical coefficient that depends on geometry; for a disk-shaped crack, $c = \pi/2$. The parameter K measures the ease of breaking the bonds holding mineral grains together as a function of distance. Measured values of K are known in part for sediments (Johnson et al., 2002) but not for igneous systems, although the experiments of Philpotts et al. (1998, 1999) implies that it is non-zero even in systems that are mostly liquid. Also, the $\alpha^{-\frac{1}{2}}$ dependence of the tensile strength implies that larger bubbles can grow by this mechanism easier that smaller bubbles. The evolution of the bubble growth by matrix fracture is shown schematically in Figure 5.5. As bubbles become increasingly elongated, the hydrostatic/lithostatic load can be transferred from the lower part of the bubble to the top, further enhancing the upward-migrating fracture front (Figure 5.5d).

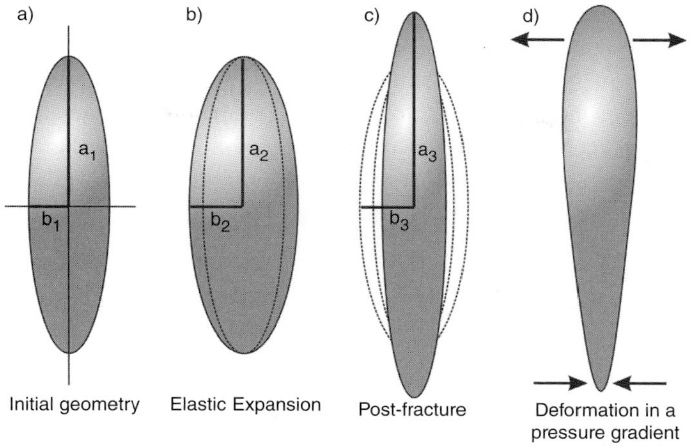

Figure 5.5 Schematic diagram illustrating bubble growth by tensile fracture. (a) Initial geometry of a vertical elongated bubble. (b) Elastic expansion of the bubble against the matrix leads to modest growth in the width of the bubble. (c) Eventually the matrix fails, leading to the bubble becoming yet more elongated. (d) For a bubble deforming in a pressure gradient, the bubble can transfer higher pressure from the bottom of the bubble to the top, leading to continued fracture at the top. Modified after Barry et al. (2010) and Algar et al. (2011), with permission.

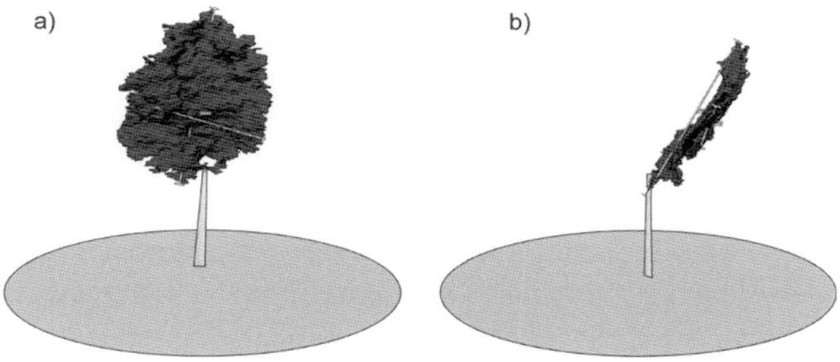

Figure 5.6 'Cornflake' bubbles formed by the injection of gas into mud. X-ray computed tomography (CT) scans of a high aspect ratio 'cornflake' bubble formed by injection of air into cohesive sediments, the latter of which have been removed from the scans. a) face-on and b) side view. The bubble is approximately 22 mm in diameter but only 0.7 mm wide with a volume of 0.3 cm^3. Post at the bottom of the bubble is the gas injection tube, lines are used for image measurements. Redrawn after Boudreau et al. (2005), with permission.

Studies have been made of injected bubbles into cohesive sediments and then imaged with X-ray computed tomography (CT scans) (e.g., Boudreau, 2005). These studies have shown that the bubbles are strongly flattened into high aspect ratio shapes resembling 'cornflakes' (Figure 5.6). Boudreau (2012) suggests that the formation of these bubbles are consistent with the elastic fracture growth model.

Oppenheimer et al. (2015) noted that, in experiments involving loose packed grains, injected bubbles become overpressured and begin to deform as they displace pore liquid (mechanism 2) and are forced into the surrounding matrix. The deformation of bubbles to form long fingers (see below) can lead to coalescence of nearby bubbles, and the coalesced bubbles escapes through pathways that open and seal intermittently (mechanism 3). They also found that bubble behaviour was mainly controlled by particle interactions and less so on liquid viscosity, container geometry or bubble growth rate. Oppenheimer et al. (2015) also note that, in systems where the crystals are more tightly packed, migration is largely by matrix fracture.

5.3 Bubble Migration in a Crystal Mush

Once a bubble forms, it will want to rise by virtue of its buoyancy. For small non-wetting bubbles, those are less than typical pore throat sizes, the bubbles can move unimpeded. For a crystal mush undergoing compaction, these smaller bubbles can move with the silicate liquid; indeed, the presence of bubbles can enhance compaction by increasing the density difference between the solid matrix and the bulk density of the liquid + bubble mixture that drives compaction (Boudreau, 2016a). For bubbles larger than typical pore throat sizes, however, movement will be restrained by the capillary forces noted above.

The problem of how a vapour bubble or a more extensive fluid phase can separate from a crystal-rich silicate liquid and begin to migrate is a class of transport problems known as three-phase theory when it involves solids and two liquid/gas phases or multiphase theory when there are additional mobile phases. Briefly, the problem involves understanding the mechanisms by which a non- or poorly wetting bubble can overcome the capillary forces that prevent the bubble from moving though a crystal-liquid mush in which the pore throats are smaller than the bubble diameter. It appears in a number of geologic problems, including bubble migration in sediments, where it is important in understanding the melting of gas hydrates and methane release to the atmosphere. Multiphase transport is of particular interest in the oil and gas industry. Indeed, the means by which non-wetting oil and gas bubbles and more continuous fluids move from the source rock to reservoirs is a similar problem of understanding bubble and vapour migration through a crystal + silicate liquid mush. As for a vapour bubble exsolved from interstitial silicate liquids, where the vapour does not wet the silicate minerals, oil and gas also initially forms as discrete dispersed bubbles in formation waters, generally in low permeable sediments. Despite these obstacles, hydrocarbons manage to migrate out from their source regions and move km-scale distances in relatively short periods of time (Appold and Nunn, 2002).

Complicating the study of bubble migration is the ephemeral nature of bubbles, as they leave little evidence for their former presence if not frozen in. This is evident in lava flows, where vesicular zones will be preserved in the quickly cooled tops and bottoms of individual flows, but are largely absent from the interior; the more slowly-cooled interior allowed the lava to degas and bubbles to escape before solidifying (Sahagian et al., 2002). Even in sediments, the mechanisms of bubble migration are not well understood. In the words of Algar et al. (2011), 'Sediments are an opaque medium; it is difficult to study a bubble if you cannot see it.'

Estimates of the requirements for bubble migration in sediments using simple buoyancy mechanisms can be wildly in error. As summarized by Boudreau (2012), these require either that the bubbles must be very large (~1 m diameter) in order to rise, or that sediments must be mechanically very weak, or that bubbles would need to reach vertical connected sizes of about 30 m in order to rise by capillary invasion in mud. Neither has experimental or observational support in the sedimentary literature, which demonstrates that gas bubbles can easily move out of sediments. Similarly, while a number of theoretical studies have also suggested that bubbles are unlikely to undergo much movement in crystal-rich igneous systems (e.g., Lake, 2013), there is field and petrographic evidence for extensive bubble migration in crystal-rich magmas. Examples include Stromboli volcano, where observed degassing requires that gas moved through magma with a crystallinity of ~50% (Métrich et al., 2001; Armienti et al., 2007). Vapour loss is also inferred by the presence of vesicle-free but crystal-rich tephra from Mount Etna, Italy, Villarrica, Chile and Medicine Lake Volcano, California (Caltabiano et al., 2004; Rust and Cashman, 2004; Polacci et al., 2006; Gurioli et al., 2008). Belien et al. (2010) suggest that, unless permeable networks are formed all the way down through the crystal-rich magma, some bubbles will have to migrate through crystal-phyric magma in these systems. Rust and Cashman (2004) suggested that bubble deformation by shear and/or partial bubble collapse allows open-system degassing of magma with a vesicularity at least as low as 20 %. Vigneresse (2015) presents a number examples of gas interactions in crystal-rich granitic mushes.

The same debate crops up in layered intrusion literature. For example, a theoretical investigation by Mungall (2015) suggests that bubbles are unlikely to migrate any significant distances in a crystal mush while the system remains partially molten, even in a crystal pile undergoing compaction. He also suggested that bubble migration was similar to slug flow in that the velocity of the compaction-driven interstitial liquid is the main driving force to bubble migration. He does not consider it likely that compaction can squeeze and overpressurize any bubbles that do form owing to limited compaction at the time of bubble nucleation and growth. In contrast, Boudreau (2016a) suggested that compaction can lead to pinched bubbles becoming overpressured such that the capillary resistance to bubble migration is easily overcome. He suggested three mechanisms by which bubbles could be force to move from their nucleation sites and are illustrating in Figure 5.7: (1) Confined bubble growth: growth of a bubble in a confined space between mineral grains leads to the internal pressure of the bubble to exceed the capillary pressure required for bubble movement. This can be aided by the movement of vapour-saturated silicate liquid to the site of the growing bubble by compaction. (2) Bubble pinching: pinching of a bubble as the solid matrix compacts and squeezes the bubble again leading to the internal pressure on the bubble to exceed the capillary pressure. (3) Pressure transfer: if a bubble blocks the pores and causes a reduction of the mush permeability, then the liquid below the bubble will support more of the lithostatic weight of the mush column. This increase of the pressure on the liquid can push the bubble through a pore restriction. The amount of overpressure required is typically $<10\%$ of the hydrostatic pressure. Recent numerical

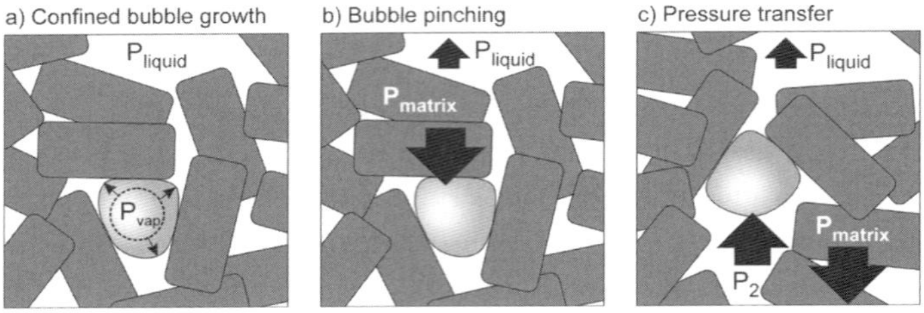

Figure 5.7 Three pressure transfer mechanisms that can drive bubble migration in a compacting mush. (a) Confined bubble growth: Bubble growth in a confined space during crystallization of interstitial liquid leads to an increase in the internal bubble pressure, P_{vap}. (b) Bubble pinching: The bubble is pinched by the collapsing solid matrix such that, over time, the bubble carries more of the matrix load (P_{matrix}) which is greater than the liquid hydrostatic pressure (P_{liquid}) on the bubble. (c) Pressure transfer: Bubbles clog the porosity, reducing permeability, resulting in the liquid below the region of reduced permeability to experience an increase in the pressure on the liquid (P_2) above the nominal hydrostatic pressure as it supports more of the lithostatic load. This locally higher pressure pushes the bubble though the matrix. Expanded after Boudreau (2016a), with permission.

modelling by Huber and Parmigiani (2018) also points to the ease of bubble migration in compacting igneous mush systems.

Related to the above is the possibility that bubbles may be pushed through a crystal-liquid mush by porosity waves. Porosity waves can form during compaction as zones of high liquid fraction (higher porosity) move up through the system These waves can travel faster than the interstitial liquid itself, much as ocean waves travel much faster than ocean water itself. Porosity waves have been suggested as a mechanism for transport of fluids through porous rock that can be much more efficient than porous media flow or even flow through fracture (Connolly and Podladchikov, 1998; Appold and Nunn 2002; Chauveau and Kaminski, 2008; Connolly, 2010). Boudreau (2016a) suggested that such waves would be effective at moving bubbles as well. Two features of the porosity wave can enhance bubble migration. First, as the high porosity front moves through, the increase in porosity allows bubbles to move upward through the more open pore space than was present initially. Second, as the wave passes through the matrix and porosity decreases, compaction can squeeze the bubbles and push them into the high porosity portion of the wave, allowing the bubbles to 'surf the porosity wave'. As the porosity wave moves through the matrix, it will collect bubbles and become more bubble-rich as it moves upward.

5.4 Bubble Coalescence and 3-Phase Flow of a Continuous Fluid

As bubbles either continue to grow as the liquid crystallizes and degases, or as bubbles migrate and locally become volumetrically abundant, they can coalesce to form a more extensive, continuous fluid. The *critical bubble percolation density* for bubbles to transition into a continuous fluid can begin when the local bubble volume fraction as a proportion of the vapour + liquid fraction exceeds approximately 0.2 (Huber et al.,

2012). For example, for a vapour-saturated crystal-liquid mush containing 70% crystals, the bubble will begin to coalesce once the bubble volume reaches about 6%. At this stage, fluids can locally fully wet the crystals. Discordant pegmatoids such as those seen in the Stillwater Complex likely formed by fluids that fully wetted the mineral grains, as is evident from the abundant fluid inclusions found in these bodies (Braun et al., 1994; Hanley et al., 2008).

Infiltration of a non-wetting fluid into a porous crystal-liquid mush is different from the movement of isolated bubbles, but is still controlled by pressure gradients, buoyancy, viscous and capillary forces. Where gravity/buoyance force can be ignored, the infiltration behaviour is largely controlled by two dimensionless parameters, the *capillary number*, Ca, and the *viscosity ratio*, M (Lenormand et al., 1988):

$$Ca = \frac{q_n \mu_n}{R^2 \sigma_{vl}} \quad \text{(Capillary number)} \tag{5.19}$$

$$M = \frac{\mu_n}{\mu_w} \quad \text{(Viscosity ratio)} \tag{5.20}$$

where the subscripts w and n refer to the invading non-wetting fluid and pre-existing wetting fluids, respectively, μ is the dynamic viscosity, σ_{vl} is again the surface tension between the two fluids, q_n the volume flux of the infiltrating fluid per unit time, and R a characteristic length scale of the system, typically taken as the average pore radius. The viscosity ratio, M, is simply the dynamic viscosity ratio of the two fluid phases. For supercritical H_2O displacing basaltic liquid, M ranges from ~5×10^{-8} to ~5×10^{-6}. The capillary number, Ca, is the ratio of the viscous to the capillary forces, and increases as the infiltration velocity increases. Because of this, flow behaviour can change as the flux rate changes.

Lenormand et al. (1988) noted that Ca and M lead to three distinct displacement regimes as a non-wetting fluid displaces a wetting fluid (Figure 5.8):

1. Stable displacement. this case is characterized by high Ca and M and infiltration of the nonwetting fluid occurs along a flat, planer front between the two liquids, with only

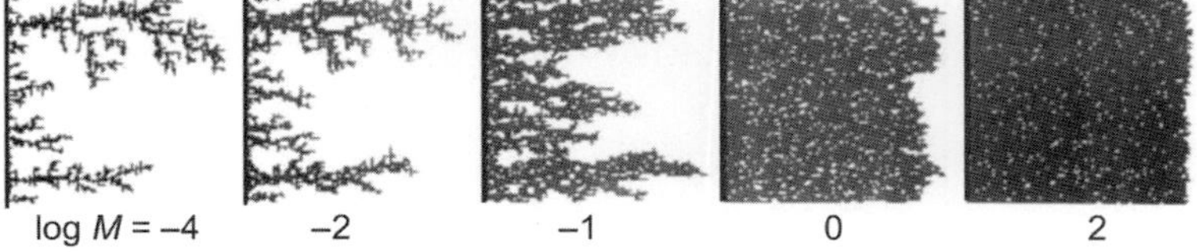

Figure 5.8 Infiltration regimes in three-phase flow. Numerical models illustrating (from left to right) the progressive change from viscous fingering, capillary fingering and stable displacement. The left-hand side of each Figure is the inflow side, and only the non-wetting phase is shown in grey. For each Figure, the capillary number, Ca, is constant and equal to 1.0, whereas the log of the viscosity ratio, log M, varies as shown beneath each Figure. From Lenormand et al. (1988), reprinted with permission.

minor irregularities at the scale of a few pores. The principal force driven fluid flow is due to the viscosity of the injected fluid and both capillary effects and pressure drop across the contact between the two fluids are negligible.

2. *Diffusion limited aggregation (DLA) viscous fingering*. At high *Ca* and for $M \lesssim 1$, viscous forces in the pore throats prevail over capillary forces and there is a distinct pressure drop across the contact between the two fluids. In this case, the infiltrating low viscosity fluid readily invades the existing fluid to produce long, dendritic, lightning bolt-like fingers that broadly develop in the direction of the pressure gradient driving fluid flow.

3. *Invasion percolation (IP) capillary fingering*. At low values of *Ca* ($\lesssim 10^{-5}$), the capillary pressure is (locally) higher than the pressure gradient that drives the fluid flow and the injected non-wetting fluid preferentially moves through the largest pores (Zhang et al., 2011). The result is that fluid fingers grow in all directions, and locally can move backwards against the general flow direction to form loops or otherwise start to fill in between the advancing fingers (Tsuji et al., 2016).

A phase map of where these flow regimes will occur in *Ca-M* phase space is shown in Figure 5.9. It is competition between the pressure gradients driving viscous flow and the capillary pressure jumps that determine if capillary or viscous fingering occurs. Furthermore, because *Ca* will increase with a higher flux rate that itself can be driven by a higher pressure gradient, viscous fingering is likely to precede hydrofracturing of the rock matrix (Løvoll et al., 2004). When *Ca* or *M* takes on an intermediate value, two displacement

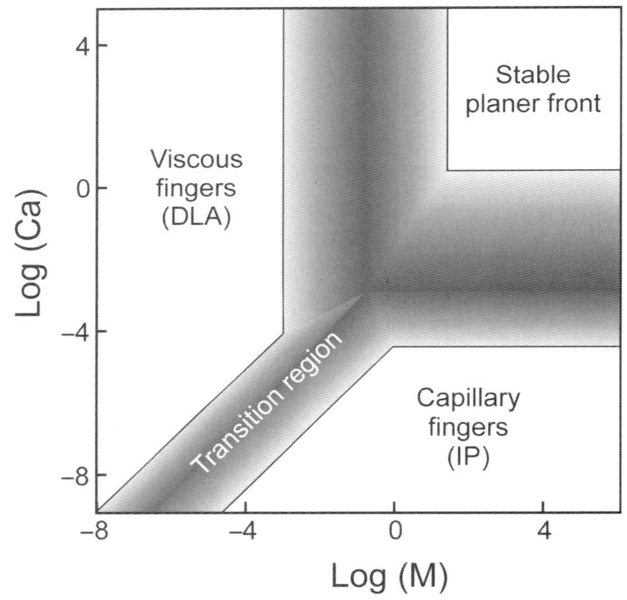

Figure 5.9 The immiscible liquid flow regimes as a function of the viscosity ratio, M, and the ratio of viscous to capillary forces, Ca. Redrawn after Lenormand et al. (1988), with permission.

patterns occur simultaneously, with the resultant pattern termed transitional or crossover. The formation of these fingers strongly influences the macroscopic, Darcy-scale fluid velocity. Fluids restricted to narrow fingers would have to have a higher velocity to equal the flux rate of a more extensively permeating fluid, which could affect the ability of the fluid to equilibrate chemically or especially isotopically with the matrix. For example, channelled fluid flow can affect the efficiency of ore element removal from source rock by hydrothermal fluids to form ore deposits. Fluids that are channelled into fingers as they move through a source rock will only see some fraction of the ore elements potentially available to be mobilized (Huber et al., 2012).

Berkowitz and Ewing (1998) noted that buoyance effects are also important where the two liquids have different densities. They suggested a modification of the Lenormand et al. (1988) Figure by inclusion of the *Bond number*, *Bo*, which is the ratio of the hydrostatic pressure drop over a pore to the capillary pressure:

$$Bo = \frac{\Delta \rho g R^2}{\sigma_{vl}} \quad \text{(Bond number)} \tag{5.21}$$

where $\Delta\rho$ is the density difference between in two fluids, g the gravitational constant, and the others as previously defined. The Bond number can be added as a vertical axis to the flow regime plot of Lenormand et al. (Figure 5.9), where the two component Lenormand plot crosses the vertical axis at $Bo = 0.0$. By convention, positive values of *Bo* will stabilize a planar infiltration front whereas negative numbers lead to a greater tendency to form fingers. The main effect on the plot of Figure 5.9 is that the region of stable infiltration in the upper right-hand corner increases or decrease as the Bond number is positive or negative, respectively. An example for the former would be a denser silicate liquid moving up and displacing a lighter, overlying interstitial silicate liquid, whereas a low density hydrous fluid migrating into overlying silicate liquids would be an example of the latter.

Further complicating the situation is the possibility that the infiltrating fluid can react with the existing solid + liquid matrix. An obvious case is the situation in which migrating vapour moves into a silicate liquid + solid assemblage that is not yet vapour-saturated. This can lead to both the infiltrating vapour to redissolve into the silicate liquid and increase the amount of liquid by the fluxing effect of the added volatiles (see Chapter 9). Alternatively, hotter fluids could move into a cooler mush and induce melting, a situation that has been explored by numerical simulations by Parmigiani et al. (2011). In addition to the three dimensionless parameters *Ca*, *M* and *Bo*, they introduced two more dimensionless numbers for *reactive flows*. The first is the *Stefan number*, *St*, which gives the ratio of the amount of enthalpy associated with the dissolution or melting of the matrix to the amount of enthalpy stored in the system. The second is a *diffusivity ratio*, ξ which is the ratio of reactant diffusivity for the two fluid phases:

$$St = \frac{c\Delta T}{L_{fus}} \quad \text{(Stefan number)} \tag{5.22}$$

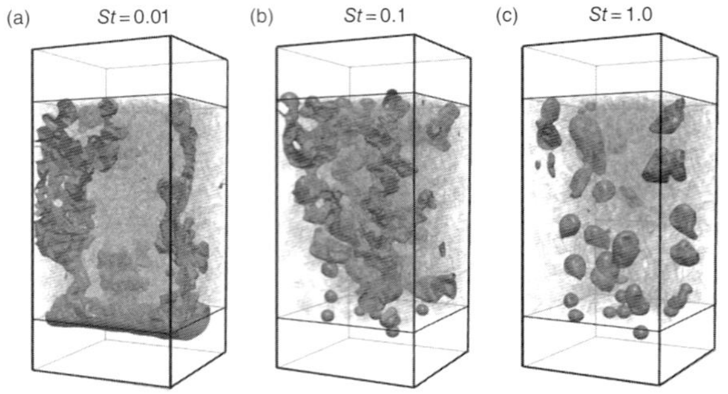

Figure 5.10 Numerical models of buoyant, non-wetting fluid migration in a 3-phase system as a function of melting efficiency, *St* (Stephan number = 0.01, 0.1 and 1.0). a) At low *St*, well-defined and stable channels of non-wetting fluid this is moving up from a vapour-rich porosity trap just above the base. b) As the melting efficiency increases (increasing *St*), the channels become less stable. c). For *St* = 1.0, the fluid flow breaks up, forming isolated bubbles. The light grey is the translucent solid fraction; note the decrease in crystallinity in a–c as the lower part of the column undergoes increased partial melting. From Parmigiani et al. (2011), reproduced with permission.

$$\xi = \frac{D_n}{D_w} \quad \text{(Diffusivity ratio)} \tag{5.23}$$

For the above, $\Delta T = T_0 - T_m$ is the temperature difference between the infiltrating non-wetting vapour and the melting temperature of the matrix, c is the specific heat, L_{fus} is the latent heat of fusion, and D_n and D_w are the diffusivities of a reactive component (e.g., H_2O) in the non-wetting and wetting liquids, respectively

An example of the numerical modelling by Parmigiani et al. (2011) is shown in Figure 5.10. The competition between the injection rate of non-wetting buoyant fluid (expressed in the capillary number, *Ca*), buoyancy (expressed by the Bond number, *Bo*), and superheat (expressed in the Stefan number, *St*) controls the stability of capillary channels. Their numerical experiments suggest that channels of the non-wetting phase are stable for low values of the Stefan number or low buoyancy differences, but that flow tends to break into slugs or bubble at higher values. Moreover, the dimensionless numbers describing the balance between injection of non-wetting fluid, buoyancy and capillary forces vary temporally and spatially because of localized melting. Thus, *Ca* decreases with time locally because of the evolution of the channel radius owing to melting of the matrix. An example for reactive viscous fingering developed in the Skaergaard intrusion is shown in Figure 5.11.

Channelled flow of bubbles can occur in inflated pahoehoe lava flows (Figure 5.12). These vertical, pipe-like bodies are bubble-rich and contain a fractionated silicate liquid. Fowler et al. (2015) suggests that a positive feedback can develop in a degassing mush column by which rising buoyant reactive bubbles and fluids cause higher local porosity. This in turn allows more evolved silicate liquids to move up into hotter parts of the crystal pile, lowering liquidus temperatures and promoting yet higher porosity (and more effective compaction).

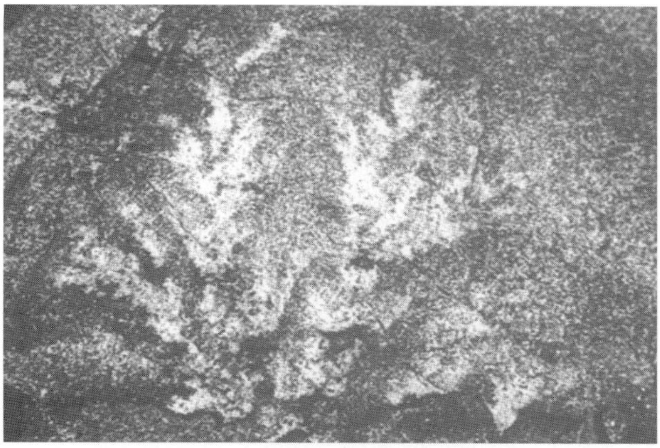

Figure 5.11 Photographs of replacement anorthosite from the Skaergaard intrusions. Dendritic anorthosites fingers are developed within the plane of layering adjacent to trough bands from Upper Zone A. The branches are oriented up-dip; note mafic haloes at base. Scale of photograph is ~1.5 m. From Sonnenthal (1992), reproduced with permission.

5.5 Lateral Migration of Fluids

The discussion above, particularly for buoyant fluids, broadly assumes that fluids will move vertically upwards. However, layered intrusion are, by definition, texturally and modally complex rock, and locally fluids may come into a zone faster than they leave such that fluids will build up over time. Various layer-parallel pegmatoids likely formed by such a mechanism, of which the Merensky pegmatoid is perhaps the most well-known example. More generally, where the layering is not horizontal, bubbles and fluids may move laterally as well as upwards (Figure 5.11). This is seen in regional fluids flow in metamorphic rocks, where fluid flow is in part controlled by lithology changes (metamorphic banding) as well as by faults and fractures (e.g., Lackey and Valley, 2004). Similarly, the role of structures in the concentration of gas and oil also demonstrates the lateral movement of fluids (e.g., Woods and Norris, 2010). Models of fluid flow at subduction zones also demonstrate considerable up-dip flow parallel to the descending lithospheric slab (e.g., Wilson et al., 2014). By analogy, any fluids generated by the dehydration of the country rocks beneath the Bushveld Complex that did not move up into the intrusion would have to have moved laterally. Evidence for this is seen in Bushveld-age Mississippi Valley-Type mineral deposits in the Transvaal Sequence distal from the Bushveld Complex (Gleason et al., 2011).

There has been very little work on this in layered intrusions, but there is evidence for such directed fluid flow. During compaction of the crystal-liquid mush, minerals can develop a pronounce foliation of tabular minerals such as plagioclase (e.g., Meurer and Boudreau, 1998b). If the rocks are not horizontal, then this foliation of individual grains would preferentially direct bubbles and more extensive fluids up-dip. This could also occur in modally layered rocks is which some layers are more permeable than others. If not horizontal, the less permeable layers can direct fluids up-dip. An example is seen in the

Figure 5.12 Vesicular cylinders in a basalt flow, Blackhead, Northern Ireland. (a) Arrows mark vertical, vesical-rich cylindrical pipes. (b) Outcrop surface approximately normal to the pipes, showing pothole-like structures resulting from preferential weathering. From Fowler et al. (2015), reproduced with permission.

Platinova Au-PGE reef of the Skaergaard. As noted in Chapter 2, the zone is characterized by a series of closely spaced mineralized layers that are upturned along the margins of the intrusion. Furthermore, the Au is particularly enriched along the margins, such that the ore zone has been described as stack of gold-rimmed plates (Nielsen et al., 2015). The Au distribution has been attributed to metal redistribution by hydrothermal fluids, which implies a lateral, up-dip migration of Au-bearing fluids along the margins (Andersen, 2006).

Other examples are seen in both the Munni Munni Complex of Australia and the Great Dyke of Zimbabwe, PGE-and base metal sulfide mineralization is concentrated along the contact between the underlying ultramafic rocks and the overlying gabbroic rocks. The contact is not flat, but it and the igneous layering is downwarped toward the centres of each

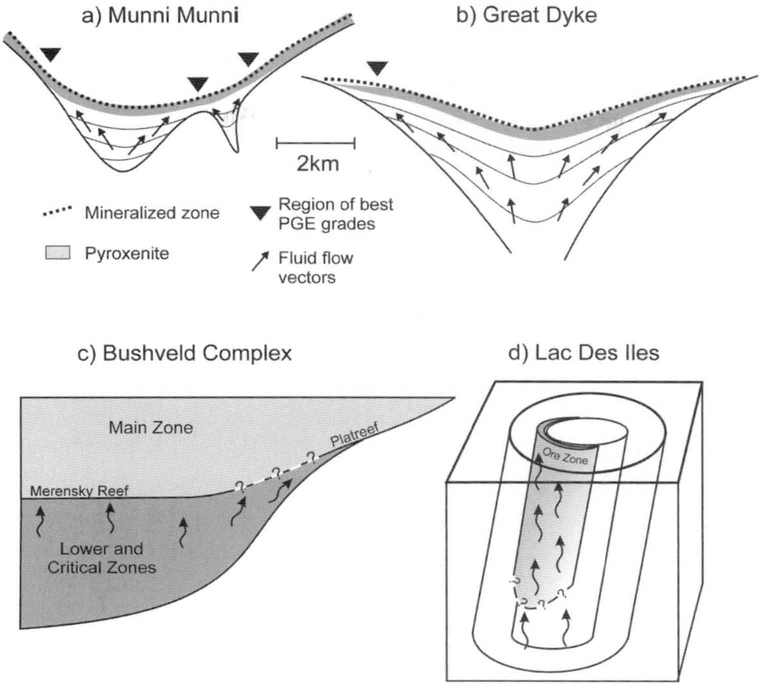

Figure 5.13 Examples of directed fluid flow in layered intrusions. (a–b) Simplified cross sections of the Muni Munni and Great Dyke showing locations of the main PGE-sulfide zones at the boundaries between the ultramafic and gabbroic sections of each intrusion. Triangles mark the regions that have the highest PGE grades, which tend to be up-dip as defined by the local layering attitudes. Arrows denote suggested fluid flow patterns as influenced by the layering. Redrawn and modified after Barnes et al. (1990), with permission. (c) A cartoon of the Platreef as an up-dip extension of the Merensky Reef as mineralizing fluids migrate upward to the region where the Bushveld rocks onlaps the country rocks. (d) Schematic view of the arc-like ore zone of the Mine Block Intrusion, Lac des Iles, showing preferential migration of ore fluids along the upper spine of the plunging intrusion. After Schisa et al. (2015), with permission.

intrusion, presumably during compaction (see Chapter 11 for more details). In the case of the Munni Munni, a similar downwarping occurs where a presumed feeder dike enters the chamber, resulting in layering being draped over a country rock high between the feeder dike and the main magma chamber. In both intrusions, the higher grade mineralization occurs along the higher, up-dip regions, including at the crest between the Munni Munni magma chamber and its feed dike (Figure 5.13a, b). The location of the higher grades is consistent with a focused fluid flow leading to mineralizing fluids to be preferentially directed up-dip.

The relationship between the Platreef and the Merensky Reef of the Bushveld Complex is uncertain. However, it is observed that the ore grade of the Platreef increases as one moves away from the contact and into the Bushveld proper. This allows the open pit mining of the Platreef to deeper levels. A hydromagmatic model would suggest that the Platreef is similar to the situation seen the Great Dyke and Munni Munni intrusions in that fluids from a compacted central part of the intrusion are moving laterally up-dip but diminish as the fluids migrate farther from the thicker parts of the intrusion (Figure 5.13c).

Finally, the Mine Block Intrusion of the Lac des Isles is a concentrically zoned, pipe-like intrusion that plunges steeply to the southwest. It is badly altered in the upper portions but becomes less so at depth. It hosts a Pd-rich ore zone that is mainly concentrated along an arc in the southwestern portion of the intrusions. Mineralization is largely confined to a varitextured plagioclase amphibolite termed the 'varitextured gabbro' but relic textures and deeper development shows the original rock was a varitextured norite unit. Mineralization terminates against an inner leucocratic unit termed the 'east gabbro' but has a CIPW normative mineralogy suggesting the original protolith was a leuco(gabbro)norite. The presence of pegmatoidal textures, the high Cl concentration in apatite in the fresher rocks, the unusually high Pd/Pt ratio of 12.3, and the very high Pd/Ir ratios (4,500–65,000) of the ores has led to a number of petrogenetic models range from those that are to those that suggest a role by fluids infiltrating across a range of conditions, from magmatic down to greenschist-grade temperatures (e.g., Talkington and Watkinson, 1984; Somarin et al., 2009; Barnes and Gomwe, 2011; Boudreau et al., 2014).

The discovery of unaltered mineralized norites found at the deeper levels has cast doubt on the role of amphibolite-greenschist alteration as the main process of ore deposition and instead likely resulted in the degradation of an originally igneous/hydromagmatic ore assemblage (Boudreau et al., 2014). To explain the arc-like distribution of the ore along one side of a nominally concentric zoned intrusion, Schisa et al. (2015) proposed a hydromagmatic model in which igneous fluids preferentially migrated upwards along the southwestern spine of the plunging intrusion. Fluid migration was restricted to a fluid-saturated annular region between the solidified contact zone at the margin of the intrusions and a hotter, vapour-undersaturated central core (Figure 5.13d).

6

Halogens in Layered Intrusions

6.1 Introduction

The study of igneous fluids and their role in the crystallization of intrusive magmas is complicated by the fugitive nature of igneous fluids and the difficulty of deciphering the evidence for them. This is especially true for the relatively low-pressure degassing of basaltic magmas, for which magmatic temperatures are above the stability of most hydrous minerals. In addition, the high abundance of carbon species (CO, CO_2) and chlorine in the magmatic fluid phase can lower the thermal stability of common hydrous minerals such as the amphiboles, but can lead to the precipitation of graphite (e.g., Elliot et al., 1982; Ballhaus and Stumpfl, 1985, 1986; Mathez et al., 1989) and high-temperature carbonate minerals (e.g., Aird and Boudreau, 2013). Nonetheless, the presence of hydrous phases can be combined with the local occurrence of fluid inclusions and hydrous silicate liquid inclusions (Chapter 7) to make inferences of volatile behaviour in igneous systems.

As described in Chapter 3, chlorine is broadly *vapourphile* and will preferentially fractionate into a vapour phase if present. In contrast, although a volatile element, fluorine is a *magmaphile* element and prefers to remain with the liquid during magma degassing. The absence of hydrous minerals until late in the crystallization of most mafic magmas would suggest that there should be no Cl-F fractionation in the absence of a fluid phase. Conversely, Cl-F variations in the absence of hydrous minerals can be used to infer the gain or loss of a volatile component.

Ideally, estimates of halogen concentrations in the parent magma (prior to saturation of the magma in a halogen-bearing phases) could be made using whole-rock analysis, assuming that the nominally anhydrous minerals contain no significant halogen concentrations and that the bulk analysis represents the concentrations of these elements in trapped (and now solidified) interstitial liquid. However, the low residual porosity most rocks in layered intrusions, the ease by which whole-rock halogen abundance may be affected by post-magmatic, low-temperature deuteric or later metamorphic fluids, and the possible incorporation of halogens into secondary minerals such as serpentine and chlorite all make the interpretation of whole-rock concentrations problematic. However, there are analyses that demonstrate that whole-rock abundances can have unusually high Cl concentrations that are also seen in the halogen-bearing minerals (e.g., Cawthorn and Poulton,

1988; Wilson, 2012). With these exceptions, however, halogen geochemistry is best determined by the compositions of unaltered halogen-bearing igneous minerals.

Where they do occur on the liquid line of descent of the parent magma, halogen-bearing minerals are typically only observed as liquidus minerals late in the crystallization sequence in most mafic and ultramafic layered intrusions. For example, in the Skaergaard intrusion, all of the liquidus minerals are nominally halogen-free until the magma became saturated in apatite during the crystallization of Upper Zone b (UZb). It is estimated that approximately 88% (McBirney, 1975) to about 94% (Wager and Brown, 1968) of the Skaergaard magma crystallized prior to the formation of UZb. Similar estimates have been made for the Kiglapait intrusion (Huntington, 1979) and the Bushveld (Cawthorn and Walsh, 1988), suggesting that apatite saturation after approximately 90% crystallization is common for many intrusions. Hence, halogen concentrations in a fractionating magma can be directly estimated from mineral chemistry only during the last ~10% of the crystallization history. This is in contrast with more siliceous liquids, in which apatite and other hydrous minerals may be saturated throughout crystallization.

This does not mean that igneous hydrous minerals are absent in layered intrusions, only that those that do occur tend to be relatively minor phases that crystallized from residual 'trapped' liquids. It is also observed in intrusions such as the Skaergaard intrusion and the Bushveld Complex that apatite is the only hydrous mineral to precipitate as a liquidus mineral along the liquid line of descent of the parent magma. That is, although minerals such as biotite and amphibole are common interstitial minerals, particularly in the ultramafic section of these intrusions, they are not observed as part of the primary mineral assemblage, even in the most evolved rocks. It is apparent that the local bulk composition effects are more important than the liquid line of descent in determining the presence or absence of these minerals.

A more serious complication is that the incorporation of Cl and F in mica and amphibole is a strongly function of crystal composition (e.g., Munoz and Swenson, 1981; Volfinger et al., 1985). This results in Mg-rich compositions to exclude Cl (*Mg-Cl avoidance*) and the Fe-rich compositions to exclude F (*Fe-F avoidance*). In contrast, halogen substitution in apatite at high temperatures is ideal or nearly so and natural apatite compositions span the complete range of OH-F-Cl end-members (e.g., Tacker and Stormer, 1989; Zhu and Sverjenski, 1992). As detailed below, the more chlorine-rich apatite compositions are most common in the lower sections of many intrusions where the bulk assemblage is relatively Mg-rich and the mica and amphibole exclude Cl. In contrast, the more F-rich apatite compositions occur in the more evolved, low *mg#* rocks in the upper parts of most intrusions where the mica and amphibole exclude F. For example, Boudreau et al. (1986) and Willmore et al. (2000) note that late, interstitial biotites in the Stillwater and Bushveld complexes have F concentrations ranging from 0.02 to 0.93 wt.% and Cl from 0.03 to 0.61 wt.%; similar ranges were also seen in late interstitial amphibole. The higher range of Cl is equivalent to mole fraction of the chlor-biotite component of about 6%. Compared with those from the Skaergaard and Kiglapait, the Stillwater and Bushveld biotites are enriched in Cl, but not F. Overall, however, biotite and amphibole in layered

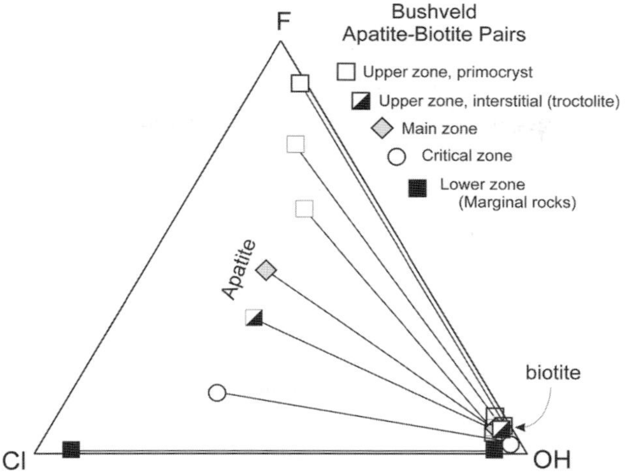

Figure 6.1 Compositions of co-existing apatite-biotite pairs from the Bushveld Complex. Redrawn after Willmore et al. (2000), with permission.

intrusions show a very limited incorporation of Cl and F (most would be considered hydroxybiotite) as compared to that seen in the co-existing apatite (Figure 6.1).

Because of these complications with micas and amphiboles, apatite is the more useful indicator of relative halogen variations in layered intrusions. Apatite is a common mineral found throughout all layered intrusions, it is stable at magmatic temperatures, and it does not readily re-equilibrate at low temperatures in the absence of extensive whole-rock alteration (but see below). As noted above, apatite is usually the first, and locally the only, halogen-bearing igneous mineral to crystallize, occurring as a late-crystallizing mineral in even the most compositional primitive rocks (Boudreau and McCallum, 1989). However, on the negative side, apatite can be sensitive to halogen loss when analysed by electron microprobe (e.g., Stormer et al., 1993).

6.2 Apatite Occurrence and Compositional Trends in Layered Intrusions

As noted above, in rock crystallized from evolved magmas (>~90% crystallization), apatite can precipitate as a liquidus mineral in modal abundance of several percent. In less evolved rocks, the abundance of apatite as an interstitial mineral is broadly a function of the amount of residual liquid. Widely spaced apatite grains can be difficult to locate in rocks with low bulk rock incompatible trace element abundances, particularly in the more primitive rocks found low in the stratigraphic section. Also, Meurer and Meurer (2006) note that late liquid migration induced by compaction can produce widely variable amounts of apatite in multiple thin sections cut from the same sample.

Apatite most commonly occurs as small, euhedral, prismatic grains with a maximum length of several hundred microns. Where the apatite is large relative to the confined space

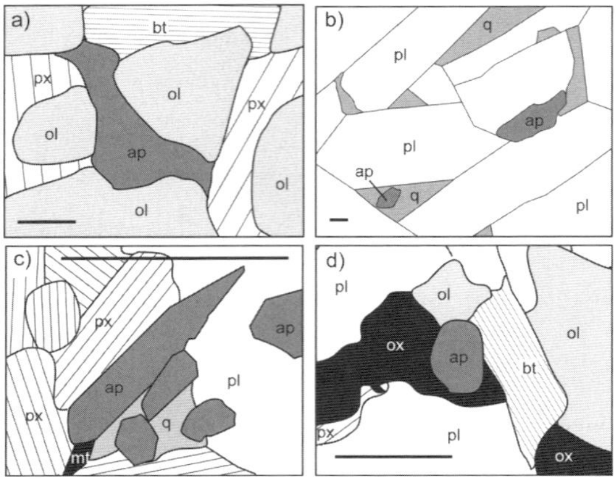

Figure 6.2 Examples of apatite occurrence in the Stillwater and Bushveld Complex. (a) Interstitial chlorapatite and associated biotite and pyroxene in a poikilitic harzburgite from the Peridotite zone of the Stillwater Complex. (b) Apatite and quartz interstitial to plagioclase from a Middle Banded series Anorthosite of the Stillwater Complex. (c) Primocryst apatite from the Upper zone of the Bushveld Complex. (d) Apatite in oxide olivine gabbro from the Upper zone of the Bushveld Complex. Abbreviations: ap = apatite, bt = biotite, px = pyroxene, pl = plagioclase, ol = olivine, ox = Fe-Ti oxide, q = quartz. Scale bar in each is 100 microns. Figures (a) and (b) redrawn after Boudreau and McCallum (1989), with permission.

in which it grew, it can form irregular, angular grains defined by the interstitial space between olivine, pyroxene and plagioclase (Figure 6.2) and be difficult to distinguish from anhedral plagioclase. Drinkwater et al. (1990) suggested that early and late-crystallizing interstitial apatite from the Forrestal section of the Dufek intrusion can be distinguish by habit, the former being euhedral and the later more distinctly anhedral. Apatite also occurs in late-stage segregations (e.g., granophyric segregations and pegmatoids) and in cross-cutting secondary replacement bodies (e.g., dunite pipes crosscutting pyroxenites).

The more F-rich apatites are relatively resistant of low-temperature alteration and weathering. This is why fluoride is commonly added to drinking water and toothpaste; the conversion of hydroxyapatite to fluorapatite makes the tooth enamel more resistant to bacterially produced acid. In contrast, Cl-rich apatites can readily lose their Cl. These chlorapatites that are rich in REE will also precipitate REE-rich phosphates such as monazite during alteration (Figure 6.3). The ease by which the Cl-rich apatite can be altered implies one should be careful assessing halogen trends in those intrusions that are extensively altered, as noted below.

Extensive stratigraphic analyses of apatite are available for a number of intrusions. These include the Bushveld Complex (Willmore et al., 2000), the Stillwater Complex (Boudreau and McCallum 1989), the Skaergaard intrusion (Nash, 1976; Brown and Peckett, 1977; Sonnenthal, 1992), the Munni Munni Complex (Boudreau et al., 1993), the Great Dyke (Boudreau et al., 1995), the Penikat Complex (Halkoaho and Alapieti, 1993), the Mine Block Intrusion of the Lac des Iles Complex (Schisa et al., 2015) and the Dufek intrusion

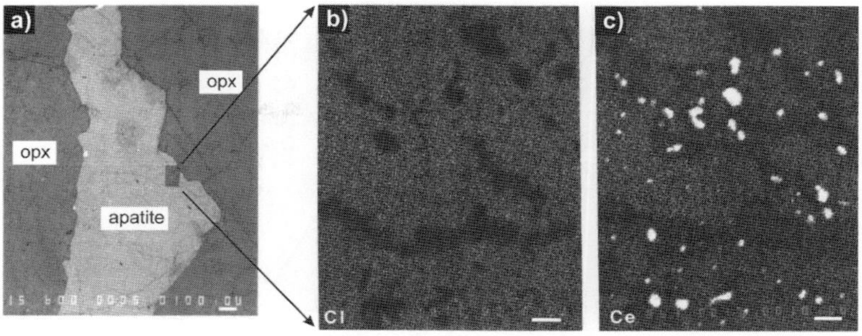

Figure 6.3 Alteration of chlorapatite from the J-M Reef of the Stillwater Complex, Montana. (a) Backscattered electron image of chlorapatite interstitial to orthopyroxene in harzburgite from the J-M Reef, Minneapolis Adit area. Although the rock is relatively fresh with little alteration evident in the orthopyroxene, the slightly darker grey patches within the chlorapatite are regions of alteration to hydroxyapatite. The bright spots that occur along the grain margins and as trails within the chlorapatite are mainly monazite and minor Fe-oxide. Scale bar is 100 microns. (b) and (c) X-ray maps of Cl and Ce, respectively, for the small shaded region shown in (a), and for which the scale bars are 10 microns each. The Figures illustrate the partial replacement of REE-rich chlorapatite with hydroxyapatite (darker grey areas in Cl plot) and monazite, (Ce, La, Nd, Th)PO4 (bright spots in Ce plot). From Boudreau and McCallum (1990), with permission.

(Drinkwater et al., 1990). A comparison of a number of intrusions shows that apatite compositions in them span almost the complete range of Cl-F-OH substitutions (Figure 6.4). However, apatite in the majority of intrusions is predominantly fluorian, with the chlorapatite component usually <40 mol.% and more commonly less than about 20 mol.%. Furthermore, apatite compositions overall become poorer in Cl and richer in F in the more evolved rocks, although this general compositional trend may be erratic and some intrusions may contain stratigraphic sections where Cl increases with stratigraphic height (Figure 6.5). The Cl-poor nature of apatite in the more evolved rocks mimics the typically Cl-poor apatite compositions that are characteristic of granites (Webster and Piccoli, 2015)

Exceptions to the general Cl-poor character of igneous apatite occur in the Stillwater and Bushveld complexes. In these two intrusions, for which stratigraphic coverage is relatively complete, the stratigraphic trend of the Cl/F ratio is characterized by an initial rapid increase in the Cl/F ratio in the toward 90+% chlorapatite compositions as one goes from the basal rocks into the ultramafic section. The apatite in these Cl-rich sections typically contains <0.4 wt.% F and >6.0 wt.% Cl (Figure 6.5) (Boudreau and McCallum, 1989; Willmore et al., 2000) and mimics the high Cl seen in bulk rock analyses (e.g., Wilson, 2012). The Cl-rich apatite occurs in rocks that are otherwise texturally 'typical' rocks from these intrusions; they are not at all pegmatoidal, for example.

In the Stillwater Complex, a change from high-Cl to more F-enriched apatite compositions occurs within Olivine-bearing zone I (OB-I), the host zone of the platiniferous J-M Reef. Although apatite compositions are variable above the J-M Reef, F-enriched compositions predominate. However, even above the J-M Reef, apatite in many rocks has Cl concentrations that are typically higher than are observed in apatite from other intrusions.

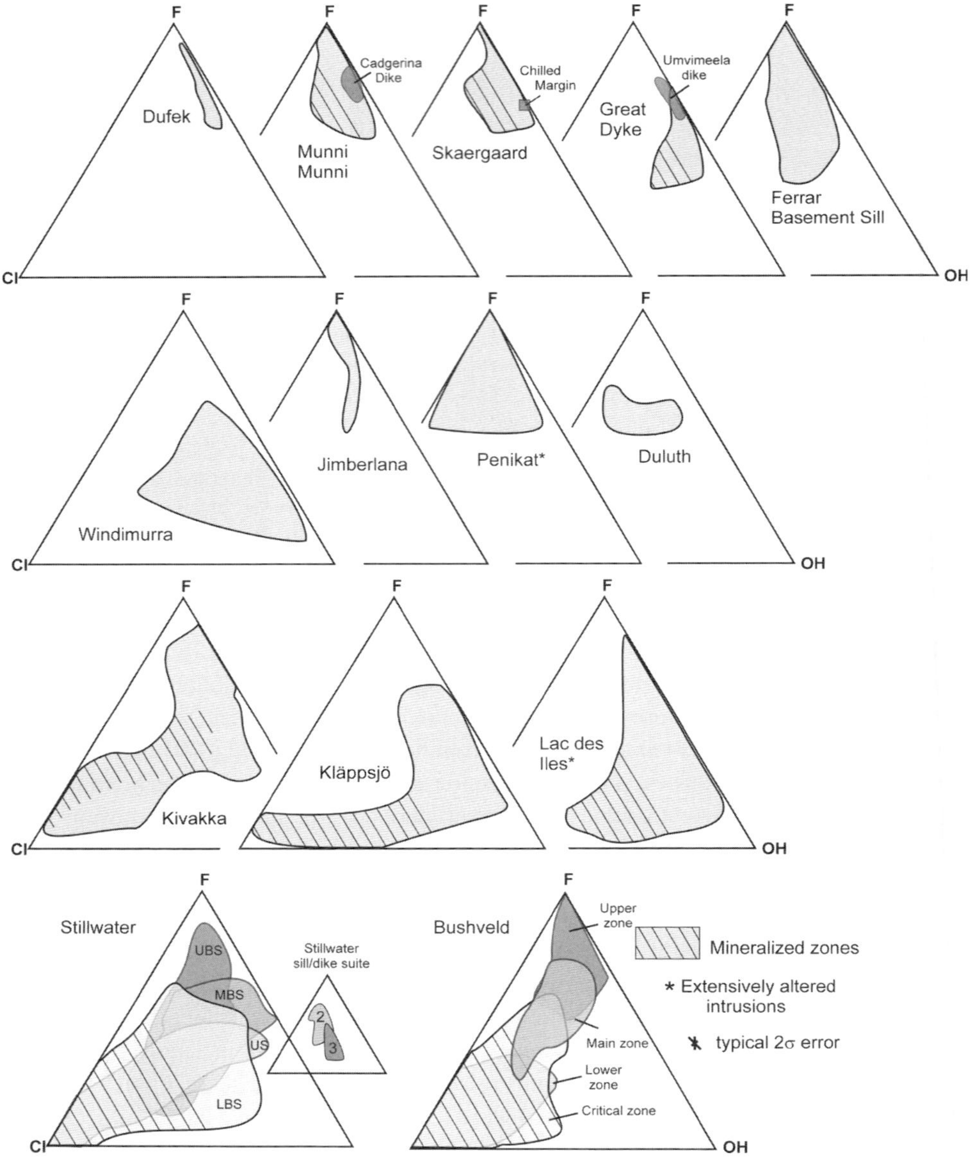

Figure 6.4 Compositional ranges of apatite from a variety of layered intrusions, from sources noted in text. The Penikat intrusions is extensively altered throughout and the Lac des Ilse is partly altered in the rocks near the surface but become fresher with depth. Ruled red regions show range of apatite compositions associated with PGE mineralization in those intrusions/sections that are relatively unaltered. The apatite from the Cadgerina Dike of the Munni Munni Complex, the Umvimeela Dike of the Great Dyke and the Stillwater Sill Dike suite are apatite from sills and dikes associated with these layered intrusion, whereas the Skaergaard chilled margin apatite is from the chilled margin of that intrusion.

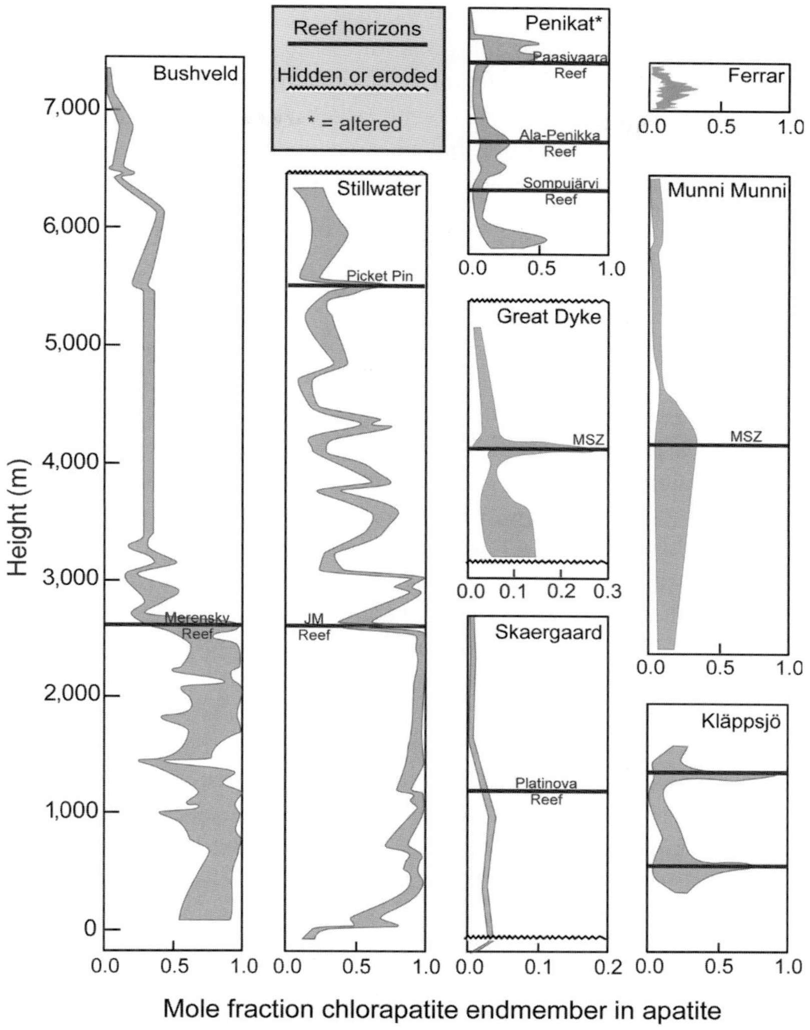

Figure 6.5 Stratigraphic trends of the mole fraction Cl/(Cl+F+OH) in apatite from a number of layered intrusions. Note the expanded x-axis scale for the Cl-poor Skaergaard and Great Dyke. The Penikat is partially altered throughout and the effect of this alteration on apatite is uncertain. From sources listed in the text.

In the Bushveld Complex, apatite from the Lower and Critical zones is also unusually Cl-rich, whereas the Main and Upper zones contain more fluorian apatite (Figure 6.5) (Boudreau et al., 1986; Boudreau and Kruger 1990; Boudreau et al., 1993b; Willmore et al., 2000). Apatite from within and beneath the Merensky Reef (including the UG2) contain >40 mol.% chlorapatite component, with many samples containing >80 mol.% chlorapatite. These compositions are similar to those of the lower part of the Stillwater

Complex. However, within approximately 60 m above the Merensky Reef, in the lower part of the Main zone, the values fall to as low as 20 mol.% chlorapatite component, which are more typical of the maximum values seen in other intrusions. Once apatite becomes a liquidus mineral in Upper Zone c (UZc), the general stratigraphic trend of decreasing Cl/F ratio with stratigraphic position continues such that apatite in the uppermost ferrogabbros is at or near end-member fluorapatite (Boudreau et al., 1993b; Willmore et al., 2000).

Besides the Stillwater and Bushveld complexes, other ultramafic-mafic layered intrusions that contain unusually Cl-rich mineral assemblages include the Koitelainen and Keivitsa-Satovaara complexes of northern Finland where chlorapatite is associated with PGE-sulfide mineralization as described in the brief report of Mutanen et al., (1988). Also, although most of the Penikat intrusion is not unusually Cl-rich, Halkoaho and Alapieti (1993) reported local occurrences of Cl–rich compositions associated with PGE mineralized zones. This is a problematic intrusion, as the rocks are variable altered. Although textures may be preserved, olivine is invariable completely altered and pyroxene and plagioclase less so. As noted above, Cl-rich apatite can be readily replaced by hydroxyapatite under these conditions. This is also observed in the Mine Block Intrusion that hosts the Lac Des Iles Pd deposits, where the rocks near the surface are extensively altered but become fresher with depth. Apatite from mineralized, unaltered rocks of the Mine Block Intrusions typically have higher Cl concentrations than those from altered mineralized rocks (Schisa et al., 2015). This intrusion is also unusual in that it is a ~vertical pipe-like body with a radial stratigraphy, with mineralization occurring as a vertically elongated, crescent-shaped zone along the south and western parts of the intrusion.

Although stratigraphic coverage is limited, apatite from the Windimurra intrusion of Western Australia tend to fall in intermediate- to hydroxyapatite-rich compositions, the latter associated with samples containing interstitial amphibole. This mineralogical association suggests that the interstitial liquids of the Windimurra were (at least locally) more OH-rich than most layered intrusions. Because these OH-rich samples are near faults and discordant zone boundaries, it is not thought that these OH-rich compositions are necessarily related to primary magmatic processes (Boudreau et al., 1993b). High OH compositions are also characteristic in the few analyses of apatite available from the layered Mt. Thirsty and Ora Banda sills of Western Australia (Boudreau et al., 1992). The stratigraphic coverage from the Dufek intrusion of Antarctica is limited to the more evolved rocks from the Forrestal section and are characterized by >60 mol.% fluorapatite and <10 mol.% chlorapatite compositions.

Considerable variation in the composition of different grains from individual thin sections is observed in all intrusions. Absolute halogen concentrations from grain to grain in the same thin section may vary by as much as 20 mol.% or more. In the more extreme cases, positive correlations are observed between Cl and Na and REE concentrations in apatite (e.g., Boudreau and McCallum, 1990; Drinkwater et al., 1990). The within-sample variation suggests that significant fractional crystallization of interstitial liquid occurred on the scale of an individual thin section as apatite grains were isolated either in isolated pocket of residual liquid or by entrapment in growing silicate minerals. The lack of significant compositional zoning in individual grains, however, suggests that the isolated grains homogenized during slow cooling.

6.3 Associated Chilled Margin and Sill/Dyke Compositions

Relative halogen abundance in the parent magmas potentially can be estimated from the composition of apatite in both chilled margins and in sill and dyke rocks that are temporally and spatially related to specific layered intrusions. These sills and dykes may be used where there is other evidence to suggest that these rocks were formed from the same magma(s) that formed the associated layered intrusions. However, analysis of sills and dykes in general, and not necessarily contemporaneous with any given intrusion, can suggest just how common high-Cl magmas might be.

Compositions of apatite from the chilled margins of the Skaergaard intrusion and from sill/dike suites associated with the Stillwater Complex, the Munni Munni Complex of Western Australia and the Great Dyke of Zimbabwe are shown in Figure 6.5. Sills and dikes that intrude beneath the Stillwater Complex have been subdivided into six groups on the basis of petrographic and geochemical criteria by Helz (1985); Helz considered the group 2 and group 3 rocks to be the most probable candidates for the Stillwater parent magma(s). The Cadgerina dike is an apparent feeder dyke to the Munni Munni Complex, a 2.9 Ga. intrusion situated in the West Pilbara block. The Umvimeela dike is a thin, unlayered intrusion coeval with the Great Dyke that intrudes parallel with it along much of its length. (See Chapter 11 for more details of the geology of these latter two intrusions)

In general, apatite from these rocks contain less Cl than can be found in the associated intrusions, although those from the Stillwater sill/dike suite are higher than are chilled margins and sill/dike rocks associated with more Cl-poor intrusions. However, even the most Cl-rich of the Stillwater sill/dike apatite is not particularly rich in Cl in comparison with the chlorapatite in the layered sequence: the Stillwater sill/dike apatite contain <3.0 wt.% Cl and >1.2 wt.% F (Boudreau and McCallum, 1989). Insofar as the apatite of the sill/dike rocks is representative of at least some of the possible parent magma compositions, there is no evidence that any of the possible initial Stillwater and Bushveld magmas were as Cl-rich as is preserved in the layered intrusion apatite. If magmas with unusually high Cl concentrations were responsible for the Cl-rich assemblages of the Stillwater and Bushveld Complexes, these magmas have either not been sampled or recorded in the sill/dike suite. To a lesser degree the same is true for the dikes associated with the Munni Munni Complex and the Great Dyke: the Cl content of apatite from ultramafic cumulates from both intrusions is typically higher than that in the associated dikes. This evidence has been used to infer that the higher Cl seen in the intrusions was the result of either equilibration with a Cl-rich fluid and/or upgrading by a zone-refining type mechanism (see below). In modern magmas, mafic magmas can contain up to ~1 wt.% Cl (Chapter 3), and high Cl/F weight ratios are associated in general with high overall volatile contents, suggesting that the Stillwater Complex and other high-Cl layered intrusions crystallized from magmas that may have contained >1 wt.% water (Boudreau et al., 1997).

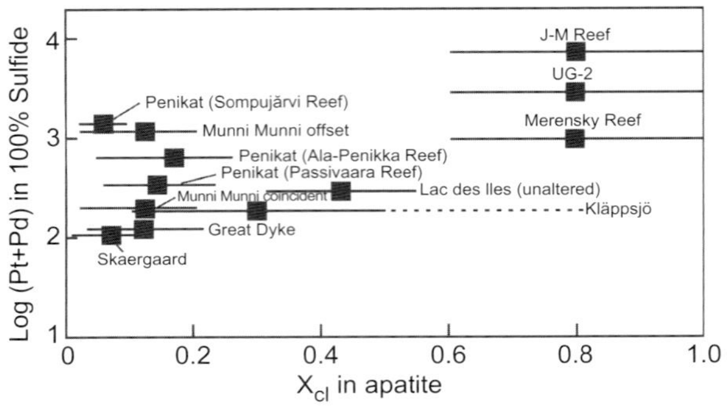

Figure 6.6 Plot of the log (Pt + Pd) in 100% sulfide fraction vs. chlorapatite component in and below the various reefs as shown. The Penikat Complex is described as partially altered throughout and the effect of this alteration on apatite composition is uncertain. Modified and expanded after Schisa et al., (2015), with permission.

6.4 Chlorine and PGE Ore Grade

The high Cl in apatite associated with PGE-sulfide zones, particularly those of the Stillwater and Bushveld complexes, is one of the stronger lines of evidence for a fluid transport of the ore component. A plot of PGE tenors in 100% sulfides shows a generally positive correlation with the Cl/Cl + F) ratio of associated apatite (Figure 6.6). One has to be careful not to over-interpret this plot, as metal tenor can be affected by such processes as later sulfide loss and chromatographic separation effects noted in Chapter 10. Similarly, the Penikat Complex contains high-grade PGE horizons, but it is extensively altered which can affect apatite Cl concentrations as noted above. Overall, however, the broad correlation of high PGE tenors associated with high Cl in associated apatite is consistent with experimental work demonstrating that Cl is important for the transport of the PGE by fluids. As summarized in Chapter 4, the PGE, even the relatively insoluble IPGE (Ir, Ru, Os) can have appreciable solubility in high-temperature, Cl-rich fluids. It also supports the suggestion of Boudreau (1993) that the halogen contents of apatite is an important exploration tool for the PGE in layered intrusions.

6.5 Apatite-Biotite Geothermometry

Zhu and Sverjensky (1992) have formulated an apatite-biotite F-OH geothermometer that takes into account the Mg/Fe dependence on the incorporation of F in biotite. F-OH exchange between apatite and biotite can be expressed in the following end-member exchange reaction:

$$\underset{\textit{biotite}}{1/2KMg_3AlSi_3O_{10}F_2} + \underset{\textit{apatite}}{Ca_5(PO_4)_3(OH)} = \underset{\textit{biotite}}{1/2KMg_3AlSi_3O_{10}(OH)_2} + \underset{\textit{apatite}}{Ca_5(PO_4)_3F} \quad (6.1)$$

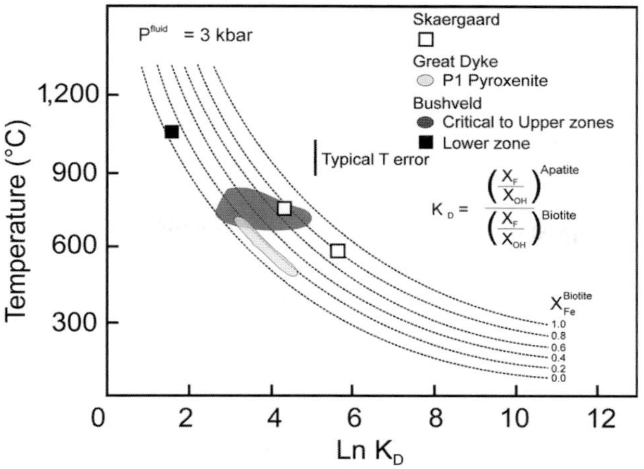

Figure 6.7 Calculated apatite-biotite OH-F exchange temperatures for the Bushveld Complex, Skaergaard intrusion, and the Great Dyke P1 pyroxenite unit. Data from Willmore et al. (2000) and Sonnenthal (1992). Redrawn after Boudreau and Hoatson (2004), with permission.

For the above reaction, one can define a partitioning coefficient as follows:

$$K_D = \frac{\left(\frac{X_F}{X_{OH}}\right)^{apatite}}{\left(\frac{X_F}{X_{OH}}\right)^{biotite}} \tag{6.2}$$

where X is the mole fraction of the subscripted component in either the biotite or apatite. Data from apatite-biotite pairs from six samples from the Skaergaard, Great Dyke and Bushveld Complex shown plotted along with the geothermometer of Zhu and Sverjensky in Figure 6.7.

For the Bushveld samples, the calculated temperatures do not change significantly for biotite compositions ranging from phlogopite- to annite-rich, the average temperature being about 750 °C. For the Great Dyke samples, the ranges shown are the total ranges observed, the spread being most strongly controlled by the low (and analytically variable) F concentrations in biotite and its effect on the value of K_D. The average temperature of the four samples from the Great Dyke P1 pyroxenite zone is 564 °C. This compares with indicated blocking temperatures on the order of 600–700 °C for olivine and 750–850 °C for pyroxene calculated using Fe/Mg partitioning between these two minerals and chromite from ultramafic sequence cumulates from the Great Dyke (Wilson, 1982).

Although Ti dependence possibly affects the calculated temperatures and is not accounted for in the Zhu and Sverjenski (1992) formulation, these temperatures do agree with the textural evidence that most samples were not affected either by low-temperature reequilibration nor the spotty alteration evident in some samples. Also, the samples fall within the region where relatively small variations in halogen concentrations result in comparatively large difference in calculated temperature. For example, reequilibration on cooling from 1,050 to 750 °C will approximately double the value of K_D. However, it is also observed that apatite in biotite-bearing samples is about the same composition as that in biotite-free systems from equivalent rock units. This suggests that apatite compositions change only modestly by exchange with biotite on cooling and that the variation in the partitioning are accomplished primarily through changes in the otherwise low amounts of halogens incorporated in biotite in these rocks.

6.6 Chlorine Isotopes

Chlorine isotopes can potentially elucidate the source of the Cl in layered intrusions as ^{37}Cl and ^{35}Cl are not strongly fractionated by magmatic and high-temperature hydrothermal processes (Liebscher et al., 2006). Furthermore, the mantle is isotopically heavy, whereas the near-surface reservoirs (oceans, atmosphere and crust) are isotopically light (e.g., Magenheim et al., 1995). In regard to the Bushveld and Stillwater Complexes, the data is conflicting. For the Bushveld, Willmore et al. (2002) noted that most samples overlap the range of values for pristine MORB $\delta^{37}Cl$ of 3–7‰ relative to sea water (95% confidence limit error is 0.2‰) (Figure 6.8). This contrasts with distal metasedimentary floor rocks beneath the Bushveld complex that have $\delta^{37}Cl$ = ~0‰. Rocks from the marginal sill and the Platreef tend to be intermediate between these two endmembers. They suggested that the Cl isotope composition and high Cl/F ratio is therefore a fundamental characteristic of the Lower and Critical Zone magmas and not derived from infiltration from the country

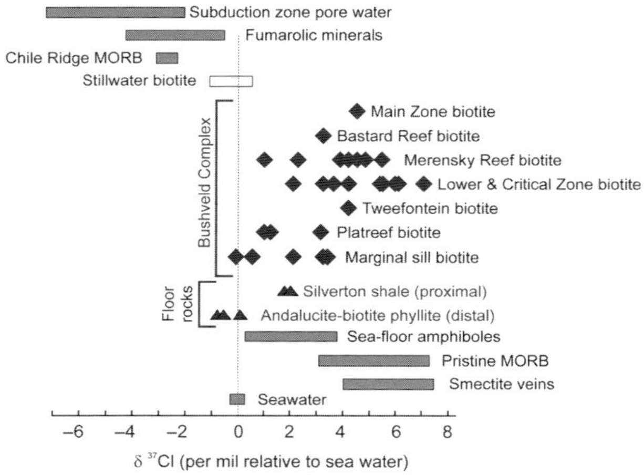

Figure 6.8 Comparison of stable Cl isotopic compositions of the Bushveld (diamonds) and associated rocks (triangles) with those from other locations, including the Stillwater Complex. From Willmore et al. (2002), reprinted with permission of publisher.

rock. These results, the high Cl/F ratio and the boninitic, arc-like character of Bushveld magma, are consistent with hydration partial melting of a depleted source in a subduction-related setting with Cl supplied either by the dehydration of a subducted slab or a 'wet' mantle hotspot, both with a heavy Cl signature. In contrast, Boudreau et al. (1997) noted that biotite from the Ultramafic series and from the J-M Reef of the Lower Banded series of the Stillwater Complex have a crustal $\delta^{37}Cl$ values ranging from 0.27‰ to –0.93‰ and average –0.18‰. They suggested that the Stillwater source rock was contaminated by crustal fluids during subduction.

6.7 Halogen Partitioning between Apatite, Silicate Liquid and Volatile Fluid

The partition of the halogens between a silicate liquid, apatite and vapour has been the subject of a number of recent studies. In general, experimental studies indicate that the affinity for Cl decreases in the following sequence, aqueous fluid > apatite ≥ silicate liquid > other hydrous minerals (biotite and amphibole) (e.g., Brenan, 1993; Sato et al., 2005; Mathez and Webster, 2005; Webster et al., 2009; Doherty et al., 2014; McCubbin et al., 2015). Briefly, the formation of chlorapatite-rich apatite can involve either high Cl concentrations in a liquid or vapour phase in equilibrium with apatite, or relatively low concentrations of the OH- and F-components. In addition, partitioning of Cl between apatite and a silicate liquid is complicated by the fact that the silicate liquid will saturate in a Cl-bearing fluid phase prior to the liquid saturating in the more Cl-rich apatite compositions.

This point is illustrated in Figure 6.9, which shows portioning of Cl and F between apatite, liquid and fluid. Figure 6.9a shows the portioning behaviour of Cl between apatite and silicate liquid for liquids ranging from basaltic to rhyolitic in composition. Using the basaltic composition partitioning of Figure 6.9a as an example, it is observed that Cl partitioning shown Henry's law behaviour up until about 4 wt.% Cl in the liquid, at which point the portioning into apatite strongly increases relative to the liquid. As summarized by Webster et al., this is the result of Cl reaching the solubility limit in the silicate liquid, after which point added Cl goes into a vapour phase or apatite. The effect is even more extreme for more siliceous liquid compositions, as are also plotted on Figure 6.9a. Webster et al. suggest that the highest concentration of Cl in a given magmatic fluid will typically be when the fluid first exsolves because the Cl content is at is maximum just prior to fluid formation and because the largest value of $D_{Cl}^{fluid/liquid}$ occurs when the fluid first exsolves. Candela (1986), however, suggests that maximum Cl concentrations are a function of the bulk D_{Cl} for systems that are crystallizing anhydrous solids as well as mixed volatile fluid. For example, systems that are saturated early in relatively low solubility fluids (in CO_2-rich magmas, for example) may continue to have Cl increase in the magma during degassing simply because the bulk D_{Cl} is less than 1.0.

In contrast, F portioning between silicate liquid and apatite also increases monotonically with increasing F concentration in the liquid (Figure 6.9c), a result of the preference of apatite to incorporate F over Cl and the higher solubility of F in silicate liquids. Also, partitioning of Cl between apatite and vapour is linear throughout the range of

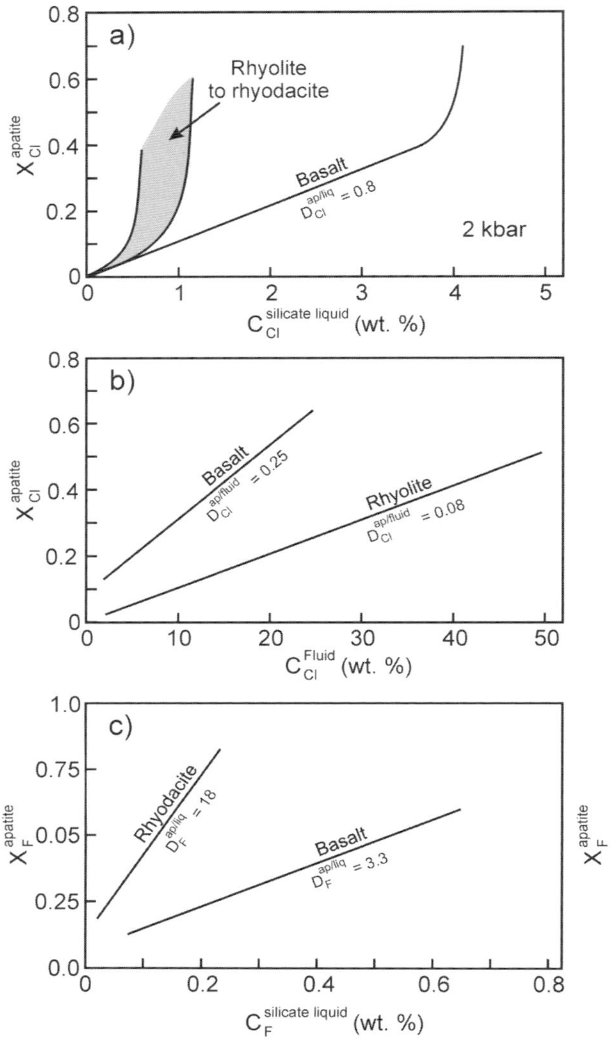

Figure 6.9 Summary of experimentally determined partitioning data as a function of melt composition for (a) Cl between apatite and silicate liquid, (b) Cl portioning between apatite and fluids and (c) F partitioning between apatite and silicate liquid. All experiments shown at 200 MPa. Chlorine generally prefers the silicate liquid and especially the fluid over apatite, whereas fluorine partitions more strongly in favour of apatite than the silicate liquid. Redrawn after Webster et al. (2009), with permission.

experimental conditions with liquids of a fixed composition (Figure 6.9b). In general, Cl preferentially partitions into a silicate liquid or especially a vapour phase in preference to apatite, whereas F preferentially partitioning into apatite over a silicate liquid or vapour. The partition coefficients shown in Figure 6.9 are broadly similar to those of Li and

Hermann (2015); in general, experimental values of $D_{Cl}^{fluid/liquid}$ for basaltic systems are <10. Experimental determinations of the partitioning of F between silicate liquids and fluids are complicated by analytical problems involving determination of F in the fluid. However, most experiments suggests that F concentrations in aqueous and aqueous-carbonic fluids at magmatic conditions preferentially prefer the silicate liquid with $D_F^{fluid/liquid} < \sim 1.0$ but some experiments have resulted in measured $D_F^{fluid/liquid}$ values > 100 (Webster et al., 2009).

Because of these problems with the simple apatite-liquid distribution coefficients for the individual anions Cl^-, F^- and OH^-, others have suggested that an exchange coefficient is found to be much less variable (e.g., McCubbin et al., 1105; Li and Hermann, 2015). For example, consider the following reaction between silicate liquid and apatite:

$$\underset{apatite}{\text{OH-apatite}} + \underset{liquid}{Cl^-} = \underset{apatite}{\text{Cl-apatite}} + \underset{liquid}{OH^-} \tag{6.3}$$

One can define an exchange coefficient, K_D, as follows:

$$K_{DCl-OH}^{apatite-liquid} = \frac{X_{Cl}^{Apatite} \cdot X_{OH}^{liquid}}{X_{OH}^{apatite} \cdot X_{Cl}^{liquid}} \tag{6.4}$$

Li and Hermann (2015) summarized a number of studies and found broadly applicable values for the exchange coefficient as follows: $K_{DCl-OH}^{apatite-liquid} = \sim 10$, $K_{DF-Cl}^{apatite-liquid} = \sim 5$, and $K_{DF-OH}^{apatite-liquid} = \sim 50$.

For subsolidus (liquid-absent) temperatures, Kusebauch et al. (2015) have suggested that apatite-fluid distribution coefficients for the halogens show systematic preferences, and although apatite-fluid exchange can be complex, they suggest useful approximate values for the apatite/fluid distribution coefficients are $D_F \cong 120$, $D_{OH} \cong 100$, $D_{Cl} \cong 2.3$, $D_{Br} \cong 0.045$, $D_I \cong 0.0025$ for crustal pressures and temperatures of 400–700 °C. The strongly incompatible nature of Br and especially I suggest why these two halogens are rarely reported in microprobe analyses of apatite.

Few basaltic liquids *sensu stricto* are saturated in apatite; as noted above, about 90% crystallization is required for a basaltic liquid to reach apatite saturation. Given that apatite commonly occurs with interstitial quartz in layered intrusions, experiments in the more Si-rich systems are perhaps most applicable to the crystallization of apatite from interstitial liquids. An example of the calculated liquid in equilibrium with interstitial apatite is shown in Table 6.1, in which the MELTS program (Ghiorso et al., 1994; Ghiorso and Sack, 1995) was used to model the equilibrium assemblage of a Bushveld orthopyroxenite. Assuming that the initial bulk orthopyroxenite can be considered a mixture of orthopyroxene and a trapped liquid component, then modelling the cooling of the bulk rock will allow one to find the composition of the interstitial liquid at apatite saturation. In this example, apatite saturation occurs when the bulk rock is ~94% crystallized, at which point the residual liquid has the composition of an andesite.

Table 6.1 *Results of MELTS calculation of the equilibrium silicate liquid at apatite saturation in a solidifying pyroxenite*

	Bulk Pyroxenite[1]	Equilibrium liquid at apatite saturation 1,025 °C, 2 kbar
SiO_2	53.35	61.2
TiO_2	0.29	1.16
Al_2O_3	4.87	14.7
Fe_2O_3	1.13	0.48
Cr_2O_3	0.67	0.04
FeO	11.32	2.31
MnO	0.27	2.84
MgO	22.95	2.07
CaO	4.17	6.43
Na_2O	0.65	3.46
K_2O	0.15	2.25
P_2O_5	0.10	1.31
H_2O	0.10	1.70
total	100.0	100.0

Note: Bulk orthopyroxenite taken from Eales (1994), analysis 12, 0.10 wt.% H_2O added for apatite saturation. Equilibrium liquid calculated using MELTS as described in text.

6.8 Causes of Halogen Variations in Apatite

Prior to the separation of a halogen-bearing phase from the crystallizing magma chamber, there should be little if any variation in the Cl/F ratio of the interstitial liquid nor of the interstitial apatite as a function of stratigraphic height. That is, although the concentrations of both Cl and F will increase during fractional crystallization of anhydrous minerals, and apatite crystallization will occur earlier in the crystallization of the interstitial liquid in the more evolved rocks, the Cl/F ratio will be fixed and the apatite that does precipitate should reflect this constancy. (The exception is the trivial amount that may partition into the nominally anhydrous minerals such as olivine, pyroxene and plagioclase). If this were generally true, one could use the apatite compositions up to the point it becomes a liquidus phases in the main magma body to infer the halogen abundances of the parent magma. However, from the discussion above, most intrusions exhibit an overall decrease in the Cl/F ratio of interstitial apatite as the rocks become more evolved. For example, in the Stillwater Complex apatite shows almost the full range of F-OH-Cl variations without apatite or any other hydrous minerals becoming a liquid phase in the main magma body in the exposed stratigraphic section. Imposed on this general trend are local sharp reversals or regions where the Cl/F ratio may increase as a function of stratigraphic position.

The problem of understanding the cause of halogen fractionating trends is further illustrated by the trends in apatite once it becomes a liquidus mineral. The decreasing

Cl/F ratio of apatite observed in the Upper zone of the Bushveld Complex and other intrusions in which the magma became saturated in apatite would suggest that Cl is behaving as a more compatible element than F, assuming apatite crystallization alone was fractionating the halogens. As summarized above, however, F preferentially partitions into apatite in preference for Cl, and fluorapatite preferentially precipitates even where the siliceous fractionates are enriched in Cl (e.g., Anderson, 1974). Hence, the decrease of the Cl/F ratio observed in the apatite-saturated rocks of the Upper zone of the Bushveld Complex is the reverse one would expect if apatite alone were fractionating Cl and F, and suggests that some additional mechanism is dominating the fractionation of the halogens even after apatite becomes a liquidus mineral.

The mechanisms that have been proposed to explain this fall into several categories: 1) Variations in initial parent magma Cl/F ratios. 2) Trapped liquid shift effects. 3) Temperature/pressure dependency of apatite/liquid partition coefficients. 4) Vapour separation/degassing of the magma chamber. 5) Vapour refining in a crystal mush. 6) Halogen incorporation into nominally anhydrous minerals. 7) External source for the Cl. These are discussed in some detail below.

6.8.1 Parent Magma Cl/F Variations

Taken at face value, the limited sill/dike and chilled margin evidence would suggest that the Stillwater and Bushveld parent magmas were enriched in Cl relative to the magmas that formed other intrusions. AS noted previously, the parent magmas to these two intrusions have been described as light REE-enriched, Ti-poor, siliceous high magnesium basalts (SHMB), of which boninites would be the modern day equivalent (e.g., Sun et al., 1989). Boninites crystallize abundant pyroxene, have late plagioclase saturation, and can have olivine and chromite as phenocrysts and are generally considered to have formed by the hydration melting of a previously melted mantle source in a back-arc setting (e.g., Crawford et al., 1989).

Hamlyn and Keays (1986), suggested that SHMB would be excellent source magmas for PGE deposits such as the J-M and Merensky reefs. They suggested that some residual sulfide remains in the mantle during an initial or 'first-stage' partial melting event. Mid-ocean ridge basalt (MORB) is an example of such a PGE-poor first-stage melt, as this magma typically contains less than 1 ppb Pd. A second melting event produces a relatively Mg-rich and Ti-poor magma, and the residual sulfides and their contained PGE are completely incorporated into the melt to produce a S-poor but PGE-fertile magma. Again, the modern equivalent would be boninites, which typically contain about 15 ppb Pd and are sulfur-undersaturated at the time of eruption.

Boudreau et al. (1997) suggested that the second stage mantle hydration melting event postulated for boninite formation (e.g., Crawford et al., 1989) would also add Cl and could explain the unusually high Cl of the Stillwater and Bushveld magmas as seen in the sill/dike suite of the Stillwater Complex, for example. The crustal Cl isotopic character of the Stillwater would be consistent with this interpretation if the light Cl came from a descending slab dewatering the light Cl fraction. The heavier Bushveld Cl would not directly support this, unless the fluxing agent was from a deeper mantle source or from a descending slab that had previously lost its light Cl fraction at a shallow level. Although this would explain some Cl-enrichment, no chilled margin or sill/dike rocks have apatite as Cl-rich as seen in the layered sequences of these two intrusions.

6.8.2 Trapped Liquid Effects

Cawthorn (1994) suggested variable amounts of apatite and trapped liquid might explain variations in Cl/F ratios. In brief, he suggests that apatite in interstitial liquids initially crystallizes as an F-rich apatite, and that Cl-rich apatite forms mainly by reequilibration with relatively large amounts of trapped interstitial liquid. This process is analogous to that by which early liquidus mafic minerals may become more Fe-rich on reequilibration with interstitial liquids during solidification (see Chapter 11). However, the abundance of apatite typically correlated with bulk rock incompatible element abundances, implying that, with the exception of the more evolved rocks, apatite itself is a product of the crystallization of the interstitial liquid. Although as suggested above some variation is likely as the crystallizing interstitial liquid becomes isolated at different times during solidification, there is no evidence that this in itself can produce unusually Cl-rich apatite, and that increasing amounts of interstitial liquid only produce more apatite.

6.8.3 Temperature and Pressure Effects

The possibility that temperature is an important control of halogen concentrations in apatite was considered by Boudreau and McCallum (1989), Boudreau et al. (1993a) and Boudreau and Simon (2007). If it is assumed that the crystallizing interstitial liquids follow a similar liquid line of descent in regards to apatite saturation as does the parent magma, then apatite would have begun to crystallize at about the same temperature throughout any given intrusion. In addition, the initial interstitial apatite to crystallize at any level should have the same composition as the first cumulus apatite to precipitate. Because this is inconsistent with a large temperature range in apatite equilibration, temperature effects alone would not be expected to be the major cause of apatite compositional variation; hence apparent variations most likely reflect actual halogen activity variations. Once apatite became a liquidus phase, however, apatite in more evolved rocks would generally reflect lower temperatures of crystallization.

Boudreau et al. (1993) illustrated the range in either temperature or halogen activity that would be required to explain the observed range in apatite compositions in the Munni Munni Complex, a Cl-poor intrusion with limited Cl/F variation. Because data on the behaviour of F and Cl in basaltic liquids and their partitioning into apatite at low to moderate pressures as a function of temperature are not available, Boudreau et al. (1993) calculated the effect of temperature and of variations in activity of F and Cl by considering equilibrium between apatite and an aqueous fluid, for which thermodynamic data are available. Although the activity of the halogens in a silicate liquid is not explicitly expressed by such a treatment, it is clear that an increase in the activity of either halogen in any phase must be seen by that component in all other phases as well.

Consider the exchange reaction between apatite and an aqueous fluid in which HCl and HF are present as neutral species:

$$\underset{apatite}{Ca_5(PO_4)_3(F)} + \underset{fluid}{HCl^\circ} = \underset{apatite}{Ca_5(PO_4)_3(OH)} + \underset{fluid}{HF^\circ} \tag{6.5}$$

Halogen substitution in apatite is close to ideal above 500 °C (Tacker and Stormer, 1989). For a fixed activity of Cl and F in the fluid, the equilibrium exchange constant as a function of temperature, *T*, is given by the following equation (relevant thermodynamic data from 500 to 1,000 °C at 2.5 kbar are from Zhu and Sverjensky 1991 and Johnson et al., 1992, and fitted to a linear equation in 1/*T*):

$$\log K_D = \log\left(\frac{X_{Cl}^{Apatite} \cdot a_{HF^\circ}^{fluid}}{X_F^{Apatite} \cdot a_{HCl^\circ}^{fluid}}\right) \tag{6.6}$$

$$= \frac{1447}{T\ (Kelvin)} + 0.1823 \tag{6.7}$$

In Equation 6.6, $X_{Cl}^{Apatite}$ and $X_F^{Apatite}$ are the mole fractions of the end-member chlorapatite and fluorapatite component of apatite, respectively, and $a_{HCl^\circ}^{fluid}$ and $a_{HF^\circ}^{fluid}$ are the activities of the neutral species HCl° and HF° in the fluid, respectively. The relationship between fluid composition and apatite composition is given by:

$$\left(\frac{X_{Cl}^{Apatite}}{X_F^{Apatite}}\right) = K\left(\frac{a_{HCl^\circ}^{fluid}}{a_{HF^\circ}^{fluid}}\right) \tag{6.8}$$

Similar equations can be defined for OH-F and HCl-H_2O exchange between apatite and fluid. For example, Piccoli and Candela (1994) have shown for the HCl-H_2O exchange,

$$\underset{apatite}{Ca_5(PO_4)_3(OH)} + \underset{fluid}{HCl^\circ} = \underset{apatite}{Ca_5(PO_4)_3(Cl)} + \underset{fluid}{H_2O^\circ} \tag{6.9}$$

Then, for the fluid fugacities $f_{HCl^\circ}^{fluid}$ and $f_{H_2O^\circ}^{fluid}$ and assuming ideal mixing in apatite (i.e., $a_i = X_i$) at magmatic temperatures, one can define the exchange constant as:

$$\log K_D = \log\left(\frac{X_{Cl}^{Apatite} \cdot f_{H_2O^\circ}^{fluid}}{X_{OH}^{Apatite} \cdot f_{HCl^\circ}^{fluid}}\right) \tag{6.10}$$

$$= \log\left(0.04661 + \frac{2535.8}{T} - \frac{0.0303(P-1)}{T}\right) \tag{6.11}$$

For the above, *T* and *P* are in Kelvin and bars, respectively.

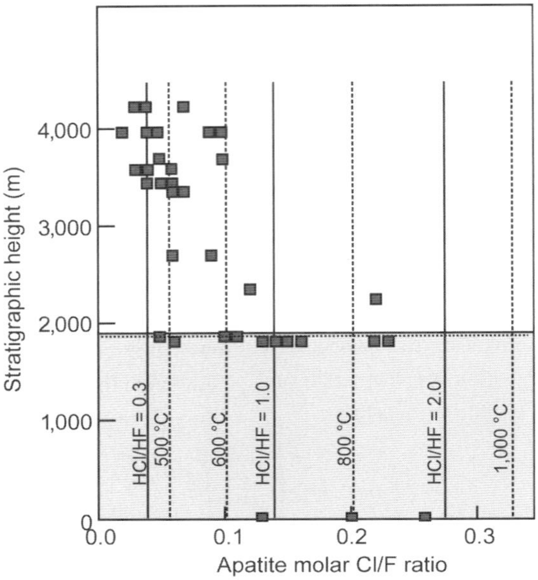

Figure 6.10 Molar Cl/F ratio of apatite from the Munni Munni Complex plotted as a function of stratigraphic height, illustrating the effects of variable HCl/HF activity ratio, temperature and degassing on halogen variations. Dashed vertical lines are calculated compositions of apatite equilibrated with a fluid at 1,100 °C that has variable HCl/HF activity ratios as noted. The solid lines are calculated compositions for apatite equilibrated with a fluid with a fixed HCl/HF activity ratio of 3.0 and at the different temperatures noted. The horizontal dashed line just below the ultramafic zone–gabbroic zone boundary is the location of the main PGE-sulfide zone. Redrawn after Boudreau et al. (1993a), reprinted with permission of publisher.

Shown in Figure 6.10 are calculated apatite compositions that would result from changing the Cl/F ratio of a fluid at a fixed temperature of 1,100 °C (dashed lines) or from changing the temperature at a fixed HCl/HF activity ratios of 3.0 (solid lines). Superimposed on this plot are stratigraphic data on halogen contents of apatite from the Munni Munni intrusion of Western Australia. This plot suggests that the observed variation in apatite compositions may be caused by changing either the HCl/HF (and H_2O/HF, although not shown) ratio by approximately an order of magnitude, or by different temperatures of equilibration which may vary by 500 °C. For high-Cl intrusions such as the Bushveld and Stillwater Complexes, in which Cl/F variations are at least an order of magnitude larger than for the Munni Munni intrusion, these ranges would have to be similarly higher. As noted above, it is unlikely that this extreme range in temperature of apatite crystallization occurred, and hence variations reflect true fractionation trends. Similar modelling has been applied to the Great Dyke (Boudreau et al., 1995), the Ferar Basement Sill (Boudreau and Simon, 2007) and the Lac des Ilse (Schisa et al., 2015).

Pressure can also have a strong control halogen partitioning into apatite. As summarized by Webster and Piccoli (2015), Cl partitions more strongly in favour of apatite as pressure decreases from 2,000 to 275 bars at 850 to 950 °C in rhyolitic melts. For the thicker intrusions such as the Bushveld Complex, there could be about 2 kbar pressure difference between top and bottom and thus pressure could be a potentially important control of apatite compositional variations. However, the expected pressure effect on apatite composition is just the opposite of what is observed in most intrusions in which the more Cl-rich apatite occur in the lower, higher pressure sections. It is clear that other processes must dominate over the pressure effects.

6.8.4 Vapour Separation and Degassing of the Magma Chamber

The separation of a volatile fluid from a crystallizing magma can fractionate the halogens as noted above. In this regard, there is abundance evidence that mafic magmas can begin to degas while still at deep crustal or even upper mantle pressures. For example, it has been estimated the extensive S loss in the Kilauea magma system is the result of volatile loss beginning at a depth of ~35 Km (Gerlach and Graeber, 1985). This is particularly enhanced if the deeper magma chambers can communicate with shallower magma chamber or even long-lived lava lakes. Degassing rates in hot spot magmatism can be substantial; the summit of Kilauea emits from 150 to 2,000 tonnes/day SO_2 since the onset of the current eruption began in 1983 (Mather et al., 2012). The Kilauea summit magma chamber is relatively shallow, 1–2 km beneath the summit caldera. For the past several years, this chamber has been open to the surface, feeding a persistent lava lake in the Overlook crater inside the Halemaumau crater, which is itself inside the larger Kilauea summit caldera. Eruptions from this lava lake are uncommon, and instead most of the degassed magma moves about 18 km down the East Rift system to erupt at Pu‘u ‘Ō‘ō. Gerlach and Graber (1985) calculated that roughly half of the original S concentration of about 1,500 ppm is lost from the magma in the summit chamber as the magma begins to degas starting at a depth of ~40 km. This estimate was made prior to the formation of the Overlook Crater lava lake: a surface opening would enhance low-pressure S loss and the formation of which is coincident with the higher SO_2 flux rates.

This S loss to shallow degassing is consistent with the observations of Dixon et al. (1991), who suggested that shallow degassed magma can mix with deeper magma storage to depths of 5 km and can erupt volatile-poor basalts on the sea floor. The loss of S to shallow degassing and remixing with deeper magma chamber is typical of other non-subduction volcanic systems reported by Shinohara (2008), who suggested that the vertical mixing can occur between the surface to as deep as 10 km. Loss of H_2O from Kilauea magmas was also attributed to a similar drainback mechanism (Wallace and Anderson, 1998)

As summarized above Cl is highly soluble in aqueous and aqueous-carbonic fluids, and hence, it partitions in favour of fluids relative to silicate liquid (i.e., $D_{Cl}^{fluid/liq} > 1$) at shallow-crustal pressure and temperature conditions. In contrast, portioning of F between silicate liquid and fluid generally prefers the liquid, with (i.e., $D_{F}^{fluid/liq} < 1.0$). Thus, the general stratigraphic trend of decreases in the Cl/F ratio in the more fractionated rock, as evident in all intrusions, can be caused by loss of Cl during degassing that accompanies

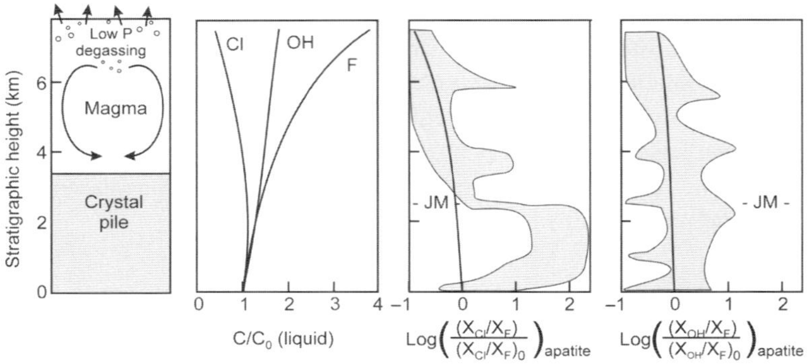

Figure 6.11 Cartoon of low-pressure degassing of a crystallizing magma (left). Calculated variations in Cl, F and OH concentrations trapped in pore liquids as a function of stratigraphic height in a single fractionating 10 km thick magma (centre) and equivalent Cl/F and OH/F variations in apatite (right), all normalized to initial concentrations at the base of the intrusion. From Boudreau and McCallum (1989), with permission.

fractional crystallization. A model for the general evolution of the halogen chemistry of layered intrusions is illustrated in Figure 6.11. As the magma crystallizes, Cl is preferentially lost to a separating fluid. This fluid most likely separating at the very top of the chamber where pressure on the magma column is low. Fluorine, which does not appreciably partition into a fluid phase, continues to increase in the magma during fractionation. Liquid trapped at different stratigraphic levels would then have initial Cl/F ratios defined by the degree Cl and F had fractionated prior to being trapped; the overall effect would be that interstitial liquids would have a progressively lower Cl/F ratio as a function of stratigraphic height.

This low-pressure degassing model is supported by the observation that, in most layered intrusions, liquidus halogen-bearing minerals other than apatite are rarely observed. That is, biotite and amphibole are rarely observed as liquidus minerals even though they are relatively common interstitial minerals, particularly in the lowermost parts of an intrusion. In addition, even relatively dry basaltic liquids such as MORB may contain as much as 0.5 wt.% H_2O (Chapter 3). This amount of water would be enough to crystallize about 25% of the entire solid assemblage as amphibole if the water were not lost prior to amphibole saturation. Hence, both the loss of Cl and the lack of hydrous cumulus minerals are consistent with low-pressure degassing and loss of volatile components from the magma as it cooled and crystallized.

6.8.5 Crystallization of Interstitial Hydrous Minerals

The fractionation of Cl and F can be complicated when the liquid becomes saturated in halogen-bearing minerals in addition to a fluid phase. For amphibole and micas, which have strong crystal-chemical controls on halogen substitution, one would expect this to strongly control halogen partitioning between these minerals and silicate liquid. For example, in experiments at 1.5 GPa and 1,050 °C, Hauri et al. (2006) measured F and Cl partitioning between amphibole, mica and silicate liquid and report $D_F^{amp/liquid} = 0.85$, $D_{Cl}^{amp/liquid} = 0.046$, $D_F^{mica/liquid} = 1.46$ and $D_{Cl}^{mica/liquid} = 0.056$. At 0.5 GPa, 1,000 °C, the partition coefficients were similar: $D_F^{amp/liquid} = 0.1.19$, $D_{Cl}^{amp/liquid} = 0.038$, $D_F^{mica/liquid} = 2.54$ and $D_{Cl}^{mica/liquid} = 0.054$. For the lower pressure results, F is preferred over Cl by a factor of 31 in amphibole and 47 in mica. Given that the silicate liquids were relatively Mg-rich mantle melt analogues, the strong preference of F over Cl would be expected by the Mg-Cl avoidance rule for these minerals.

For the Mg-rich rocks in the lower sections of most layered intrusions, this preference for F in micas and amphiboles can lead to an increase in the Cl content of the liquid relative to F. It is even possible that crystallizing these minerals can reversing the preferential loss of Cl to an exsolving vapour if the mass fraction of halogen-bearing minerals crystallizing exceeds the effect of fluid on the bulk distribution coefficient. Considering the case in which the weight fraction of fluids separating from a crystallizing silicate liquid, w_{fluid}, is equal to 0.01 (appropriate for a relatively CO_2-rich fluid), and the weight fraction of biotite separating, $w_{biotite}$, is 0.10. If one assumes Cl and F biotite/liquid partition coefficients at 0.5 GPA Of Hauri et al. (2006) noted above and vapour/liquid distribution coefficients of $D_{Cl}^{fluid/liq} = 10$ and $D_F^{fluid/liq} = 0.1$, then the bulk distribution coefficients for Cl and F are given by

$$\begin{aligned} \mathrm{D}_{Cl}^{bulk} &= w_{fluid} \cdot D_{Cl}^{fluid/liquid} + w_{biotite} \cdot D_{Cl}^{biotite/liquid} \\ &= 0.01 \cdot D_{Cl}^{fluid/liquid} + 0.01 \cdot D_{Cl}^{biotite/liquid} \\ &= \ 0.11 \end{aligned} \tag{6.12}$$

and

$$\begin{aligned} \mathrm{D}_{F}^{bulk} &= w_{fluid} \cdot D_{F}^{fluid/liquid} + w_{biotite} \cdot D_{F}^{biotite/liquid} \\ &= 0.01 \cdot D_{F}^{fluid/liquid} + 0.01 \cdot D_{F}^{biotite/liquid} \\ &= 0.25 \end{aligned} \tag{6.13}$$

In this case, although both elements behave incompatibly, F is removed from the liquid in preference to Cl. Thus, the saturation of a liquid in one or more halogen-bearing minerals can reverse the decrease in the Cl/F ratio caused by fluid loss alone, and can lead to the precipitation of minerals with higher Cl/F ratios in the later stages of crystallization. However, as noted above in most rocks, apatite is the principle or only halogen-bearing mineral and is present in rather modest proportions; hence, its crystallization may not significantly affect the halogen concentrations of the fractionating fluid-saturated liquids. For example, crystallization of apatite as a cumulus mineral did not reverse or even significantly change the general stratigraphic trend of decreasing Cl/F in the Bushveld Complex.

The nominally anhydrous minerals olivine, pyroxene, and plagioclase that are the major minerals in layered intrusions can incorporate small amounts of H_2O and the halogens at magmatic temperatures and especially at mantle pressures (e.g., Peslier, 2010; Joachim et al., 2015). Mathez and Webster (2005) suggested that H_2O (and presumably F) preferentially partitioned into pyroxene relative to Cl in the orthopyroxenites the Stillwater and Bushveld this might be the source of some of the high Cl apatites of these two intrusions. However, a correlation between Cl-enrichments and pyroxenites in layered intrusions is not generally observed, but more focused studies may find otherwise.

6.8.6 Vapour Refining in a Crystal Mush

As noted above, fractionation of hydrous minerals can lead to increases in the Cl/F ratio. However, the rapid rate of increase seen in the base of the Stillwater and Bushveld Complex and the lack of liquidus hydrous minerals imply that halogen-bearing mineral fractionation could not have involved fractionation of the entire magma chamber (e.g., Boudreau and McCallum 1989). Hence, one must look for mechanisms that operate in more limited volumes. McBirney, (1987) suggested that incompatible elements may be strongly enriched in the interstitial liquid by a process he terms constitutional zone refining (see Chapter 9). In this process, the interstitial liquid is enriched in incompatible elements as crystallization continues. This enriched liquid is then transported upward into overlying crystal-liquid mush by interstitial liquid convection or compaction of the crystal pile; alternatively, upward enrichment of incompatible elements can occur by diffusion of these elements within the interstitial liquid as they move down a concentration gradient. Because the displaced, enriched liquid is out of equilibrium with the mush assemblage it moves through, the solids must react with this liquid. For the case in which the enriched element is a fluxing agent such as H_2O, it can lead to partial melting of the existing minerals. Because the zone over which the crystal mush is solidifying is small relative to the entire chamber, the enrichment processes will occur on a scale of relatively short stratigraphic length. Crystallization of halogen-bearing minerals that preferentially incorporate F relative to Cl will lead to Cl-enrichment in the interstitial liquid. The wide variations in Cl/F in apatite within individual samples noted previously is evidence for *in-situ* fractionation of the halogens. The Cl-enriched liquid can then migrate upward into hotter cumulates which have not yet begun to crystallize apatite, perhaps driven by compaction of the crystal pile. This results in an elevated Cl/F ratio in the apatite that precipitates higher in the stratigraphic sequence.

The observation that strong Cl-enrichments in MORB and ocean island basalts can be caused by the assimilation of concentrated brines derived from seawater (e.g., Coombs et al., 2004) suggests that a zone-refining process involving Cl-rich fluids can operate for the halogens and was explored by Boudreau and McCallum (1989). Fluid separating from interstitial liquid will preferentially incorporate Cl relative to F. This fluid may then migrate upward to redissolve in hotter interstitial liquid which have lower volatile contents and have not yet become fluid saturated. This dissolution will result in elevating the Cl/F ratio of the stratigraphically higher interstitial liquid as well as the hydrous phases that eventually separate from this liquid (Boudreau et al., 1986) (Figure 6.12). This process was called 'vapour driven' constitutional zone refining by Boudreau and McCallum (1989) to

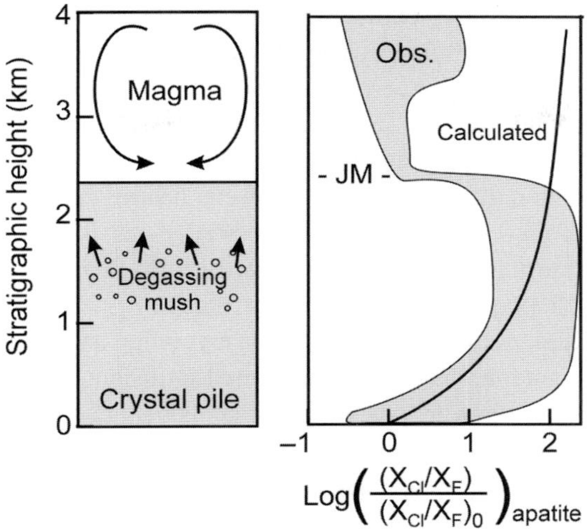

Figure 6.12 Cartoon of a vapour-refining front moving through a solidifying cumulus pile (left). Plot on the left shows the trend of observed apatite Cl/F compositions (shaded) and a calculated zone-refining trend, both normalized to 'average' group 2 and 3 sill/dyke compositions. The calculated refining trend fails to match the observed trend at and above the level of the platiniferous J-M Reef. After Boudreau and McCallum (1989), with permission.

emphasize the importance of a volatile-rich fluid phase, although the effects are similar to those described by McBirney (1987). Indeed, the observation noted above that Cl in a silicate liquid may separate a Cl-rich fluid at concentrations too low to produce Cl-rich apatite is one of the lines of arguments of Mathez and Webster (2005) that the Cl-rich rocks of the Stillwater, Bushveld and other intrusions were saturated in a Cl-rich fluid.

This process was modelled by assuming that the fluid that separates from one zone redissolves and enriches the immediately overlying zone in volatile components. This process then repeats itself in the next zone. The result is a progressive increase in both the amount of fluid that eventually separates at each zone (and a consequent earlier fluid-saturation during the crystallization of the interstitial liquid at each zone) and an increase in the Cl/F ratio with stratigraphic height. The reversal in the Cl-enrichment that occurs at the level of the J-M Reef is explained as the point at which the fluid migration front reached the top of the crystal pile and was diluted as it mixed with the overlying magma. This fluid escape and mixing may have been localize at pothole structures that occur at this level (e.g., Boudreau, 1992).

6.8.7 External Sources for the Halogens

One of the characteristic features of the two most Cl-rich intrusions, the Stillwater and Bushveld complexes, is that both have extensive metamorphic aureoles several km thick underlying them in which the rocks have undergone extensive dehydration and even partial melting. For the Stillwater Complex, they contact aureole largely affected biotite schist to produce a pyroxene-cordierite hornfels. The crustal isotopic characteristics of the Stillwater Complex noted above would support an external source for the Cl. For the Bushveld Complex, the sedimentary footwall rocks were not thermally metamorphosed prior to the intrusion of the Bushveld magma, and thus formation of the aureole could have generated significantly larger amounts of fluid as they were heated than the Stillwater aureole. However, the light crustal signature of the footwall rocks is distinct from the heavier isotopic Cl observed in the Bushveld layered series. It is clear that more work could be done to clarify the role played by external fluids, if any.

The above summarizes several mechanisms involving halogen-bearing minerals or fluids that can fractionate the halogens in a layered intrusion. These mechanisms can be chamber-wide (as in the case where the magma chamber is fluid saturated), or can be more local in effect (as perhaps in the case of zone-refining models). It is likely that several mechanism may operate independently to give a rather complex distribution of the halogens that can be decoupled from the major element trends. In general, however, it appears that most intrusions become fluid saturated at least at some point during their crystallization such that Cl is either lost from the main magma body (presumably at the top of the chamber where pressure is lowest) or separates from the interstitial liquid as the mush solidifies and migrates through the crystal pile. In regards to the latter, it appears that apatite for which the mole fraction of the chlorapatite component is >10 of mol.% requires the presence of a Cl-rich fluid.

7
Melt and Fluid Inclusion Evidence

7.1 Introduction

Silicate liquid and volatile fluid inclusions are the most direct evidence of melt and fluid compositions. They are found in many igneous rocks from a variety of environments, from the common occurrence of melt inclusions in olivine from hot spot and mid-ocean ridge volcanism to fluid inclusions in minerals from arc volcanism. In layered intrusions, however, inclusions that directly sample the magma that they precipitated from are rare, and descriptions of fluid inclusions are limited to just a few studies. In part, this may represent a lack of curiosity; those who have gone looking have generally found them. Where they do occur they provide support for liquid and fluid compositions deduced from mineral assemblages and experimental studies.

7.2 Silicate Liquid Inclusions

Silicate liquid inclusions can become entrained in the host mineral during original growth or during post-accumulation recrystallization events within the crystal mush. Given the common occurrence of melt inclusions in volcanic rocks, the paucity of inclusions in layered intrusions has not been addressed. The slow growth of the original mineral can limit inclusion trapping, although it does not seem to be a problem for inclusions found in large volcanic olivine crystals that presumably grew over a long time span. Textural re-equilibration in the crystal pile (crystal aging, compaction-driven recrystallization or even perhaps mineral exsolution) may lead to the loss of pre-existing inclusions in some instance but promote the trapping of interstitial liquids in other case.

7.2.1 Silicate Liquid Inclusions in the Stillwater Complex

Chromite containing polymineralic assemblages that are interpreted to have crystallized from a trapped silicate liquid are perhaps the most commonly reported host. A typical example is described by Spandler et al. (2005) from the Stillwater G chromitite and is shown in Figure 7.1. The ferromagnesian inclusion minerals are close to the pure Mg end-member compositions owing to Mg-Fe exchange with the host chromite during cooling.

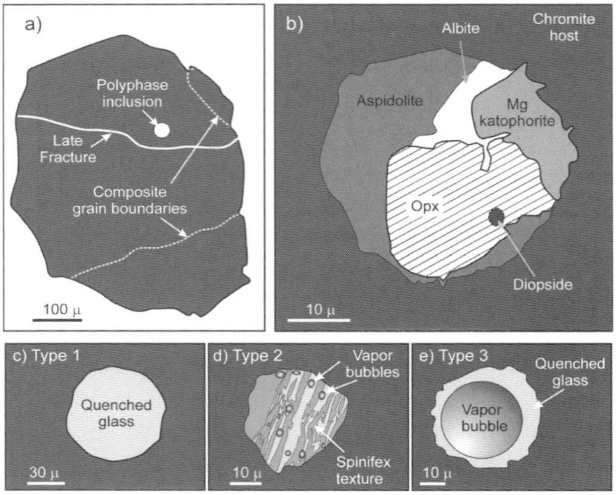

Figure 7.1 Examples of polyphase inclusions in chromite from the G chromitite cyclic unit, Stillwater Complex, Montana. (a) Chromite grain with composite grain boundaries and polyphase inclusion (shown in red). (b) Detail of the inclusion shown in (a). (c–e) Three types of inclusions observed after reheating to 1,450 °C: (c) Type 1, quenched glass; (d) Type 2, olivine-spinifex texture (e) Type 3, vapour-rich. All three types can contain a vapour bubble, but their abundance and size increases from types 1–3. Redrawn after Spandler et al. (2005), with permission.

The mica and amphibole in the inclusions are significantly more Na rich than interstitial amphibole and phlogopite from the surrounding rocks (Page and Zientek, 1987). Spandler et al. attempted to homogenize the inclusions by heating them to 1,300–1,450 °C, although complete remelting and homogenization only occurred at the higher temperature and all types could contain one or more vapour bubbles.

Average compositions for the three types of homogenized melt inclusions are listed in Table 7.1. A missing but non-trivial component is H_2O, given the abundance of Na-phlogophite in the inclusions that contain in excess of 4 wt.% H_2O. The high Fe is anomalous in that that the daughter minerals are Mg-rich, which Spandler et al. attribute to extensive exchange of Fe, Mg and Mn between the melt inclusions and chromite host during the reheating. The inclusions are otherwise characterized by a nominally low SiO_2, very low CaO and high Na_2O contents. Cl likely was affected by Cl loss to the vapour bubbles.

Spandler et al. suggested that the compositions (corrected for the Fe-enrichment) are consistent with variable degrees of mixing between a high-Mg basaltic parental magma and a Na-rich trondhjemitic melt. The trondhjemite was suggested to derive from partial melting of mafic or metasedimentary country rocks. However, Boudreau (2015) noted that the metamorphic aureole adjacent to the Stillwater complex was dehydrated to a pyroxene hornfels and that any partial melts would be unlikely to contain much of a volatile component.

Table 7.1 *Average H_2O-absent homogenized melt inclusion compositions in Stillwater G chromite. Data from Spandler et al. (2005).*

Type	1	2	3
SiO_2	47.5	43.3	39.8
TiO_2	1.12	1.12	0.81
Al_2O_3	10.1	8.65	7.27
FeO	18.0	24.6	24.2
MnO	0.29	0.36	0.42
MgO	15.9	17.1	16.3
CaO	0.95	0.79	9.44
Na_2O	5.28	4.11	2.90
K_2O	0.23	0.34	0.18
Total	99.4	100.4	100.2
Cl (ppm)	750	320	530

Note: Average Cl is a minimum concentration, as some analysis list Cl as not detected and are assumed to be 0.

The relative enrichment in Na_2O and Cl suggests these components could have been introduced by saline fluids described in Chapter 6. This point is illustrated by a simple MELTS calculation to demonstrate the effect of adding a H_2O + NaCl volatile component to an initially dry olivine + liquid assemblage. In this case the model composition is made by combining 10% of the Stillwater gabbronorite sill liquid composition from Helz (1995) with a typical Stillwater Peridotite zone olivine in the weight proportions of 10% liquid and 90% olivine. This bulk composition is listed in Table 7.2 as 'Bulk liquid + solid'. The calculated liquid composition at 1,250 °C and 200 MPa pressure is listed as the 'Model dry liquid' and the liquid composition after adding 3% of a vapour composed of 85% H_2O and 15% NaCl to the bulk is listed as '3% vapour added, liquid'.

Compared with the dry liquid, and with the exception for the increase in Na_2O which is introduced by the vapour, the changes result from the flux melting of the olivine to increase FeO, MnO and MgO and decrease in SiO_2 (modestly), Al_2O_3, CaO and K_2O. Comparison with the Type 1 remelted inclusion of Spandler et al. (2005) (Table 7.1) shows the observed inclusions to be higher in MgO + FeO than the model liquid and likely represents some melting of the host spinel as they were reheated to 1,450 °C. In addition, the inclusions will become enriched in the Mg/(Mg + Fe) ratio by re-equilibration with the host chromite on cooling (e.g., Melcher et al., 1997). The Type 1 inclusions also have markedly lower in CaO than the model liquid, although the Type 3 inclusions have higher CaO. The overall similarity, however, suggests that the inclusions are simply trapped hydrated interstitial liquids.

Table 7.2 *Hydration of interstitial liquid, MELTS model.*

	Bulk liquid + solid[1]	Model dry liquid[2]	3% vapour added, liquid[3]
SiO_2	40.57	53.6	52.7
TiO_2	0.10	1.16	0.50
Al_2O_3	1.52	15.1	11.6
FeO	14.5	8.00	12.0
MnO	0.20	0.12	0.23
MgO	41.46	8.36	8.99
CaO	1.00	9.83	7.80
Na_2O	0.18	2.93	5.23
K_2O	0.03	0.54	0.36
total	99.5	99.6	99.4
% Liquid	–	6.1	9.6

Notes: 1. Bulk composition produced by mixing 10% Stillwater gabbronorite sill composition (Helz, 1995) with 90% olivine. 2. Calculated MELTS liquid composition for the bulk composition (1) at 1250 °C and 200 MPa. 3. Composition of MELTS liquid after addition of 3% vapour containing 85% H_2O and 15% NaCl at 1,250 °C and 200 MPa, recalculated as dry for comparison with the homogenized inclusions of Spandler et al. (2005) in Table 7.1.

7.2.2 Other Examples of Silicate Liquid Inclusions

Similar Mg, Na and H_2O-rich inclusions in chromite and other minerals are reported from the J-M Reef of the Stillwater complex and the Merensky Reef of the Bushveld complex (e.g., Boudreau, 1988; Li et al., 2005) (Figure 7.2). Boudreau (1988) suggested that the J-M Reef examples were entrapped during the formation of chromite by the incongruent flux melting of a Cr-bearing (gabbro)norite protolith. For the Merensky Reef samples, Li et al. (2005) interpreted the hydrous liquids trapped in chromite to have resulted from variable degrees of mixing between the magma on the floor of the chamber and Na-K-rich fluids expelled from the underlying crystal pile that both caused dissolution of orthopyroxene and leaving the magma saturated only in chromite.

Hydrous mineral-rich inclusion have also been reported from the Western Laouni intrusion of the Southern Hoggar, Algeria (Lorand and Cottin, 1987) and the Early Palaeozoic ophiolite sequence of the Kempirsai Massif, Kazakhstan (Melcher et al., 1997). Lorand and Cottin (1987) argued against trapping during the initial precipitation of the chromite based on the overabundance of Na-amphibole, Na-phlogophite and other incompatible-rich minerals. They instead suggested that the inclusions trapped interstitial liquid during a prolonged period of crystal aging/annealing (e.g., Hulbert and Von Gruenewaldt, 1985; Waters and Boudreau, 1996). Further, they suggested that the trapping was enhanced by fluid fluxing of the original assemblage, as is suggested here for the G chromitite samples.

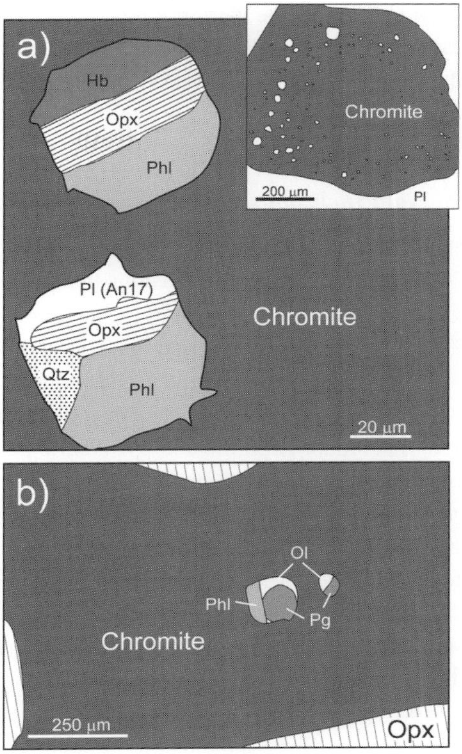

Figure 7.2 Other examples of polymineralic inclusions in chromite, interpreted to have crystallized from hydrous melt inclusions. (a) Assemblages in chromite inclusions from the Merensky reef. Redrawn after Li et al. (2005), with permission. (b) Assemblages in chromite inclusions from the JM Reef, Stillwater Complex. Redrawn after Boudreau (1988), with permission. Abbreviations: Hb = hornblende, Ol = olivine, Opx = orthopyroxene, Pg = pargasite (Na-rich hornblende), Phl = phlogopite, Pl = plagioclase, Qtz = quartz.

Loferski and Arculus (1993) report multiphase clinopyroxene-ilmenite-apatite inclusions in plagioclase from the thick anorthosite units and other rocks of the Middle Banded series of the Stillwater Complex. They noted that the unusual assemblage, including a distinctive Mn-rich ilmenite and baddeleyite (ZrO_2), indicates formation from an unusual liquid. They suggested that small globules of an immiscible liquid, enriched in Mg, Fe, Ca, Ti, P, REE, Zr and Mn, exsolved from the anorthosite parent liquid, became trapped in the plagioclase, and later crystallized to form the inclusions.

7.3 Fluid Inclusions

Fluid inclusions can preserve direct evidence for the presence of a fluid. However, to date there are only a handful of studies of fluid inclusions in layered intrusions. With few exceptions, the fluid inclusions that are found are hosted by pegmatoids that form at some later time than the host rock. Thus, the period of early bubble generation and fluid movement is not likely sampled and must be inferred. Nonetheless, the broad features of fluid compositional evolution during solidification of layered intrusions is beginning to be understood from those that have been report.

7.3.1 Fluid Inclusions in the Stillwater Complex

An example of the variety of fluid inclusion types that can occur in layered intrusions is the study of Hanley et al. (2008) of fluid and melt inclusions that occur below the J-M Reef of the Stillwater Complex. They found a variety of fluid types, ranging from NaCl-dominated halide melts (>82 wt.% $NaCl_{equiv}$) to compositionally complex Na-Ca-K-Fe-Mn-Ba-Si-Al-Cl brines (28–79 wt.% $NaCl_{equiv}$) that were trapped along with a moderate density carbonatitic fluid (CO_2 + CH_4). These where hosted in rocks ranging from fresh dunite from the ultramafic zone to discordant gabbroic and pyroxenite pegmatoids. In detail, they observed six fluid and melt inclusion types.

Type 1: polyphase brine inclusions. The Type 1 inclusions contain an aqueous liquid phase, a small vapour bubble and up to 16 solid phases (Figure 7.3a–d). They occur in quartz and olivine. In the olivine, Type 1 inclusions occur along healed fracture planes and are secondary in origin. In quartz, Type 1 inclusions occur with Type 2 and Type 3. They have an estimated trapping temperatures of 480–640 °C at pressures ranging from 1.5–5.5 kbar. The high solids fraction suggests that some may indeed be transitional fluids as defined in Chapter 3.

Type 2: carbonic fluid inclusions. Type 2 inclusions are mainly single-phase liquid CO_2 or critical density CO_2 liquid-gas mixtures (Figure 7.3e). They occur only in quartz and were not observed in any olivine-bearing samples. Type 2 inclusions occur coevally with Type 1 and Type 3 inclusions. Calcite and rutile are relatively common as accidentally trapped minerals and show highly variable phase proportions.

Type 3: mixed polyphase brine-carbonic fluid inclusions. The Type 3 inclusions contain variable amounts of a single-phase CO_2 and an aqueous brine with solids and were found only in the pegmatite samples (Figure 7.3f and g). The solid phases present are similar to those observed in the coeval Type 1 inclusions. Like types 1 and 2 inclusions, Type 3 commonly contain large, accidentally trapped calcite crystals. The Type 3 inclusions coexist with Types 1 and 2 inclusions in quartz. Where all three are present in graphic quartz in pegmatoid, they have an estimated trapping temperatures of 700–715 °C at pressures ranging from 4.3–5.6 kbar. In contrast, where all three are in massive quartz infilling miarolytic cavities in the centre of the same pegmatoids, estimated trapping temperatures and pressure are lower: 275–560 °C at pressures ranging from 1.4–4.0 kbar.

The classification of some calcite in the Type 1, 2 and 3 inclusions as accidental implies it began to precipitate before entrapment, consistent with the findings of Aird and Boudreau (2013) that unmixed carbonate assemblages in the Stillwater began to precipitate above the maximum carbonate solvus temperature of about 950 °C

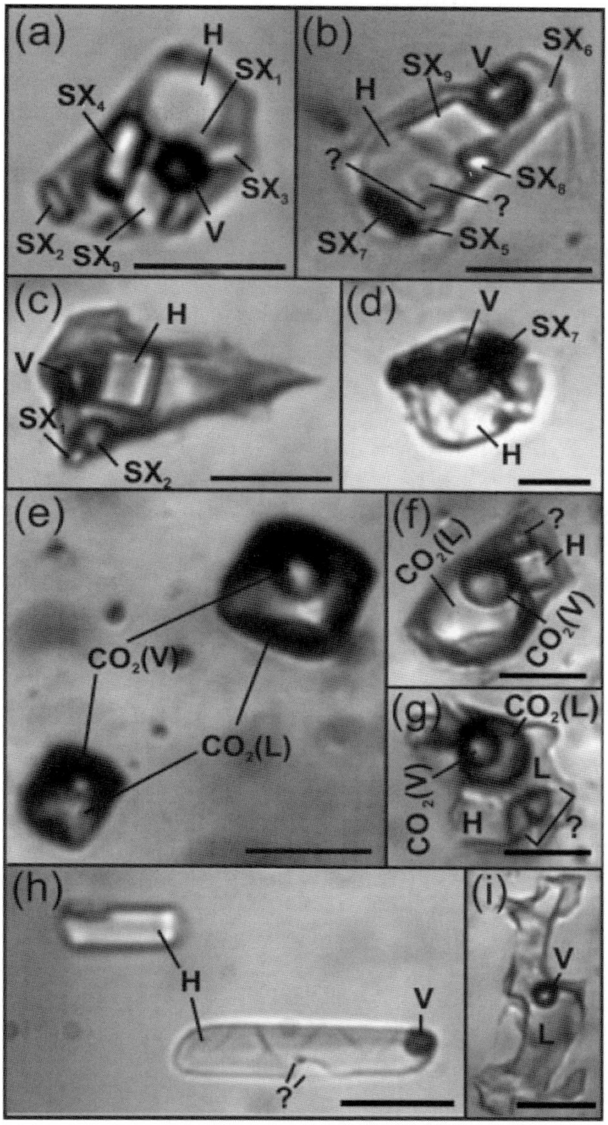

Figure 7.3 Inclusion types identified in the pegmatite and orthopyroxenite beneath the J-M Reef of the Stillwater Complex, Montana. Scale bars in all images = 10 microns. Phase abbreviations (other as noted): L = aqueous brine; V = vapour; H = halite; CO_2(L) = CO_2 liquid; CO_2(V) = CO_2 vapour; ? = unidentified solid phases. (a–d) Type 1 brine inclusions hosted in graphic-textured quartz (a and b), massive quartz from the pegmatite core (c), and olivine (d). Phases: SX_1 = K-feldspar or muscovite; SX_2 = Ca, Cl-bearing phase; SX_3 = Ba, Cl-bearing phase; SX_4 = lawrencite-molysite; SX_5 = 'chalcanthite'; SX_6 = sylvite; SX_7 = 'pyrosmalite'; SX_8 = Fe, Mn, Cl-bearing phase; SX_9 = Ca-Fe-Mn carbonate. (e) Type 2 carbonic inclusions in pegmatite quartz. (f and g) Type 3 inclusions in pegmatite quartz showing variable volumetric ratios of carbonic fluid and aqueous brine as evidence for two immiscible fluids. (h) Type 4 inclusions in olivine. (i) Type 5 late secondary aqueous inclusion in quartz. From Hanley et al. (2008), reprinted with permission.

Type 4: halide melt inclusions. Type 4 inclusions (Figure 7.3h) contain >90 vol.% halite. The halite within the inclusions does not have a cubic morphology but instead mimics the shape of the inclusion wall. They are interpreted as representing a trapped molten salt phase rather than accidentally trapped halite crystals precipitated from a halite-saturated aqueous fluid at temperatures ranging from 660–800 °C.

Type 5: two-phase aqueous inclusions. Type 5 inclusions contain an aqueous liquid phase and, rarely, a vapour bubble; they have no daughter minerals. They occur in graphic and massive quartz in the pegmatite (Figure 7.3i) and are secondary in origin. They have an estimated trapping temperatures of 125–225 °C at pressures ranging from 1.0–2.5 kbar. The low temperature estimate and simple compositions implies they formed either during initial cooling of the intrusion or during later alteration.

Type 6: silicate melt inclusions. These rare inclusions occur along growth zones in the graphic quartz. On the basis of their Si- and Al-rich composition, Type 6 inclusions are considered to be Cl-rich silicate melt inclusions (transitional fluids?) that recrystallized after entrapment.

The entrapment T-P range estimated for the various inclusion types, as well as T-P estimates from the hornblende-plagioclase geothermometer and the calcite-dolomite solvus, are shown in Figure 7.4. The P-T space and various phase fields and boundaries for the two component NaCl-H_2O system are from Sourirajan and Kennedy (1962) and Bodnar (1994). The H_2O-saturated albite-quartz eutectic melting curve is shown in bold (Holtz et al., 1992) this can be considered as a proxy for the solidus of the host rock. The dashed line field marked 'a' is the maximum solvus temperature for interstitial for calcite-dolomite

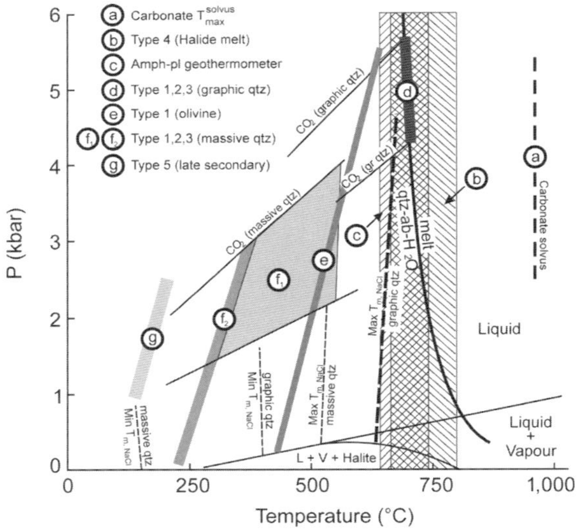

Figure 7.4 Estimation of trapping conditions for fluid and melt inclusion assemblages identified in the Stillwater Complex. See text for details. Modified after Hanley et al. (2008), with permission.

solid solutions from the Stillwater Complex (Aird and Boudreau, 2013). Field 'b' is the range of minimum entrapment temperatures for Type 4 (halide melt) inclusions in olivine. Field 'c', which overlaps that of field b, is the range in temperatures obtained from hornblende-plagioclase thermometer of Holland and Blundy (1994). Field 'd' is the estimated range of entrapment conditions for the immiscible brine-CO_2 assemblages in the graphic albite-quartz intergrowth at the centre of the pegmatite. Field 'e' is the range of isochores for Type 1 brines in olivine. Field 'f_1' is the range of entrapment conditions estimated for the immiscible brine-CO_2 assemblages in the massive quartz. Calculated isochores for Type 1 brine inclusions that are similar to pure halite melting temperatures are labelled as field 'f_2'. Field 'g' shows the estimated range in trapping conditions for late secondary (Type 5) low temperature inclusions.

Overall, the fluid and melt inclusions are interpreted to record a range of high T magmatic-hydrothermal to low T metamorphic events, with pressures ranging from lithostatic to hydrostatic. The inclusions show that residual silicate melt and two volatile phases of markedly different composition move through the crystal pile as it solidified. The system H_2O-CO_2-NaCl shown in Figure 7.5 shows that hydrosaline fluid and carbonic fluid ± halite can coexist over a wide range of compositional space at elevated temperatures and mid to upper crustal pressures (Aranovich et al., 2010). Also, as noted in Chapter 3, a similar extensive two fluid phase field exists in the H_2O-NaCl-$CaCO_3$ system, for which carbonic and hydrosaline fluids can coexist with calcite (Figure 3.7). However, Hanley et al. (2008) suggest that the hydrosaline and carbonic fluids exsolved separately from different parts of the igneous pile that used the host pegmatoid as a common pathway rather than unmixing from a single high temperature exsolved fluid.

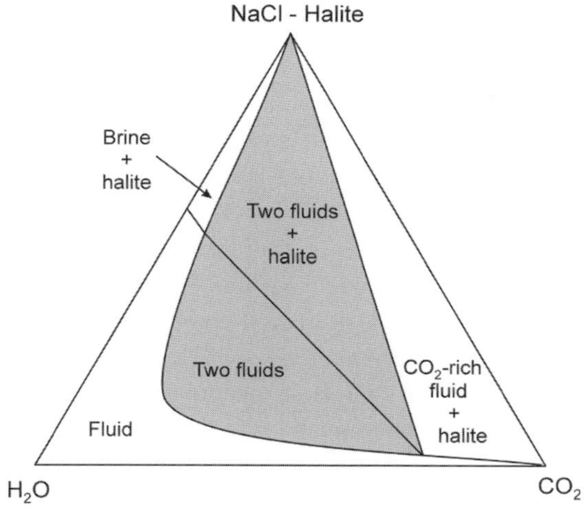

Figure 7.5 Generalized phases relationships in the system H_2O-NaCl-CO_2 for elevated temperatures (500–800 °C) and 2–9 kbar Pressure. Shaded region shows the two fluid phase field ± halite. Generalized after Aranovich et al. (2010), with permission.

They suggest that this can explain the significant variations in salinity when different assemblages are compared, even within a single crystal. In either event, the fluid inclusion data constrain only the entrapment conditions for the volatiles, not the maximum temperatures at which the volatiles existed. Thus, hydrosaline and carbonic fluids were probably present at well above the solidus when the overlying magma chamber was still crystallizing and that possibly originally exsolved as a single mixed fluid.

Hanley et al. (2008) envision that a residual silicate liquid from which the pegmatite was formed was also fluid-saturated because aqueous brine and CO_2 were trapped in the pegmatite as it crystallized. They suggest that brines, in particular, would have been highly mobile along intergrain boundaries in the underlying rock units, owing to their effective wetting properties (Brantley and Voight, 1989), as well as being enhanced by compaction of the host rocks (Meurer and Boudreau, 1998). As hydrothermal activity waned and the transition from lithostatic to hydrostatic conditions took place, brine and CO_2 continued to flow through the cooling pegmatite core. Continued quartz and calcite precipitation would occur by simple fluid cooling, decompression and decrease in fluid salinity.

7.3.2 Fluid Inclusions and Hydrosaline Melt Inclusions in Other Intrusions

The fluid inclusion reported from the Bushveld Complex are broadly similar to those of the Stillwater Complex and indicate a similar, prolonged cooling history during which the composition of the volatiles evolved from water-poor (80+ wt.% $NaCl_{equiv}$) hydrosaline melts to water-rich brines trapped with immiscible carbonic fluid. For example, Ballhaus and Stumpfl (1986) described a complex suite of H_2O-NaCI-($CaCl_2$)-CO_2-CH_4 fluid inclusions in interstitial quartz from the Merensky Reef. The earliest fluid inclusions are NaCl-rich with less than 10 vol.% water, CO_2 coexisting with a halite daughter crystal and brine, and polyphase inclusions with up to six daughter and accidental phases and high contents of divalent cations. The maximum trapping temperature is around 730°C at 4 to 5 kb pressure. Schiffries (1990) described fluid inclusions in quartz from the core of a zoned mafic pegmatoid in the upper Critical Zone in the eastern lobe of the Bushveld Complex near the platiniferous Driekop dunite pipe (Chapter 8). The inclusions range from NaCl-H_2O (70 to >80 wt.% $NaCl_{equiv}$); immiscible NaCl-H_2O-CO_2 (15–60 wt.% $NaCl_{equiv}$) and hydrated $CaCl_2$ inclusions with trapping conditions <750 °C at the same pressure range reported by Ballhaus and Stumpfl (1986). Schiffries and Skinner (1987) describe the presence of fluid inclusions with a halite daughter crystal in hydrothermal veins that cut the complex.

In the Skaergaard intrusion, Larsen et al. (1992) and Sonnenthal (1992) describe primary H_20-NaCl-CH_4 inclusions in quartz, apatite and feldspar from both gabbroic pegmatoid and in interstitial minerals trapped at temperatures between 655 and 770 °C. Wernette and Boudreau (2018) describe Cl-rich inclusions containing native silver associated with halite and sylvite hosted in pyroxene from the Skaergaard Middle Zone.

Hanley et al. (2011) report fluid inclusions from a footwall granophyre from the Sudbury Igneous Complex that they interpret to contain mixtures of a high salinity (44–70 wt.% $NaCl_{equiv}$), Na–Fe-rich aqueous magmatic fluid and a lower salinity, Ca-

and Sr-rich groundwater. They high Sr concentrations (0.18 wt.% Sr) has implications for Sr isotopic contamination by infiltrating country fluids discussed in Chapter 13.

Hanley and Gladney (2011) report one and two-phase carbonic fluid inclusions with up to 10% CH_4 and only minor H_2O in pegmatoids associated with the Pd mineralization of the Roby zone of the Lac des Iles Complex of northwest Ontario, Canada. They suggested there were two stages of pegmatite formation and fluid entrapment: (1) the crystallization of magnesiohornblende – plagioclase intergrowth at T ~650° to 850°C and (2) the crystallization of quartz at T ~535° to 650°C, both at a maximum depth of 10–12 km. Somarin et al. (2009) noted five types of fluid inclusions in the (altered) pegmatitic plagioclase from the same mineralized zone with homogenization temperature and salinity ranging from 240°C to 445°C and 15.37 to 48.52 wt.% $NaCl_{equiv}$, respectively.

Ripley (2005) reported both primary and secondary high-salinity fluid inclusions occurring in microfractures in contact aureole units of the Duluth Complex with trapping temperatures of about 500 °C. Frost and Touret (1989) reported primary and secondary inclusions in quartz from the Sybille Monzosyenite within the Laramie Anorthosite Complex with compositions ranging from CO_2–CH_4, H_2O–NaCl and silicate–oxide–chloride–carbonate melt inclusions. Trapping conditions for primary CO_2 and melt inclusions were estimated to be 950–1,000 °C and 3 kbar. Glebovitsky et al. (2001) describe secondary fluid inclusions in quartz from gabbroic pegmatites, quartz veins and sulfide-bearing alteration assemblages in the Lukkulaisvaara intrusions in Russia. Compositions ranged from H_2O, CO_2, H_2O–NaCl (~43 wt.% $NaCl_{equiv}$), CO_2–H_2O, CH_4–N_2 and CO_2–CH_4–N_2, all trapped at relatively low temperatures of 290–370 °C and 1.4–2.5 kbar.

The presence of molten salts and hydrosaline melts have been observed in a few other mafic igneous systems as well. Hanley et al. (2005b) reported inclusions of Cu-rich, NaCl melt associated with sulfide deposits at the Sudbury Igneous Complex, Ontario, Canada, and Frost and Touret (1989) reported a primary assemblage of coexisting CO_2 and Na-K-Cl melt inclusions from the Sybille Monzosyenite noted above. In non-layered intrusions igneous environments alkali halide melts have been reported in a variety of felsic and ultramafic xenoliths, as well as in kimberlites (e.g., De Vivo et al., 1995; Kamenetsky et al., 2006). Hydrous salt melts have also been observed in rocks of granulite-facies metamorphism (e.g., Newton et al., 1998).

7.4 The Origin of High-Salinity Brines and Halide Melts

The fluid and melt inclusion data show that Cl-rich fluid and silicate liquids were present late in the crystallization history of the Stillwater and Bushveld complexed, and are in agreement with predicted volatile compositions from experimental studies noted in Chapter 3 and apatite compositions discussed in Chapter 6. As summarized by Hanley et al. (2008), the fluid and melt inclusion assemblages in the Stillwater Complex share several key characteristics with those reported in other mafic-ultramafic systems (1) they contain extremely high-salinity volatiles with either dissolved or coexisting immiscible carbonic fluid; (2) the high-salinity volatiles are NaCl-dominant; (3) they occur within compositionally evolved pegmatitic phases, or in interstitial and post-magmatic quartz; (4) they were trapped at conditions at or slightly below the expected solidus temperatures for interstitial residual silicate melt.

The experimental data presented in Chapter 3 illustrates that Cl solubility varies inversely with the H_2O concentration of basaltic liquids, and that a hydrosaline fluid can only exist when the Cl concentration of the liquid is very high. In typical basaltic melts, in which Cl concentrations are <0.1 wt.% Cl, the exsolved fluid will contain <4 wt.% Cl. To achieve a high enough Cl/H_2O ratio to exsolve a high-salinity fluid, a basaltic liquid must either crystallize or react with abundant hydrous or nominally anhydrous phases that exclude Cl from their structure, or acquire an unusually high Cl content from its source region, through contamination, or by a vapour-refining mechanism.

As noted in Chapter 6, the sill and dyke compositions beneath the Stillwater Complex proper that are thought to represent parental Stillwater liquid compositions contain apatite that has anomalously high Cl/F ratios compared with most other intrusions. However, even these, which contain approximately equal amounts of the chlor-, fluor- and hydroxy-apatite components do not have the nearly pure chlorapatite compositions found in the lower parts of the Stillwater Complex. Thus while the Stillwater may have higher than average Cl concentrations, it does not appear to be anomalously enriched.

Furthermore, hydrous minerals are rare in the Stillwater Complex and would have had little influence on the volatile budget as they crystallized. It has been suggested that tens to hundreds of ppm of H_2O could partition into the nominally anhydrous minerals olivine and pyroxene but exclude Cl to produce the high Cl/H_2O ratio necessary to exsolve a hypersaline volatile phase (Mathez and Webster, 2005). However, many large mafic-ultramafic intrusions such as the Great Dyke also have extensive section of pyroxene- and olivine-bearing rock but do not contain the diagnostic mineral phases such as Cl-rich apatite expected if Cl enrichment could be generated by this process.

Boudreau and McCallum, (1989), Boudreau and Meurer (1999) and Willmore et al. (2000) have suggested that the process of *vapour refining* could lead to high Cl/H_2O and CO_2/H_2O ratios. As described in Chapter 6, Cl- and CO_2-rich vapour lost during the crystallization of the lower parts of the crystal pile can rise up and enrich hotter, vapour undersaturated liquid higher in the crystal pile in these volatile components. When these stratigraphically higher rocks become vapour-saturated, the fluid will be enriched in Cl and CO_2 over what would have originally been expected from an unmodified interstitial liquid. By this process of vapour refining, unusually high Cl concentrations can be achieved in interstitial liquids.

7.5 Ore Element Concentrations in Fluid Inclusions

As described in Chapter 8, some of the Stillwater pegmatoids can have significant bulk rock PGE-base metal concentrations that are similar to those of the J-M Reef. In this regard, Hanley et al. (2005a) note that 5% of the Stillwater brine inclusions have significant concentrations of Pt and Pd ranging from 0.4 to 4 ppm. Cu/Pt and Cu/Pd ratios in the brines range from ~20 to ~400 with Pd/Pt ratios between ~1 and ~2. These compare with typical values from the J-M Reef, which has a Cu/Pt ratio of ~400 and Pd/Pt ratio of 3.

Hanley et al. (2005c) reported laser ablation ICP-MS analyses of brine, hydrocarbon and halide melt inclusions associated with footwall-style Cu-PGE-Au deposits in the Fraser Mine of the Sudbury Igneous Complex. They contain significant concentrations of Cu (0.1–1 wt.% range), Au, Bi, Ag and Pt (all 0.1–10 ppm range). Cu/Pt and Cu/Au ratios in the inclusions are significantly lower than in the host alteration veins and adjacent massive sulfide ore veins, suggesting either (1) early Cu loss from the volatiles by chalcopyrite precipitation or (2) enhanced Au and Pt solubility relative to Cu at the temperatures of entrapment.

Hanley and Gladney (2011) noted that the carbonic inclusions of the Lac des Iles Complex described above contain estimated concentrations of Pd and Bi in the low ppm range and in the tens of ppm for the base metals. Concentration ratios for the various metals are Cu/Ni = 1.2–5.2; Cu/Pd = 36–182; Ni/Pd = 30–35. These ratios are comparable to those in the highest grade mineralized samples in the Roby zone of the Mine Block intrusion of the Lac des Iles Complex.

In general, the fluid inclusion data support the experimental evidence described in Chapter 4 that carbonic, Cl-rich fluids are particularly favourable for fluid transport of the PGE. Given that CO_2 is typically most concentrated in the first fluids to be separated from a silicate liquid on cooling and crystallizing and given the preference by Cl to partition into the fluid, this would suggest that the PGE are most readily transported by the earliest fluids to separate from the crystal mush. These H_2O-poor fluids would also have higher S concentrations that would any later, H_2O rich fluids (Chapter 3). A simple model would envision the PGE and sulfur being transported most readily during early degassing the crystal pile, where they reprecipitate PGE-rich sulfide as the fluid redissolves in fluid-undersaturated silicate liquid higher in the crystal pile. Later, the influx of more H_2O-rich fluids largely alter the originally precipitated assemblage, removing S but leaving the PGE as metal alloys as observed in the potholes of the Merensky Reef, for example. These qualitative models, however, have not been confirmed by more rigorous quantitative calculations.

8

Pegmatoids, Pipes and Potholes

8.1 Introduction

Discordant pegmatoids are common to many layered intrusions and are some of the clearest evidence from fluid flow through the solidifying rock. As noted in the previous chapter, they are the main host of fluid inclusions. Where mineralized, they also are strong evidence for fluid transport of the ore elements.

8.2 Pegmatoids and Discordant Sulfide-Bearing Bodies

Pegmatoids in the Lower Banded series below the J-M Reef of the Stillwater Complex are typical of those found in many intrusions (Figure 2.12a). They occur as pipe-like to podiform bodies up to about 0.5–1.0 m in diameter that can both cut across or run parallel to the igneous layering. They are composed of the same minerals as the host gabbronorites (plagioclase, orthopyroxene and clinopyroxene $\pm$ amphibole) and in approximately the same proportions, but with grain sizes that may be $>$10 cm in length. Some examples can have quartz- and albite-rich cores. Compared with host-rock mineral compositions, the pyroxenes of the pegmatoid have a lower *mg#* (as low as 71 vs 85 in the host rock) and the plagioclase a lower An content (58 vs. 85). This is consistent with calculated fluid compositions that are relatively rich in Fe and Na and poor in Ca and Mg. The pegmatoids are also enriched in the REE as compared with the host (gabbro)norites and anorthosites (Braun et al., 1994; Hanley et al., 2008)(Figure 8.1a).

A number of studies (e.g., Cawthorn and Barry, 1992) have suggested that the origin of the stratiform pegmatoids such as that associated with the Merensky Reef could be attributed simply to the presence of a greater proportion of residual liquid that permitted a greater period of time for maturing of crystals to larger grain size. Most other interpretations involve some role by volatile fluids to drive recrystallization and the development of coarse textures (e.g., Barnes and Campbell, 1988; Nicholson and Mathez, 1991; Hanley et al., 2008; Boudreau, 2008; Mathez and Kinzler, 2017). The discordant pegmatoids in particular are generally considered to be regions by which late evolved fluids migrated through a partly solidified crystal pile. This interpretation is supported by the fluid inclusion evidence described in Chapter 7. Boudreau and McCallum (1986), Braun et al.

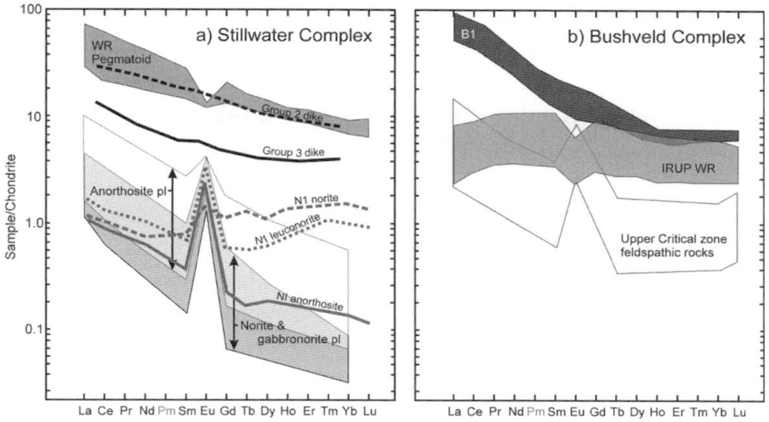

Figure 8.1 Chondrite-normalized REE whole rock concentrations in pegmatoids from the Stillwater and Bushveld Complexes. (a) Stillwater Complex, whole rock REE from pegmatoid in Gabbronorite Zone I (labelled WR pegmatoid). Also shown are whole rock data for a norite (heavy grey solid line), leuconorite (heavy grey short dash line) and anorthosite (heavy grey long dashed line) from Norite Zone I; whole rock REE in proposed Stillwater parent magmas the Group 2 dykes (solid black dashed line) and the Group 3 dykes (solid black dashed line); REE in plagioclase separates from norites and gabbronorites (shaded labelled region) and anorthosites (unshaded labelled region) from the Lower Banded series units Norite I, Gabbronorite I and Olivine-Bearing Zone I for comparison. Data from data from Hanley et al. (2008), Lambert and Simmons (1988), and Ryder and Spettel (1985). (b) Iron-rich ultramafic pegmatoids (IRUP) form the Bushveld Complex. The field labelled 'IRUP WR' is whole rock analysis from Reid and Basson (2002). Also shown is the range for a proposed Bushveld parent magma composition B1, and whole rock REE in feldspathic Upper Critical Zone rocks. Data from Maier and Barnes (2010), Reid and Basson (2002).

(1994), and Boudreau and Simon (2007) noted that the fluid channels tend to be enriched in incompatible elements and late-crystallizing phases such as quartz and suggested that the fluids migrated upward using pre-existing liquid channelways. The compositional difference between host rock and pegmatoid and the presence of abundant high-temperature fluid inclusions is consistent with pegmatoid formation by recrystallization driven by channelled fluids that were out of equilibrium with the host minerals (Braun et al., 1994; Hanley et al., 2008).

Hanley et al. (2008) presented a detailed description for the formation of the pegmatoids found from 150 to 500 m below the J-M Reef of the Stillwater Complex. They noted that they cannot represent a sample of an evolved silicate liquid derived from crystallization of the Stillwater magma. That is, although the pegmatite centre is highly evolved in terms of its major element composition compared with its GNI host rock, its incompatible trace-element composition is otherwise similar to the presumed Stillwater parent magma (Figure 8.1). They suggested instead that the albite-quartz intergrowth in the core of the pegmatite formed from a residual silicate liquid moving though the crystal pile that was also fluid saturated with aqueous brine and CO_2. This fluid-liquid mixture was channelled through the partially crystallized layered units to form the pegmatite pipes. Hanley et al. (2008) suggested that the brines, would have been highly mobile along grain boundaries in the underlying rock units, owing to their effective wetting properties (Brantley and Voight, 1989). Fluid and silicate liquid migration was also aided by crystal compaction of the crystal pile (Meurer and Boudreau, 1998).

Hanley et al. (2008) suggested that recrystallization and metasomatism of the host gabbronorite produced the coarse-textured margin zone of the pegmatite with a similar incompatible trace-element abundance patterns and trace-element ratios in the margin as the centre of the pegmatite. The pure (near end-member) albite at the centre of the pegmatite was interpreted to have formed by the compositional refining of an impure feldspar by the NaCl-rich brine: albitization of plagioclase released Ca to the fluid and allowed precipitation of calcite in the graphic quartz and entrapment of calcite within fluid inclusions. Following Newton and Manning (2002), Hanley et al. suggested that calcite precipitation in the pegmatite would be promoted by reaction of the carbonic fluid with Ca-bearing brine:

$$\underset{\text{---}}{CaCl_2} + \underset{\textit{fluid}}{H_2O} + \underset{\text{---}}{CO_2} \rightarrow \underset{\textit{calcite}}{CaCO_3} + \underset{\textit{fluid}}{2HCl} \tag{8.1}$$

and

$$\underset{\text{---}}{CaCl_2} + \underset{\textit{fluid}}{CH_4} + \underset{\text{---}}{2O_2} \rightarrow \underset{\textit{calcite}}{CaCO_3} + \underset{\text{---}\textit{fluid}\text{---}}{2HCl + H_2O} \tag{8.2}$$

Reactions such as these will increase the HCl and H_2O components. Hanley et al. (2008) also suggested that the observed increase in the CO_2: CH_4 ratio in the carbonic fluid with decreasing entrapment age indicates either an increase in the $f(O_2)$ imposed on the fluid during fluid entrapment, or precipitation of graphite from the carbonic fluid on cooling. As temperatures dropped, quartz and calcite precipitation would continue owing to decreasing solubility with cooling, decompression and decrease in fluid salinity.

Field evidence for sulfur and PGE transport through these same channelways is seen in some pegmatoids that host both disseminated sulfide to massive PGE-rich sulfide cores several tens of centimetres in diameter in both the Bronzitite Zone of the Ultramafic series and Gabbronorite Zone I of the Stillwater Complex below the J-M Reef. The sulfides are nominally composed of the same major sulfide minerals as make up the reef rocks (pyrrhotite, chalcopyrite, pentlandite) and containing platinum-group minerals such as braggite, (Pt, Pd, Ni)S, that can be large enough to be visible in hand specimens. The association of disseminated and massive PGE-rich sulfides hosted in the pegmatoid is strong evidence that the fluids can transport the ore component, and is consistent with the ore metal concentrations in fluid inclusions found in the pegmatoids (Hanley et al., 2005a; see Chapter 7).

The example of a massive sulfide core to a pegmatoid shown in Figure 2.12b occurs adjacent to a pothole-like feature in the Gabbronorite subzone of OB-I. The Pt + Pd vs Cu concentrations of two other examples of PGE-enriched pegmatoids beneath the J-M Reef are shown in Figure 8.2. The Janet 50 samples are hosted in the upper part of the Bronzitite

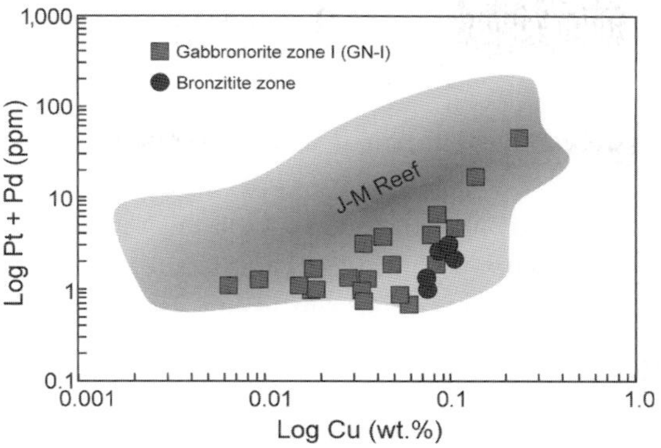

Figure 8.2 Bulk rock Pt + Pd vs Cu of mineralized pegmatoids from the Bronzitite Zone of the Ultramafic series and Gabbronorite Zone I in the Lower Banded series beneath the J-M Reef along with the compositional field of the J-M Reef (Shaded field, darker represents higher data point density). Redrawn after Zientek et al. (2002) with permission from the Canadian Institute of Mining, Metallurgy and Petroleum.

Zone near the top of the Ultramafic Series and consists of a pegmatoidal orthopyroxenite with disseminated sulphide minerals and local concentrations of graphite (Volborth and Housley, 1984). The Coors 602 samples are described as an orthopyroxene-rich pegmatoid in GN-I below the J-M Reef and a dyke-like body of orthopyroxenite that cuts the Lower Banded series (McIlveen, 1996).

Podiform sulfide-rich bodies have been described in the Lac des Iles by Duran et al. (2015, 2016). They are composed of pyrrhotite + pentlandite ± chalcopyrite + pyrite, the latter presumably the result of post-intrusion greenschist alteration that has affected much of the higher levels of the intrusion. The amount of sulfides in the pods varies from <10 to ~100 modal percent, with those with >60 mode % sulfide considered 'massive' sulfides that can be up to 162 cm thick. Although obscured by the later alteration events, the surrounding silicate minerals are not described as pegmatoidal. Duran et al. discounted the precipitation of sulfides from hydrothermal solutions under the assumption that this would have resulted in an increase of alteration in the silicate minerals present around the sulfide pods and instead suggested that they represent sulfide liquid mobilization during co-magmatic deformation. They did not consider that igneous fluids may have been initially in equilibrium with the igneous minerals as appears to be the case for the Stillwater pegmatoids.

In the Stillwater Complex, PGE-sulfides also occur as discordant disseminated sulfide-bearing 'pipes' in the uppermost 150 m of the Upper Anorthosite Zone of the Middle Banded series (Figure 2.14a) and as trails of disseminated sulfide leading upward to larger, podiform regions of disseminated sulfide (Figure 2.14), both occurring as part of the low grade Picket Pin PGE deposit (Boudreau and McCallum, 1985, 1986). The presence of transgressive zones of sulfide leading up to conformable sulfide-rich zones, the association of sulfide with late-crystallizing minerals such as apatite and quartz, and the Cu-rich nature of the sulfide assemblage have been interpreted to have formed when mineralizing solutions percolated upward through residual liquid-rich interstitial channelways during solidification of the Upper Anorthosite. The model they suggested is broadly analogous to the formation of fumarolic pipes in a cooling volcanic ash flow tuff: for both a cooling thick anorthosite unit or a cooling tuff, gas migrating from the lower part of the body becomes focused into vertical channels in the upper part of the body. The lack of PGE-sulfides in the medium-grained anorthosite member at the top of the Upper Anorthosite or in the overlying troctolite suggests that the monomineralic medium-grained anorthosite member acted as a relatively impermeable cap which trapped the migrating fluids at its lower contact. Calculated fluid compositions in equilibrium with the observed mineral assemblage show that these fluids were silica- and alkali-rich and that HCl and HF were important fluid species. Locally, fluid concentration appears to have been high enough to fully wet the silicate mineral grains to produce small (few tens of centimetre maximum diameters) pegmatoidal bodes with sulfide-rich cores that are now mostly rusted out.

8.3 Iron-Rich Ultramafic Pegmatoids and Pipes of the Bushveld Complex

The Bushveld Complex hosts a number of discordant ultramafic bodies, some of which contained significant PGE mineralization prior to being mined out in the early 20th century. As summarized by Scoon and Mitchell (2004a), the discordant bodies fall into two main types, iron-rich ultramafic pegmatites (IRUP) and magnesian dunite pipes (Figure 8.3), although Cawthorn et al. (2017) notes that the two have significant compositional

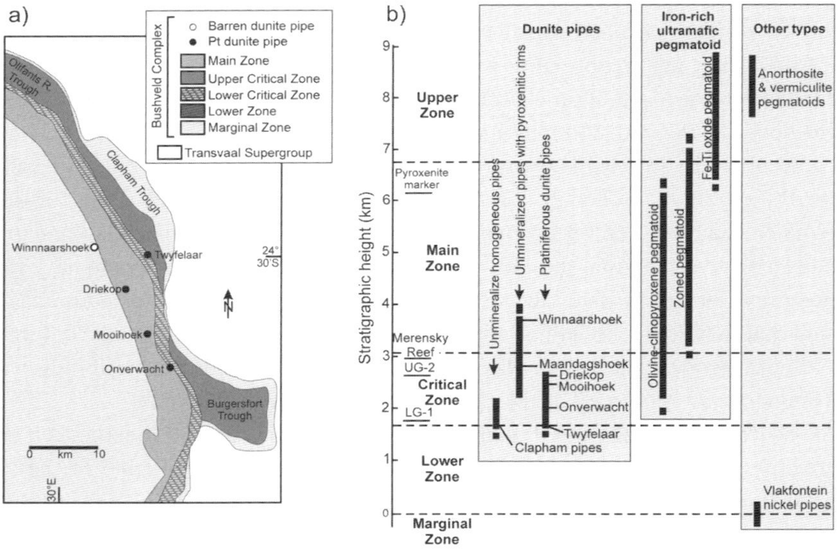

Figure 8.3 Discordant bodies in the Bushveld Complex. (a) Simplified geologic map of the Eastern Bushveld Complex, showing the location of the pipes discussed in the text. (b) Types and locations of various discordant bodies within the Bushveld Complex. Redrawn after Scoon and Mitchell (2004b), with permission.

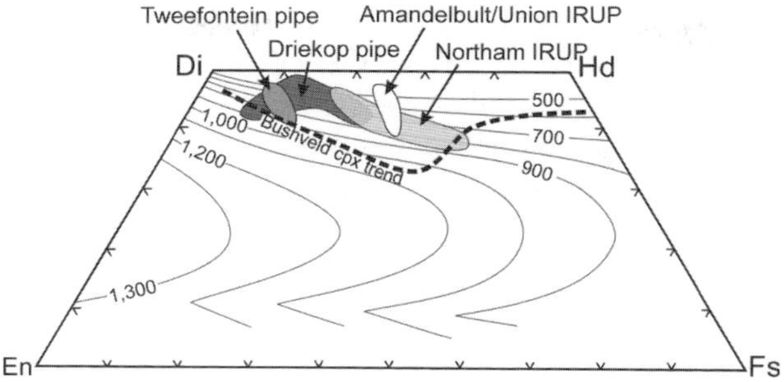

Figure 8.4 Compositions of Bushveld IRUP and pipe clinopyroxene (data from Scoon and Mitchell, 1994, 2009; Cawthorn et al., 2000; Reid and Basson, 2004), plotted on the pyroxene solvus for the two pyroxene field (after Lindsley, 1983). Line denoted 'Bushveld trend' is the range of clinopyroxene from the host Bushveld rocks (Atkins, 1969).

overlap. The dunite pipes include unmineralized and generally homogeneous pipes to those that are PGE-bearing and that can have Fe-rich cores. A few miscellaneous types include the Valkfontein nickel pipes that occur near the base of the complex in the western limb and anorthosite and vermiculate pegmatites found in the Upper Zone.

The IRUP are coarsely crystalline rocks composed largely of fayalitic olivine, augite, Fe-Ti oxides, and hornblende, and locally can form irregular sheets and lenses that pseudomorph igneous layering. They are not found in the underlying Transvaal sequence nor in the Lower Zone but are only observed starting in the Critical Zone; but even here they are not uniformly present along strike. Most studies agree that the IRUPS are largely a volume-for-volume replacement of the original rock without evidence for dilatational displacement. Clinopyroxene tends to be rich in the wollastonite component and lacking in orthopyroxene exsolution, which can allow a minimum temperature of formation of ~700 °C (Figure 8.4). Like the Stillwater pegmatoids, they are enriched in the REE relative to the host rock.

The platiniferous pipes are know from only four localities in the central area of the eastern limb of the complex: The Driekop and Mooihoek pipes are exposed in the Upper Critical Zone, the Onverwacht in the Lower Critical Zone and Twyfelaar pipe in the Lower Zone. Three of the better known pipes are shown in Figure 8.4 and have been the subject of a number of reviews by Scoon and Mitchell (e.g., Scoon and Mitchell, 2009, 2010, 2011), from which the following is a summary. In brief, the dunite pipes, both mineralized and unmineralized, can be up to several hundred m in diameter. They appear to be concentrated near faults and structural upwarps in the floor rock (Viljoen and Scoon, 1985). While typically described as 'carrot-shaped,' the down-section extent and three-dimensional shapes are not well known. Mineralization does appear to disappear with depth, however, and no similar pipes are reported in the country rocks beneath the Bushveld Complex. Modal layering in the host rocks tends to be strongly downwarped and thinned in the vicinity of the pipes, although given the evidence of replacement in and around the pipes it is unclear to what degree this 'layering' is original or related to later metasomatism in the host rock.

The host rocks tend to be composed mainly of anorthosite and leuconorite where the pipes cut Upper Critical Zone rocks, which elsewhere is more typically composed of norites and gabbronorites. The margins can locally be irregular; the upper part of the Onverwacht pipe, for example, expands outward for several hundreds of metres into the host rocks as gently dipping sheets, fingers and lenses of magnesian dunite and harzburgite to produce a mushroom-shaped body in cross section. Viljoen and Scoon (1985) note that

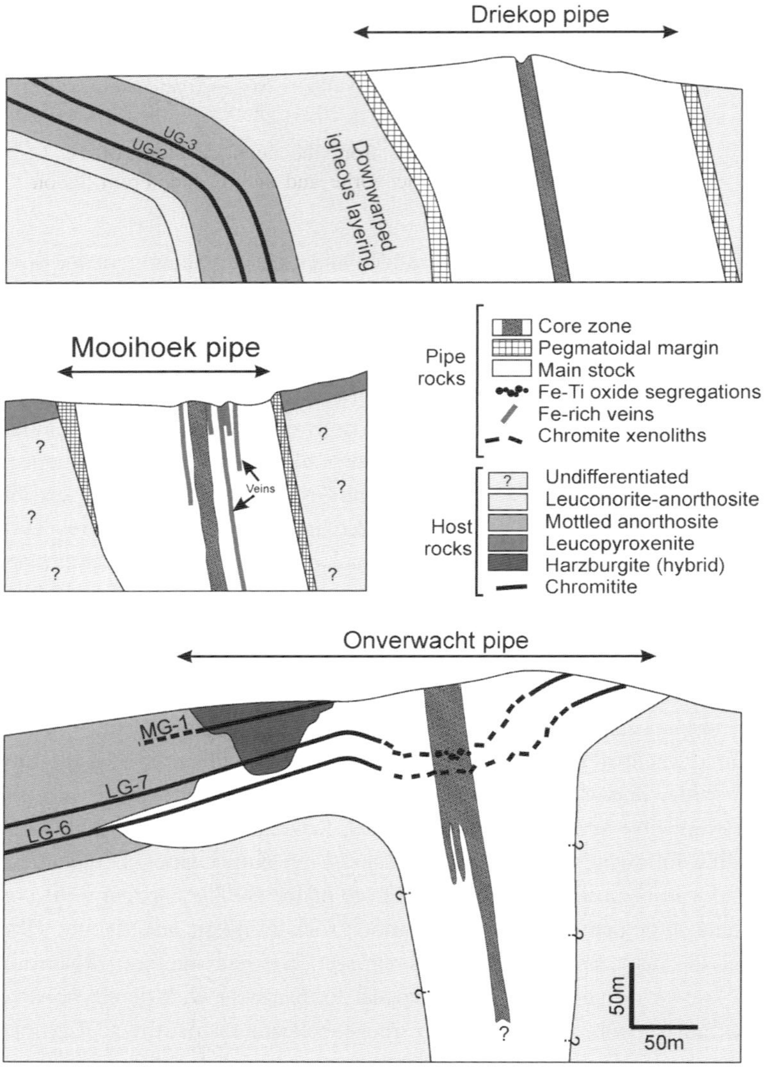

Figure 8.5 Cross sections of the Driekop, Mooihoek and Onverwacht platinum pipes, all to the same vertical and horizontal scale. Redrawn after Scoon and Mitchell (2009, 2010, 2011), with permission.

Figure 8.6 Iron-rich clinopyroxenite pegmatite from the outer envelope of the Mooihoek pipe developed within and down into the host leuconorite and anorthosite. From Scoon and Mitchell (2011) reproduced with permission.

plagioclase-rich layers are preferentially replaced with pipe assemblage relative to pyroxene-rich layers (Figure 8.6). Broken blocks of chromitite are present in some pipes and appear to define remnants of chromitite layers inherited from the host, although they are displaced to a lower stratigraphic position.

Outside of the pipe margins and cores, the pipes are largely composed of a magnesian olivine (Fo 80–85) with minor Cr-spinel; the olivine have a higher Mg# than the host rocks and have NiO contents similar to the olivine in the layered rocks of the Lower Zone. Schiffries (1982) noted that large, anhedral olivine grains can show undulatory extinction produced by kink bands that are locally recrystallized to form smaller undeformed polygonal grains. The marginal contacts with the country rocks can be more complex, with clinopyroxene-rich rocks, Fe-enrichments (olivine Fo64–44) and pegmatoidal textures typical. Cawthorn et al. (2000) note that the minor plagioclase found in the pipes are isotopically enriched as compared with the host rocks (initial $^{87}Sr/^{86}Sr$ ratio of 0.7073), but is similar to that of the Upper Zone. Stumpfl and Rucklidge (1982) report the local presence of abundant graphite and, in the Mooihoek pipes, alkali amphibole inclusions in chromite. Bulk rock compositions (Figure 8.7) illustrates the general enrichment in mafic component in the pipe, the Fe-enrichment in the core and the abundance of anorthosite and leuconorites in the immediately surrounding host rocks.

Although now mined out, PGE mineralization in the mineralized pipes was concentrated within a core zone up to 10–20 m in diameter. This preferential concentration of mineralization in the core is broadly analogous to the massive sulfide cores surrounded by pegmatoidal silicates seen in the Stillwater Lower Banded Series pegmatoids. Scoon and Mitchell (2009) note that previous workers have suggested that the Driekop core zone was composed of hortonolite, grading out to more magnesian olivine, whereas they suggest that the core composed of a more heterolithic package, with segregations of Fe-rich dunite (Fo73) and lesser Fe-rich wehrlite hosted by the magnesian dunite. The Fe-enrichment is most extreme in the Onverwacht and Mooihoek pipes. Here, coarse crystals of hornblende

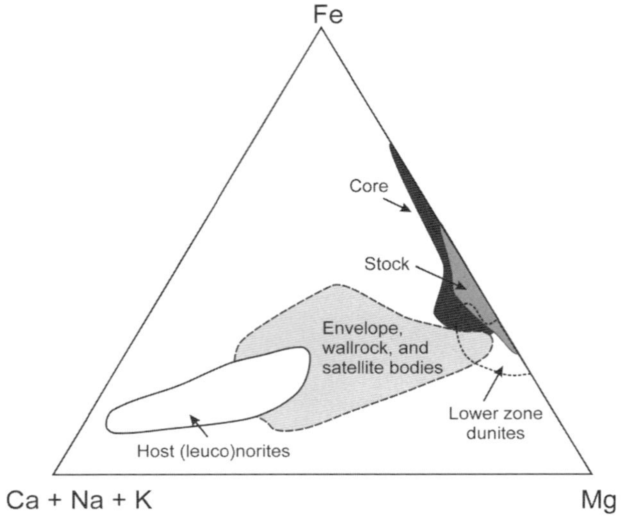

Figure 8.7 Modified AFM diagram (Na+K+Ca)-Mg-Fe bulk rock plots for the four mineralized pipes and the barren Winnaarshoek pipe. Also shown are field for host rock (Leuco) norite and Lower Zone dunite. Redrawn after Scoon and Mitchell (2004b), with permission.

and phlogopite, small grains of apatite, and metre-sized pods of Fe-Ti oxides presumably replacing chromitite were reported as prominent minerals in the core of the Onverwacht pipe (Wagner, 1929; Scoon and Mitchell, 2010). As summarized by Stumpfl (1961) and Peyerl (1982), the relative proportions of the major Driekop PGE host minerals include 50% ferroplatinum (Pt–Fe) and ferroplatinum-base metal sulfide intergrowths, 45% As and Sb minerals (e.g., sperrylite), and 5% PGE-sulfide (e.g., braggite, laurite). Mineralized veins of iron-rich dunite and wehrlite, as well as barren veins of iron-rich clinopyroxenite pegmatite, crosscut the magnesian dunite outside of the core zone.

Peyerl (1982) documented ore mineral variations that occur in the UG2 chromitite away from the Driekop pipe that appear to have been influenced by the presence of the pipe (Figure 8.8). He noted that sulfur-poor assemblages, composed of PGE-Fe alloy metals, PGE arsenide, and PGE-stibnite minerals make up most of the host minerals for the PGE within about 1 km of the pipe, similar to what is observed in the pipe itself. In contrast, farther from the pipe PGE are in a more typical reef-type Pt-Pd sulfide assemblage. The size of the PGM and the base metal sulfides also increase towards the pipe, but whether this is due to higher grade or a coarsening of the assemblage was not noted.

The barren Winnaarshoek pipe crosscuts a relatively homogenous sequence of norite and leuconorite in the Lower Main Zone, 100–300 m above the Merensky reef. While the main stock and the pegmatoidal contact with the host are similar to that of the platiniferous pipes, it does not have the Fe-rich core of the mineralized pipes. However, if the carrot-shaped nature of the central mineralized zone is typical, it may be that a potential Winnaarshoek ore zone occurred at a stratigraphically higher, but now eroded, part of the complex.

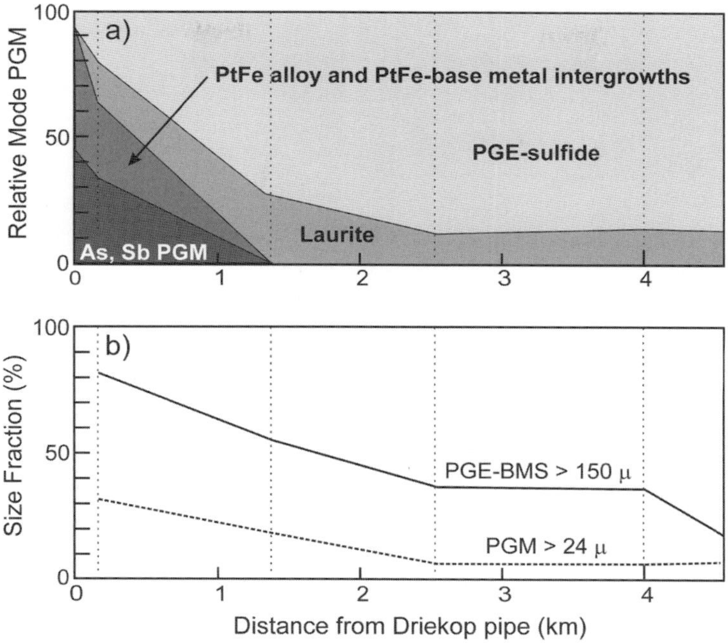

Figure 8.8 The effect of the influence of the Driekop pipe on the platinum-group minerals (PGM) of the UG2 chromitite as a function of distance from the pipe. (a) Relative mode of the various PGM. (b) Relative size fraction of discrete PGM (blue line) and PGE-base metal sulfides (red line) that exceed the listed minimum size. Constructed with data from Peyerl (1982).

8.4 Bushveld Plutonic Diatreme

Occurring in the eastern Bushveld Complex and possibly related to the dunite pipes for reasons discussed below is a 10-metre diameter diatreme (breccia pipe) that is exposed cutting and anorthosite of the Upper Critical Zone. The pipe contains blocks of Bushveld anorthosite, norite and pyroxenite derived from lower in the stratigraphy (Figure 8.9). No country rocks from below the complex are observed. The blocks are commonly elongate parallel to the axis of the pipe, and are tightly packed within an ultramafic pegmatite (IRUP) groundmass. IRUP dykes radiate outward from the pipe. Boorman et al. (2003) noted that plagioclase in the breccia pipe blocks reveal that these grains are unzoned and have higher anorthosite content (~An_{85}) than host rock plagioclase (~An_{75}). This and the presence of smaller ultramafic blocks both indicate that the breccia pipe blocks were transported considerable vertical distances. Whole-rock major- and trace-element data show enrichment in fluid-soluble incompatible elements, supporting the role of volatile fluids in the formation of the pipe. Concentric ring fractures occur in the host rock around the pipe out to a distance of several m.

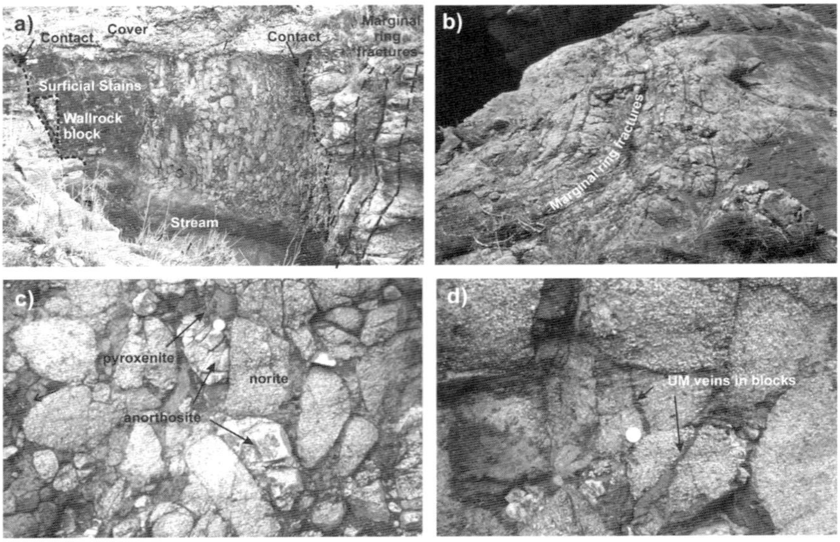

Figure 8.9 Breccia pipe from the Upper Critical Zone of the Bushveld Complex. (a) The ~10 m diameter pipe as it is exposed along the steep cliff along a dry streambed. (b) Ring fractures developed in the host rock around the perimeter of the breccia pipe. (c) Detail of the variety of rock types present in the pipe. (d) Detail of ultramafic material between and cutting some blocks. The ultramafic material is locally developed as larger pegmatoidal bodies. From Boorman et al., (2003) with permission.

Ferguson and McCarthy (1970) noted the similarities of the Bushveld diatreme with kimberlites and suggested it formed in response to fluid overpressure, similar to mechanisms proposed for classic kimberlite diatremes (e.g., Lorenz, 1975; McCallum 1985). This idea was followed up by Boorman et al. (2003), who modelled the vapour pressure increase that can occur during solidification of the interstitial liquid, and suggested that certain layers within the complex can become fluid-rich source zones that can decompress suddenly.

More recently, Boudreau (2016c) noted the similarity with km-long blowout pipes that can occur in sedimentary basins (Figure 8.10). The sedimentary pipes range from several m to several 10's of m in diameter, also can have ring fractures in the host sediments, and are believed to have formed by the sudden release of overpressured gas over a period of hours or a few days (e.g., Løseth et al., 2011). Boudreau noted that the diatreme and the larger dunite pipes are unique to the Bushveld Complex, that IRUP tend to have an enriched crustal Sr isotopic signature, and that the Transvaal sediments were unmetamorphosed prior to the intrusion of the Bushveld Magma. He extended the idea initially proposed by De Wall (1977) who suggested that the Vlakfontein nickel pipes to have formed by CO_2-rich fluid fountains released during thermal metamorphism of the footwall country rock. They modelled pipe

Figure 8.10 Sedimentary blowout pipes as analogues to the Bushveld breccia pipe. (a) Exposure of a pockmark (a pothole analogue structure) at the top of a blowout pipe and the surrounding ring fractures in the lowermost part of the Lindos Bay clay; Cape Vagia, Rhodes. (b) Detail of the surrounding ring fractures. From Løseth et al. (2011), with permission. (c) Acoustic seismic profile of seafloor pockmarks and km-scale pipes, Niger Delta. Redrawn after Løseth et al. (2011), with permission.

formation as a response to much more extensive devolatilization of the footwall country rocks than seen in other intrusions that intruded 'drier' country rock. Loading due to inflation of the Bushveld magma chamber also aided fluid movement, the system behaving as the 'world's largest whoopee cushion.' This is supported by the association of pipes with upwarps and faults in the underlying floor rocks noted above, which would have focused country fluids.

Boudreau (2016c) suggested that the apparently unique character of the Bushveld metamorphic hydrothermal aureole was responsible for the large platiniferous replacement pipes and the diatreme described below. While smaller replacement features can be found in any intrusion, the size and scale of the larger pipes and the diatreme are unique to the Bushveld system. A first-order model for the understanding of how pipes can develop during dehydration of the underlying country rock is shown in Figure 8.11. In this Figure, fluid is generated at the base of the sill by dehydration of the country rock; fluid generated within the intrusion is ignored. If the rocks are also undergoing compaction, then the lithostatic load pressing on the fluid at the base of the fluid generation zone can be transferred upward by the fluid. This can result in an overpressured situation, leading to dilation of the rocks in the upper part of the fluid generation zone. If the fluid pressure exceeds the tensile strength of the rock, the rocks will fracture and the fluid can move upward into the sill.

Because of the low density and low viscosity of the fluid, a volatile fluid is much more effective in transferring the lithostatic load upward as compared to a silicate liquid: the fluid becomes progressively more overpressured relatively to ambient lithostatic pressure as the fluid moves upward. In essence, squeezing the bottom of a toothpaste tube causes the cork at the top of the tube to pop open. As long as fluid transport can keep up with the

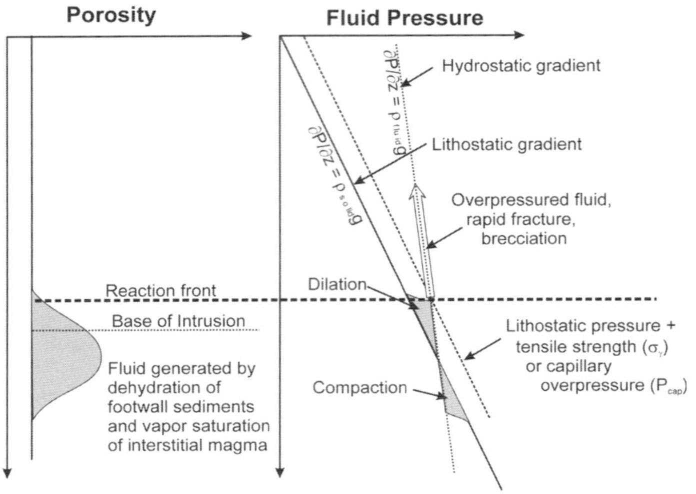

Figure 8.11 Schematic model of the compaction-driven migration of a contiguous intergranular fluid across the lower country rock-sill contact (labelled 'Base of Intrusion'). The fluid-saturated rocks are composed of a lower continuous fluid zone that includes the intergranular fluid in the country rock and a continuous fluid in the mush that exists up to the dashed line labelled 'Reaction front', where the rock will begin to fracture. Channelled movement of this fluid upward can result in the fluid becoming increasingly overpressured. Revised after Connolly (2010), with permission.

developing fracture network, the migrating fluid induces more rapid fracture and more rapid transfer as the fluid moves upward. If fluid flow is rapid enough, the result is a brecciation in the upper part of the developing pipe and presumably explosive injection of country fluid into the magma chamber. If slower, the migrating fluid can still develop a high permeable, channelized zone by which both country fluids and late igneous fluids generated by crystallization of the interstitial liquid in the mush column can migrate through the crystal pile. Such a process can be an effective mechanism by which a magma chamber can be contaminated by country fluids, and particularly in those crustal isotopes that might be especially enriched in the country fluids (Sr, Pb).

8.5 Existing Bushveld Pipe Petrogenetic Models

Models for the formation of the dunite pipes and IRUPs range from those the envision an orthomagmatic origin (e.g., Wagner, 1929; Hatch et al., 1974; Scoon, 1994; Cawthorn et al., 2000; Scoon and Mitchell, 2009) to those that interpret them as hydrothermal replacement features (e.g., Sohnge, 1963; Cameron and Desborough, 1964; De Waal, 1977; Stumpfl and Rucklidge, 1982; Schiffries, 1982) or some combination of the two (e.g., Rudashevsky et al., 1992). For example, Scoon and Mitchell (2009) suggested that the pipes formed by injections of olivine and chromite-saturated magmas injected along vertical (and presumably open) conduits. The formation of leucocratic rocks, and downwarping of the host layering is attributed to disequilibrium partial melting of the host rocks. This partial melting produced a dense, PGE- and Fe-rich liquid that then drained downward and percolated into the core zone by exploiting fractures associated with the pipe. In a variation, Cawthorn et al. (2000) suggested that the dunite pipes formed as replacements due to focused flow of magma through porous crystal mush. This calls on similarities to magma infiltration and replacement mechanisms proposed to explain secondary dunite and other replacement bodies seen in ophiolites (e.g., Kelemen et al., 1992; Batanova and Savelieva, 2009) and layered intrusion such as the Rum Complex (e.g., Bédard et al., 1988; Holness et al., 2007). They did not directly address the mineralization but favoured the hypothesis advocated by Wagner (1929) who suggested PGE in the pipes was selectively melted from earlier-formed reefs.

Scoon and Mitchell viewed the IRUPS to have formed by a separate mechanism that the pipes. They suggest that they represent the downflow of Fe-rich interstitial liquids from higher, more evolved portions of the Bushveld Complex. A problem with this is that the interstitial liquids high in the intrusion are quartz and plagioclase saturated, and the hydrous minerals are much more F-rich than is observed in the IRUPS and pipes. Cawthorn et al. (2018) suggested that the IRUPS and pipes are cumulates that formed by one fractionating magma. They noted similarities of mineralogy with the minerals precipitated from a meimechite or ankaramite magmas. These are volatile-rich, SiO_2-poor magmas that crystalize olivine and clinopyroxene early and plagioclase relatively late. They suggested that the pipes and IRUPS formed by replacement of host rock with a liquid, which then crystallized and fractionated to form the bodies as cumulates.

In contrast to these orthomagmatic models are those who have interpreted the pipes as the result of fluid activity (e.g., Schiffries, 1982; Stumpfl and Ruckledge, 1982; Boudreau, 2016c). Where the pipes intruded pyroxenites of the Lower and Lower Critical Zones, a simple silica loss reaction can explain the olivine-rich nature of the pipes:

$$\underset{\textit{orthopyroxene}}{2(\mathrm{Mg,Fe})\mathrm{SiO_3}} = \underset{\textit{olivine}}{(\mathrm{Mg,Fe})_2\mathrm{SiO_4}} + \underset{\textit{fluid}}{\mathrm{SiO_{2,aq}}} \tag{8.1}$$

Because pyroxenes can contain more Cr than can be incorporated into olivine, Reaction 8.1 can be expanded to produce the minor chromite component of the pipes (ignoring the chromitite in the large blocks inherited from the host):

$$\underset{\textit{orthopyroxene}}{2\left(\mathrm{Mg^{2+},Fe^{2+},Cr^{3+}}\right)\left(\mathrm{Si^{4+},Al^{3+}}\right)\mathrm{O_3}} \rightarrow \underset{\textit{olivine}}{(\mathrm{Mg,Fe})_2\mathrm{SiO_4}} + \underset{\textit{Cr spinel}}{(\mathrm{Mg,Fe})(\mathrm{Cr,Al})2O4} + \underset{\textit{fluid}}{\mathrm{SiO_{2,aq}}} \tag{8.2}$$

Schiffries also noted the Cl-rich nature of amphibole from the Driekop pipe suggested that aqueous chloride complexes played an important role in the transport and precipitation of the platinum-group elements, the base metals and other cations. This is consistent with the fluid inclusion evidence (Chapter 7) and the unusually Cl-rich minerals, particularly apatite, that occur in the lower part of the Bushveld Complex (Chapter 6). Furthermore,

Cl-rich fluids can carry significant Mg and Fe, the *mg#* of the olivine need not be the same as the original orthopyroxene. As the olivine in the core of the pipe is too fayalitic to have been produced by simple silica loss from the host-rock orthopyroxene, Schiffries suggested it was enriched by reaction with a Fe-rich fluid:

$$\underset{\textit{forsterite}}{Mg_2SiO_4} + \underset{\textit{vapour}}{2FeCl_2} \rightarrow \underset{\textit{fayalite}}{Fe_2SiO_4} + \underset{\textit{vapour}}{2MgCl_2} \tag{8.3}$$

Where the pipes cut plagioclase-bearing gabbronorite and norite that make up most of the Upper Critical Zone, both the magmatic and metasomatic models require this to be either eroded (by the open conduit orthomagmatic models) or replaced by olivine +/– pyroxene with the loss of plagioclase (by infiltrating silicate liquid or hydrous fluid metasomatic models). Schiffries (1982) proposed that a high-temperature chloride solution flowed through irregular channelways along a structural weakness near the axis of the pipe caused a metasomatic replacement involving the desilication of orthopyroxene to produce olivine and the complete dissolution of plagioclase. The 67% volume loss accompanying these reactions resulted in the structural collapse of the Critical zone rocks around the pipe.

The abundance of plagioclase-rich rocks (anorthosite and leuconorite) in the host rock where the pipes transgress the upper Critical Zone also has several explanations. The orthomagmatic models suggest that this was the result of partial melting of a norite with preferentially melting of the mafic component as the rocks were heated by the intruded magma, perhaps aided by kinetic effects that inhibited the melting of plagioclase. The mafic components is then presumably lost to the hotter, intruding/infiltrating magma. The metasomatic models would interpret the anorthosites and leuconorites as the compliment to the mafic component enriched in the pipes. Schiffries (1982) suggested that the plagioclase lost from the formation of the dunite pipe was moved outward to make the anorthositic host rock. An alternative model presented below suggests it was just the reverse and fluids moved into the pipe, leaching the mafic component from the host rocks and reprecititated in the ultramafic pipe.

The observation of Cawthorn et al. (2017) that the IRUP have affinities with volatile-rich meimechite or ankaramite magma leads to the question if such a liquid could have been produced by flux melting within the Lower Zone of the Bushveld Complex. For example, Green et al. (2004) suggest that ankaramite liquids can form by flux melting of a refractory lherzolitic mantle. Hydration melting of Lower Zone rock by a CO_2- and Cl-rich volatile fluid that is carrying significant Fe, Ca, alkalis, and REE moving into hotter ultramafic rocks could produce a meimechite- or ankaramite-like liquid that are characterized by relatively high concentrations of these components and low CaO/Al_2O_3 wt. ratios. This would explain the lack of evidence for these magmas in the underlying country rock and lower parts of the Layered series. The volatile enrichment would explain the coarse texture of the IRUPs, the volatile-enriched liquid would favour crystallization of olivine and clinopyroxene over orthopyroxene and plagioclase, the melt would crystallize at a much lower temperature, and otherwise rapid cooling would result in a lack of exsolution in the clinopyroxene. Why go into the mantle when you can make everything in your back yard?

8.6 Other Examples of Mafic-Felsic Mineral Segregations and Pipes

Ignoring mafic-felsic modal layering (which may itself may have developed by the processes of constitutional zone refining described in Chapter 9), podiform replacement bodies with pronounced ultramafic-felsic segregations are a common feature of layered intrusions. For example, Naslund (1986) noted podiform anorthosite-ultramafic segregations that occur beneath the Basistoppen sill. He suggested that intrusion of the Basistoppen basaltic sill into the still-hot Skaergaard ferrogabbros led to preferential melting and movement of the mafic component to produce local pods of anorthosite restites and adjacent Fe-rich melagabbros enriched in the mafic component lost from the anorthosite. However, similar felsic-mafic segregations occur in the Lower Zone (Figure 8.12) and in the Marginal Border Series (McBirney and Sonnenthal, 1990; McBirney and Creaser, 2003), where there was no source of reheating and indeed the rocks would have been cooling rapidly. Holness et al. (2007) suggested that mixing of differentially evolved interstitial liquids across a curved cotectic boundary can result in a reactive liquid saturated

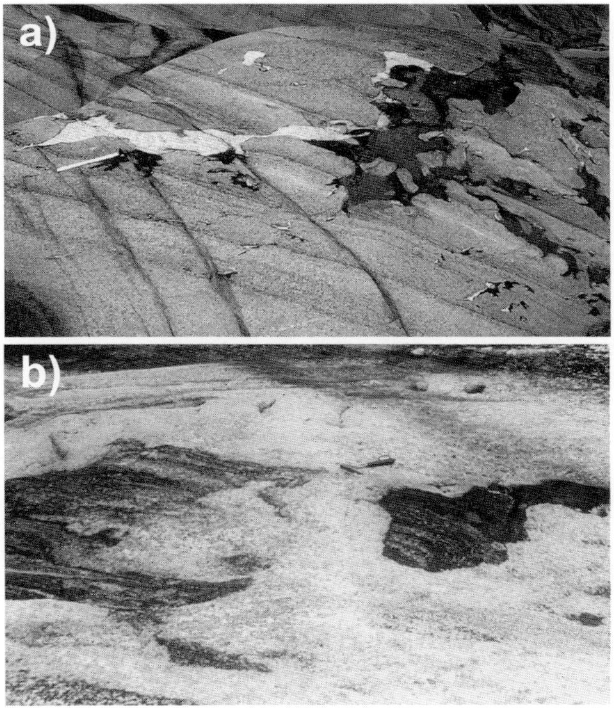

Figure 8.12 Examples of discordant mafic-felsic replacement bodies from the Skaergaard intrusion, East Greenland. (a) Anorthosite and olivine pyroxenite replacement pods crosscutting layered gabbro of Lower Zone A. (b) Anorthosite replacing parts of Lower Zone A, in which residual islands of the original gabbro have been enriched in the mafic components lost from the anorthosite. From McBirney and Sonnenthal, (1990), reproduced with permission.

in plagioclase alone. Sonnenthal (1992) suggested that the mafic/felsic segregations were due to reaction with Cl-rich fluid, but did not present details of the segregation mechanism.

Discordant and rootless bodies of troctolite and anorthosite up to several metres in size and hosted by (olivine) gabbro occur in the Middle Banded series of the Stillwater Complex. The discordant bodies are elongate perpendicular to the layering of the host cumulates with local concordant apophyses (Figure 8.14a, Figure 8.13). Plagioclase in the discordant bodies is blocky and about twice the size of that in the hosts where the plagioclase is more tabular/prismatic. The olivine forms large, lobate 'amoeboidal' grains with rounded plagioclase inclusions. The pods lack any discernible foliation despite being hosted by relatively well-foliated gabbro and have sharp margins with the host with no disruption of the host mineral foliations, implying they formed after the main compaction-driven texture of the host rock had formed. Average olivine in the host rocks is slightly more magnesian than that of the discordant bodies by about 1–2 *mg#* but NiO contents are

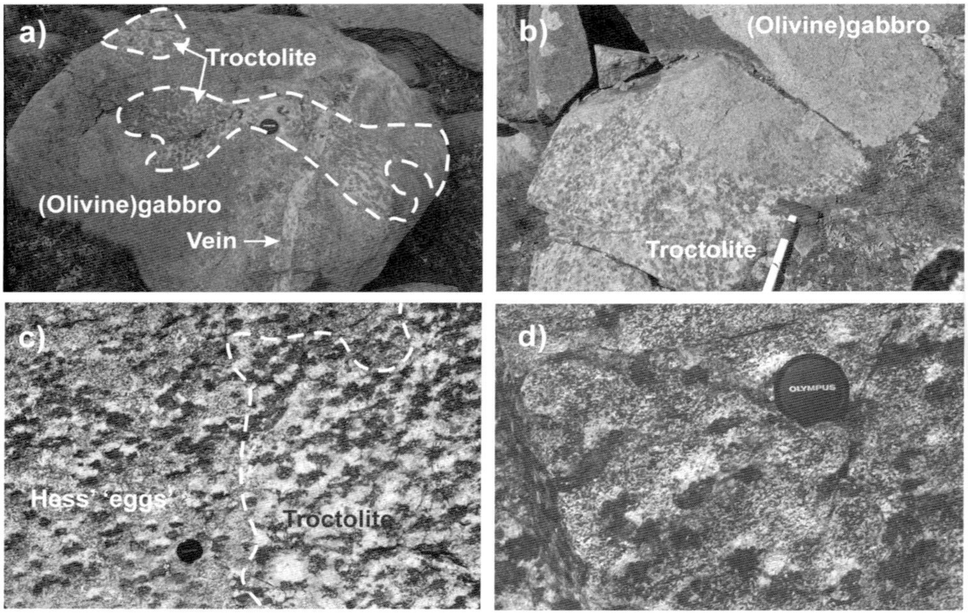

Figure 8.13 Examples of discordant troctolites from the Middle Banded series of the Stillwater Complex. (a) Discordant troctolite bodies in a boulder from Olivine-Bearing Zone III. The boulder is also cut by a large plagioclase + quartz + hornblende vein, which are otherwise rare in the complex. (b) Discordant troctolite in OB IV just below the Upper Anorthosite at the top of the Middle Banded series. Stratigraphic 'up' is towards the top of the photo. (c) Example of the 'egg' texture in an olivine gabbro described by Hess (1960), with a transition to discordant troctolite in OB IV. F) Detail of 'egg' texture; large, black olivine defines the 'yoke' in the centre-bottom of the surrounding plagioclase-rich 'egg white', both hosted by a medium-grained gabbro. From Boudreau (2016), with the permission of the Mineralogical Society of Great Britain & Ireland. A black-and-white version of this figure appears in some formats. For the colour version, please refer to the plate section.

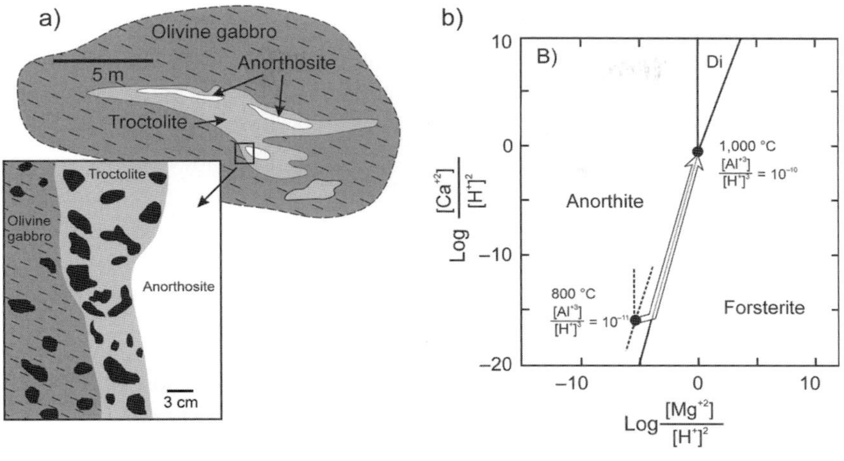

Figure 8.14 Example of a zoned discordant troctolite with anorthositic cores from the Middle Banded series of the Stillwater Complex, Montana, illustrating the selective leaching of mafic components. (a) Detail of discordant troctolite with anorthositic cores, showing the transition from a well-foliated olivine gabbro (plagioclase–olivine–clinopyroxene) to an unfoliated troctolite (plagioclase and coarse-grained amoeboidal olivine [black]) with a core of plagioclase alone (anorthosite). (b) Activity diagram showing the stability fields of anorthite, diopside and forsterite as a function of the activity ratios $[Ca^{+2}]/[H^+]^2$ and $[Mg^{+2}]/[H^+]^2$ for values of the activity ratio $[Al^{+3}]/[H^+]^3$ as shown. A fluid initially in equilibrium with the three minerals at 800 °C would find itself in equilibrium with anorthite alone at 1,000 °C and react with the solid assemblage along the open path. The resulting sequence of mineral assemblages is the same as observed in the field. Redrawn after Meurer et al. (1997), with permission.

similar, and plagioclase compositions are indistinguishable. Whole-rock major- and trace-element compositions of the discordant bodies are generally indistinguishable from layered troctolite with similar modal abundance, but the anorthositic cores are enriched in K, Na, Ba, Sr and P (Meurer et al., 1997). Irvine (1974) notes that olivine in replacement dunite bodies (Fo80) from the Duke Island Complex is only slightly less magnesian than host dunite (Fo82–91) but similar to host olivine clinopyroxenite (Fo78–84)

Meurer et al. (1997) suggested that the discordant bodies formed when cooler volatile fluids moved upward into hotter rocks and became undersaturated in silica and pyroxene components as the fluid became heated, as illustrated in the activity-activity diagram of Figure 8.14b. Reaction of the fluid produced olivine after pyroxene by preferential loss of silica at low fluid/rock mass ratios. Continued leaching at higher fluid/rock mass ratios resulted in the loss of all mafic components, leaving just plagioclase. Fluid fluxing-reaction increased the permeability along the flow path and focused the flow, allowing the original bulk compositions to be modified and leaving zoned plagioclase-rich troctolite and anorthosite replacement bodies. These bodies lack the mafic-felsic paring seen in the Skaergaard examples but illustrate that the mafic component of the host rock can be readily removed.

8.6.1 Discordant Bodies in the Mantle and Lower Crust

Discordant dunite and pyroxenite bodies are relatively common features of the mantle sections of ophiolites and they are viewed as particularly important in allowing the rapid flow of mantle partial melts as compared with more dispersed porous flow. Like the Bushveld pipes, they show evidence of a volume-for-volume replacement of host lithology and have chemically equilibrated with mid-ocean ridge basalt (*MORB*). The evidence includes remnants of original structures or protolith preserved in the replacement and the lack of evidence for dilatational opening, leading most to interpret them as metasomatic replacement bodies involving a focused reactive flow (e.g., Kelemen et al., 1995, 1997). This flow focusing is driven by the increased solubility of orthopyroxene with decreasing pressure, leading to dissolution of orthopyroxene and precipitation of olivine as mantle liquids move upwards.

As summarized by Keller and Katz (2016), volatiles are particularly important on lowering the mantle solidus and in promoting a channelled, pipe-like transport of mantle melts despite the low concentrations of volatiles in the mantle. They used a thermodynamic model for multi-component melting coupled with mass transport equations to model low-degree, volatile-rich melting for both hydrated and carbonated mantle melting. The migration of these reactive melts results in a channelled flow behaviour (pipes) that allows for efficient melt migration. Modelling of these mantle structures has become increasingly sophisticated and studies of silicate liquid and volatile fluid transport in layered intrusion could benefit from the application of these methods.

PGE pipes of the Italian Alps. Ni-Fe-Cu-PGE-enriched ultramafic branching pipes locally occur in the Italian Alps that were emplaced in a slice of lower continental crust and lithospheric mantle (e.g., Garuti et al., 2001; Locmelis et al., 2016; Sessa et al., 2017). These have roughly similar diameters as the Bushveld pipes (50–300m) and intruded a gabbroic to metasedimentary host rock; the former is enriched in plagioclase and pegmatoidal adjacent to the pipes. The pipes are composed of coarse-grained amphibole pyroxenite with cm-sized amphibole oikocrysts and mm-size olivine intergrown with phlogopite. Olivine in the pipes ranges from Fo_{72-78} and the low Ca pyroxene is Al-bearing and has an *mg#* of 75–79. Cl-F apatite, carbonate and Al-Cr-Fe spinels are some of the accessory minerals. Mineralization is characterized by aggregates, veinlets, and 1 mm–5 cm patches of sulfide composed mainly of pyrrhotite, pentlandite and chalcopyrite. Mineralization is described as variable within the pipe and most abundant within a metre of the contact with the host gabbro. This is in contrast with the Bushveld pipes where mineralization is largely in the core of the pipes. PGE are largely contained in Pt-Pd tellurides.

Because these rocks interacted with fluid at lower crustal pressures, amphibole is stable at higher temperatures than expected for the Bushveld pipes. Sessa et al. (2017) calculate olivine + plagioclase is stable to about 750 °C at 2 kbar, but breaks down to amphibole-bearing assemblages at about 950 °C at 6 kbar pressure. They suggest that the textural and mineralogical evidence is consistent with a metasomatic reaction of a peridotite protolith pipe at lower crustal conditions:

$$\begin{aligned} &1.0\text{olivine} + 0.55\text{plagioclase} + 0.23\text{H}_2\text{O} \rightarrow \\ &\quad 0.41\text{orthopyroxene} + 0.23\text{amphibole} + 0.18\,\text{spinel} + 0.07\,\text{FeMg}_{-1} \end{aligned} \tag{8.4}$$

where $FeMg_{-1}$ represents an exchange between the fluid and the product minerals. They suggest that the pipes were originally conduits for mafic magmas that became filled with olivine-rich cumulates under open-system conditions. As the system cooled down, the cumulate crystal mush was infiltrated by and reacted with volatile-rich, hydrous mafic to variably differentiated magma, possibly analogous to slab-derived melts proposed by Locmelis et al. (2016), to produce the observed replacement and ore minerals. A metasomatic origin is discounted based on the pipe mineralogy being independent of host rock lithology (gabbro or metasedimentary) and sharp contacts with the host rocks. However, as noted above, sharp contacts are to be expected in a metasomatic replacement, and the mineralogy of the replacement is controlled by the infiltrating liquid, not the protolith, except at the replacement front. Regardless of how the pipe itself may have formed, the concentration of PGE-sulfides and coarse textures to the margins of the pipe suggests that the presence of the pipes served to focus the transport of fluids from the surrounding host rock.

8.7 A Metasomatic Model for Dunite Pipe and IRUP Formation

Metasomatic zones should be broadly consistent with the solubility relations and fluid compositional controls discussed in Chapter 3. The fact that the mineral segregations observed in the Bushveld PGE dunite pipes, the ultramafic-felsic segregations seen in many layered intrusions and the Cl-rich nature of the hydrous mineral assemblages is consistent with the enhanced solubility of plagioclase, olivine and pyroxene in HCl-rich solutions. Silica has the highest solubility, followed by the alkalis, and hence can be easily lost to covert pyroxene into olivine and produce a more Ca-rich feldspar (e.g., Orville, 1972), if not complete dissolution of these minerals. Solute concentrations in saline solutions increase with both temperature and chlorinity. As noted in Chapter 3, fluid inclusions commonly have in excess of 10 wt.% Fe at 700 °C and presumably even higher amounts at higher temperatures (Yardely, 2005). More broadly, the formation of fluids enriched in Fe, Mg and Ca can occur as acidic initial solutions leach these elements from the host rock.

A revision of Schiffries (1982) metasomatic model for the formation of the dunite pipes is shown in Figure 8.15. Viljoen and Scoon (1985) noted the preference for the pipes to be located in regions of faulting and upwarping of the floor rocks, suggested that that this will focus flow of country fluids generated during dehydration for the Transvaal sediments during heating of the underlying metamorphic aureole. Using the Bushveld breccia pipe as an example, the diameter of the pipe is about the same as the mineralized cores of the dunite pipes. They suggested that fluid will move upward into the layered igneous rocks in long, narrow pipe-like channels of similar diameter. This infiltration may be passive percolation through the host rocks, such as the fluid flow that occurs during the formation of classic porphyry metal deposits. At the other extreme fluid infiltration may more rapid and explosive, leading to the formation of a diatreme. The lack of extensive metasomatism associated with the breccia pipe would suggest that it happened after much of the surrounding igneous rock had degassed (Figure 8.14a).

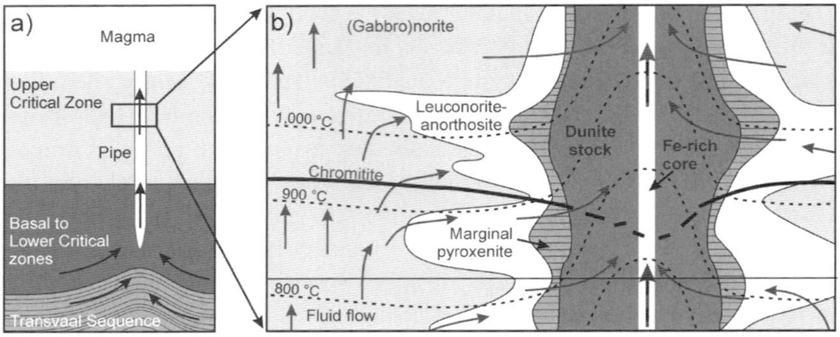

Figure 8.15 Cartoon of the formation of a Bushveld platinum dunite pipe (no scale). (a) Initial formation of a narrow, 10–20 m diameter blowout pipe by release of overpressured fluids from both the crystal pile and from dehydration of Transvaal Sequence sedimentary rocks in the metamorphic aureole beneath the complex that are focused by upwarps in the footwall. (b) Detail of the replacement in the Upper Critical Zone. Original pipe serves as a permeable chimney for Cl- rich country and magmatic fluids to move up and out of the crystal pile, leaching PGE and mafic components from the host leucocratic rocks and re-precipitation the mafic and ore components in the cooler stock.

The initial upward advance of a developing pipe is driven by fluid overpressure, but this is a transient condition occurring at the advancing front. Once formed, a pipe becomes a region of higher permeability than the surrounding host rocks that persists as long as there is a fluid supply. The consequence of this is that the pipe forms a high-permeability channel that allows Cl-rich, interstitial igneous fluids from the surrounding environment to move both laterally towards the pipe and upward to drain the crystal + liquid + vapour mush (Figure 8.14b). The upward movement allows the fluid the see hotter rocks, where it becomes heated and reacts with original host norite, dissolving the mafic and ore element component, resulting in the formation of anorthosite and leuconorites. As the magmatic vapour moves towards the pipe core it will encounter an environment that is cooling from fluids moving up through the centre of the pipe from the colder, deeper parts of the intrusion. The results of fluid cooling is to expand the pyroxene and olivine fields relative to plagioclase drive the solute-rich fluid to precipitation of olivine ± chromite and resorb plagioclase to form the main dunite stock, and the precipitation of more Fe-rich minerals and ore component in the yet cooler core. Precipitation of the latter may be aided by mixing with yet cooler residual deeper Bushveld and country fluids moving upward in the core region of the pipe. The observation of Viljoen and Scoon (1985) that plagioclase-rich layers of the host rock are preferentially replaced with mafic minerals relative to pyroxene-rich layers is consistent with the anorthosites being regions of higher fluid flux. The ore may have originally precipitated as sulfide but later loss of S produced the predominance of S-poor PGE minerals and alloys. Because of extensive volume changes, ring fractures and faulting in and around the core can produce other vertical pathways for fluid movement and mineral precipitation to produce a more complex geometry.

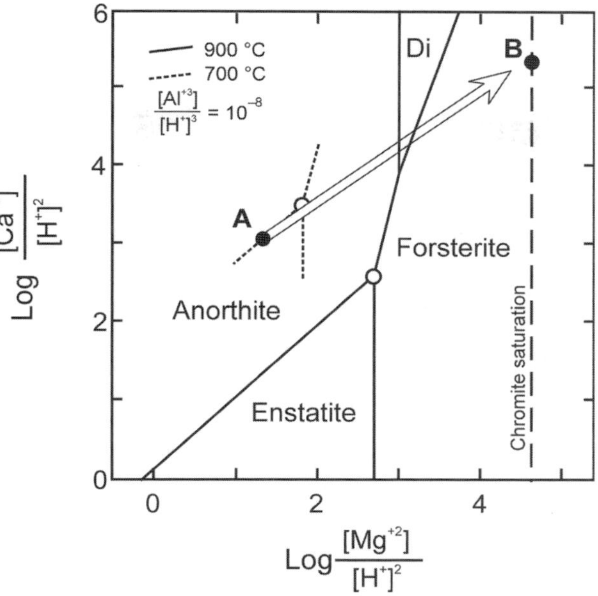

Figure 8.16 Activity diagram for the system Ca-Al-Mg-Si-O-H-(Cr). Stable mineral fields shown at 900 °C, 2 kbar and the phase boundaries around the En-An-Fo triple point at 700 °C as calculated with the program SUPCRIT (Johnson et al., 1992), 2 kbar total pressure, solid minerals balanced on silica, and chromite saturation shown schematically. Activity of the components are denoted by the square brackets; e.g., $\left[Ca^{+2}\right]^2 = (a_{Ca^{+2}})^2$. Stability fields are calculated based on $[Al^{+3}]/[H^+]$ activity ratio = 10^{-8}, all other phases assumed pure. A solution saturated in anorthite + enstatite at 700 °C at point 'A' will be saturated in anorthite alone at 900 °C. Mixing with a fluid in the core region at point 'B', undersaturated in anorthite but saturated in forsterite and a chrome spinel, will produce a rock sequence ideally anorthosite–pyroxenite–dunite.

An activity-activity diagram for the major minerals related to the formation of the Bushveld pipes is shown in Figure 8.16. A fluid initially in equilibrium with a norite (plagioclase + orthopyroxene) at 700 °C would fall along the dashed opx-pl saturation boundary shown at point 'A'. Heating of this fluid to 900 °C will cause the phase boundaries to shift to the solid lines, and the original fluid will now be in the plagioclase saturation field. Reaction of this fluid with the opx-pl assemblage will drive the fluid to become enriched in the mafic components. Changing physical conditions (e.g., moving towards a cooling core of a pipe) can reverse the situation and the fluid can now find itself oversaturated in olivine (± chromite) at point 'B'. This is a somewhat oversimplified model fails to include time-dependent changes in temperature or fluid composition, but it does explain the generally observed rock sequence of anorthosite-pyroxenite-dunite.

The data of Peyerl (1982) described above is consistent with the pipes having influenced the PGE mineralogy of chromitite out to at least 2 km from the centre of the pipes. He suggested that the changes were related to the migration of volatiles *outward* from the

Driekop pipe. The model presented here, however, would suggest that it was the inflow of fluids towards the high-permeability pipe that resulted in both the core PGE enrichment and the modification of the host chromitite layers as the fluids were increasingly concentrated around the pipes. In regards to the mineralization, a simple mass balance calculation illustrates that the host rocks can readily be the source for the PGE in the pipes. The average Pt concentrations from the Lower Critical Zone and Upper Critical Zone rocks, excluding the chromitites where Pt is generally enriched, average 40 and 119 ppb Pt, respectively (summary of Barnes and Maier, 2002a). Assuming that as little as 5 ppb was, on average, scavenged from a 1 to 2 km circular region and concentrated into a 10 m pipe core, the core would contain from 50 to 200 ppm Pt.

8.8 Potholes

Potholes are transgressive, bowl- to funnel-shaped regions with circular to elliptical cross sections where parts of the magmatic stratigraphy are missing or replaced (Figure 8.17). They are associated with zones platinum-group element (PGE) mineralization in both the Bushveld Complex (the Merensky Reef and the UG2 chromitite layer) and the Stillwater Complex (the J-M Reef) (e.g., Ballhaus, 1988 and earlier references therein; Turner et al., 1985). Potholes may be several hundred metres wide and typically range from a few metres to several tens of metres deep. For example, the Merensky reef in the northern part of Winnaarshoek area is disrupted by both the Winnaarshoek pipe and a pothole several hundred m wide. The pothole removes the Merensky and Bastard units such that the Main Zone overlies the Merensky Footwall cyclic unit. Potholes contain abundant pegmatoidal material and, in Merensky potholes, they locally contain abundant graphite (Ballhaus and

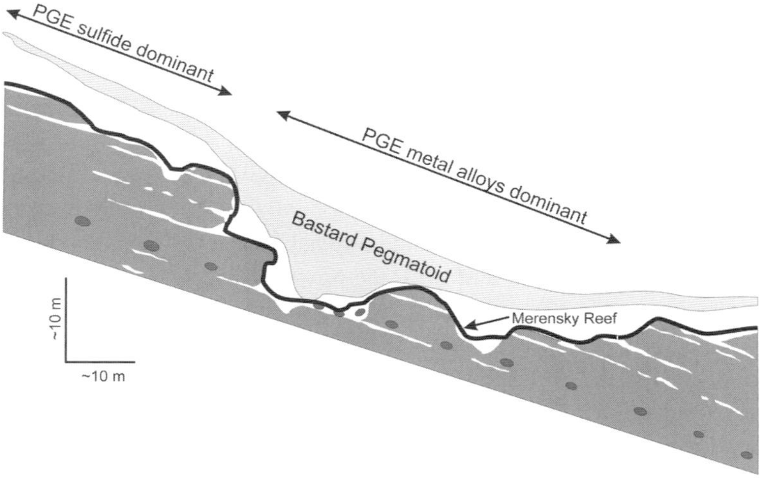

Figure 8.17 Cartoon of a typical pothole one the Merensky Reef. Redrawn after Ballhaus (1988), with permission.

Stumpfl, 1985; Buntin et al., 1985). Kinloch (1982) and Kinloch and Peyerl (1990) have noted that Pt–Fe alloys tend to predominate in the Merensky potholes while Pt sulfides are the dominant PGE phase in normal Merensky Reef. The pothole region is also characterized by abundant leuconorite, in contrasts with the modally variable rocks seen elsewhere.

Smith and Basson (2006), using a statistical study of Merensky reef pothole shapes a the Northam Platinum Mine, found the average aspect ratio of potholes increases as the size increases, which they suggested was the result of merging of potholes that were forming along a linear trends presumably controlled by the underlying structure. Chitiyo et al. (2008), on a similar size-distribution study of both Merensky Reef and UG2 potholes at the Rustenburg Platinum Mine noted that the UG2 is more densely potholed, but those of the Merensky Reef are, on average, larger.

Buntin et al. (1985), in a study of the intrinsic oxygen fugacity found the pothole bottoms are characterized by being more reduced ($f(O_2) = 10^{-9} - 10^{-12}$ at 1,150 °C), inorganic carbon concentration is up to 14 times greater, and the $FeO_{total}/(FeO_{total} + MgO)$ is higher than in adjacent normal Merensky Reef. Reid et al. (993) suggested $\delta^{18}O$ values in the rocks below potholes cannot distinguish possible presence of magmatic vapour where the rocks are unaltered. Cawthorn and Poulton (1988) looked at incompatible element concentrations beneath both potholed and normal reef noted that P_2O_5 was similar in both environments, but that nominally more mobile elements such as K, Rb and Cu are enriched and more variable beneath potholes. Cawthorn and Barry (1992) noted that the potholes associated with the UG2 have no geographic or structural relationships with the Merensky potholes. They suggested that comparison of geochemical profiles through normal and potholed UG2 chromitite show no evidence of fluid channelling as has been documented through Merensky potholes.

Theories on the origin of potholes in the Bushveld Complex have been summarized by Ballhaus (1988) and Smith and Bisson (2006), among others. As for theories for the PGE mineralization, these range from wholly magmatic to models that view them as vapour escape structures. Strictly magmatic models include suggestions that potholes formed by the scouring of loosely consolidated cumulates by magmatic convection currents (e.g., Schmidt, 1952) or scouring by and deposition from plagioclase-rich slurries (Maier et al., 2016). Others have included a thermal erosion component (e.g., Irvine et al., 1983; Campbell, 1986; Smith and Bisson, 2006), although these usually require a pre-existing depression on which to enlarge. Other interpretations have suggested they are related to structural slumping event within the crystal pile (Carr et al., 1994). All of these tend to ignore the evidence for fluids and rarely explain the different ore mineralogy in the pothole environment.

Models involving a migrating vapour include that of Kinloch (1982), who suggested the preference of PGE alloys in potholes is the result of sulfur loss to late fluids moving through the potholes. Elliott et al. (1982) and Buntin et al. (1985) noted the apparent changes in oxygen and sulfur fugacity in the vicinity of potholes as well as the presence of discordant pegmatoids which occur below some potholes. They have suggested that the potholes are the plutonic equivalent of fumarolic zones from which C-H-O-S fluids were

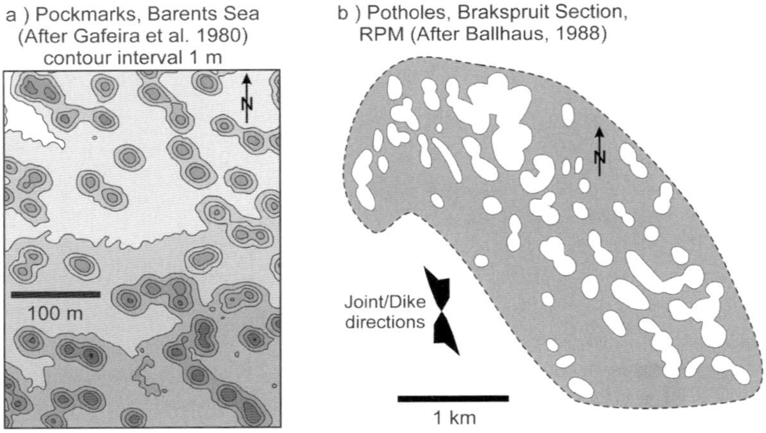

Figure 8.18 Seafloor pockmarks as an analogue for Bushveld potholes. (a) Plan view of pockmarks on the sea floor from the Barents Sea, Norway. Redrawn after Gafeira et al. (2018), with permission. (b) Plan view of the distribution of potholes from the Rustenburg Platinum Mine. Note difference in scale. Redrawn after Ballhaus (1988), with permission.

injected into the magma chamber. In a hybrid magmatic-vapour model, Ballhaus (1988) has suggested that potholes are sites of non-deposition of cumulate crystals caused by fluid introduction into the chamber at pothole locations which lower liquidus temperatures and prevent cumulates from crystallizing.

Boudreau (1992) noted that morphologically similar features form in the sedimentary environment on the sea floor, where they are known as *pockmarks* and are formed by gas escaping from gas-rich sediments (Figure 8.17). As shown in Figure 8.18, sedimentary blowout pipes can form pockmarks where they reach the seafloor. By analogy with theories on the formation of pockmarks, Boudreau proposed that the development of igneous potholes require the formation of a volatile fluid overpressure and fluid channelization within the crystal pile. Boudreau suggested that it is the presence of an overpressured fluid that initiates pothole development by the formation of dome-shaped instabilities in the upper parts of the cumulate pile and also forces interstitial liquid and fluid upward. Calculations by Boorman et al. (2003) show that these overpressures are readily developed during crystallization and degassing of interstitial liquids.

These models generally envision the replacement of a generally coherent crystal $\pm$ liquid mush with liquid. However, the leucocratic nature of the pothole rocks and evidence for replacement of pyroxene-bearing rocks with anorthosite suggests that they could be subliquidus replacement structures. In Olivine-Bearing Zone II (OB-II) below the Lower Anorthosite Zone of the Stillwater Complex is a pothole-like structure. The troctolite-anorthosite assemblages defines OB-II, where isolated pods of troctolite are hosted in a discordant anorthosite (the 'pillow troctolite' of Hess, 1960), both of which crosscut well-foliated gabbronorite (Figure 8.20). The troctolite pods/pillows can have cores of

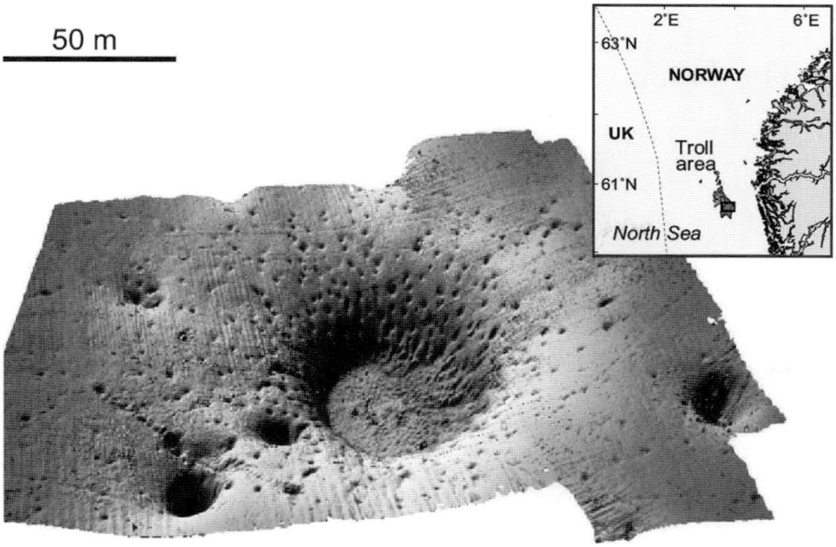

Figure 8.19 Sea floor pockmarks as analogues to pothole structures in layered intrusions. Example of high-resolution bathymetric profile of a 15 m deep pockmark with a flat, circular central region the North Sea area. From Mazzini et al., (2017), with permission.

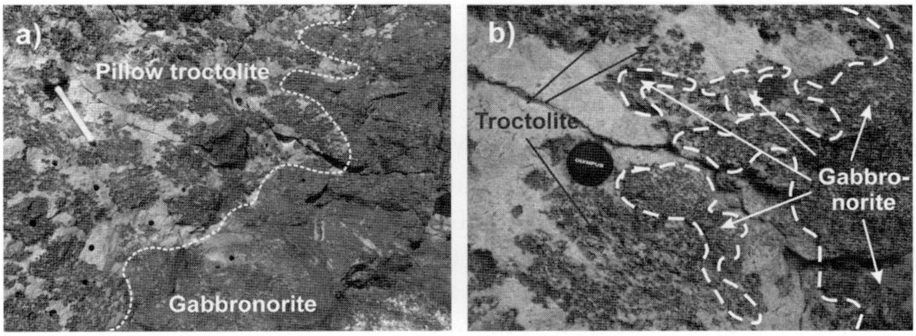

Figure 8.20 Photographs of examples of discordant troctolite in pothole-like structure from OB-II of the Stillwater Complex. (a) Discordant 'pillow troctolite' (terminology of Hess, 1960) of OB-II crosscutting foliated gabbronorite host. Stratigraphic "up" is towards the top of the photo. (b) Detail of transition from the host gabbronorite to the troctolite pillows. From Boudreau (2016b), with the permission of the Mineralogical Society of Great Britain and Ireland.

gabbronorite, the foliation of which can be rotated relative to that of the surrounding gabbronorite. The lithologic relationships are similar to the replacement troctolite-anorthosite bodies that occur elsewhere in the Middle Banded series of the Stillwater Complex, and both have been attributed to a replacement origin (e.g., McCallum et al., 1977; Meurer et al., 1997).

9

The Effects of Volatiles on Mineral Stability and Volatile Fluxing

9.1 Introduction

Fluids and the volatile content of magmas and their solid crystallization products can influence the final solid by a number of mechanisms. Going from high to low temperatures, these effects can be summarized as follows:

Liquidus phase boundary shifts. Volatiles can have a direct influence on the crystallization of magma by changing the liquidus temperature and shifting cotectic boundaries. In addition, the gain or loss of sulfur as it partitions between a fluid phase and the magma can induce or delay sulfide saturation.

Volatile Fluxing. Along with decompression (adiabatic) melting, flux melting is one of the two main ways by which melting is induced in the mantle and is responsible for the production of volatile-rich magmas at subduction zones. However, it is often overlooked as a mechanism to produce modal variations in layered intrusions. Within a crystal-liquid mush, the addition of volatiles as they move into hotter rocks can induce remelting of the initial assemblage. Where modest, the gain or loss of volatiles can lead to favouring one mineral over another and result in modal layering. Where melting is more extensive, incongruent melting can result in both the formation of olivine and chromite after Cr-bearing pyroxene.

High temperature subsolidus reactions. High temperature subsolidus heating of vapour as it migrates into hotter rocks can lead to the fluid becoming undersaturated in silica and other components. Reaction of this vapour with the host rock can produce a number of secondary replacement features such as finger structures and discordant bodies. Examples were discussed in Chapter 8.

Low temperature reactions. As the rocks cool, fluid interaction results in more typical metamorphic assemblages. In this case, it can be difficult to discern alteration that is the result of cooling primary magmatic fluids with those that may infiltrate during a later metamorphic event.

Excluding perhaps the lowest temperature alteration reactions, olivine and chromite are two minerals whose occurrence in layered intrusions can be formed by all of these processes. There are a number of mechanism that can lead to the precipitation of olivine and chromite in a layered intrusion as either primary or secondary (replacement) minerals. (1) An influx of an olivine ± chromite-saturated primitive liquid. (2) A mixing event involving the resident magma and a more primitive liquid that leads to the magma to be saturated in one or both minerals. These first two are common explanations for the reoccurrence of these minerals in the stratigraphic sequence. (3) Hydration of liquid in the chamber and expansion of the olivine and chromite phase fields phase field. This mechanism is a variation on the magma mixing model but involves vapour-magma mixing and is discussed in more detail below. (4) Incongruent melting by heating. Injection of a hotter magma can heat the crystal pile, potentially leading to the formation of olivine and chromite by incongruent melting of Cr-bearing pyroxene. (5) Incongruent flux melting by the addition of volatiles. In this case, addition of H_2O to a pyroxene-bearing assemblage can again lead to incongruent melting, leaving olivine and chromite are restite (residual solid) minerals. Hydration melting of the mantle wedge is the current paradigm for the production of arc-related magmas in subduction zones. (6) Metasomatic reactions involving volatile leaching of the more soluble components such as silica to leave olivine and chromite as a relatively non-soluble residue as discussed in Chapters 8 and 12. The first three mechanisms involve compositional changes in the liquid alone, the rest involve reactions in the crystal pile.

9.2 The Effects of Volatiles on Liquidus Phase Relationships

Volatiles, especially the more soluble components such as H_2O, can have a profound effect on the crystallization behaviour of magmas. Despite this, the initial volatile composition magmas are generally not well known, and many estimates of parent liquid compositions are tabulated as dry. This practice of ignoring volatile concentration follows from the difficulty of estimating the original magma volatile content, even if these liquids are otherwise well preserved as chilled margins or associated sills and dikes. The best estimates of the volatile contents of basalts are done using analyses of trapped glass inclusions (e.g., Lowenstein, 1995), which are rarely found in the slowly cooled rocks associated with layered intrusions (but see Chapter 7).

The effect of bulk H_2O content on mineral stability in a basaltic and intermediate liquid compositions has been discussed by Kushiro et al. (1968), Kushiro (1975), Eggler and Burnham (1973), Ford et al. (1972), Cawthorn (1976), and Nicholson and Mathez (1991), Almeev et al. (2007, 2012), among others. The primary effect of the addition of H_2O is to lower the liquidus temperature and, for partially molten systems, the amount of liquid will increase at any given temperature. An empirical equation for the effect of the concentration of H_2O in a basaltic liquid, $C_{H_2O}^{Liquid}$, on temperature suppression relative to a dry liquid, $\Delta T\ (°C) = T^{dry} - T^{wet}$, has been presented by Médard and Grove (2007):

$$\Delta T(°C) = 39.2\left(C_{H_2O}^{Liquid}\right) - 2.842\left(C_{H_2O}^{Liquid}\right)^2 + 0.07722\left(C_{H_2O}^{Liquid}\right)^3 \quad (9.1)$$

Almeev et al. (2007) suggest an even simpler expression for MORB olivine-saturated liquidus suppression at 200 MPa:

$$\Delta T(°C) = 39.2(C_{H_2O})^{0.73} \quad (9.2)$$

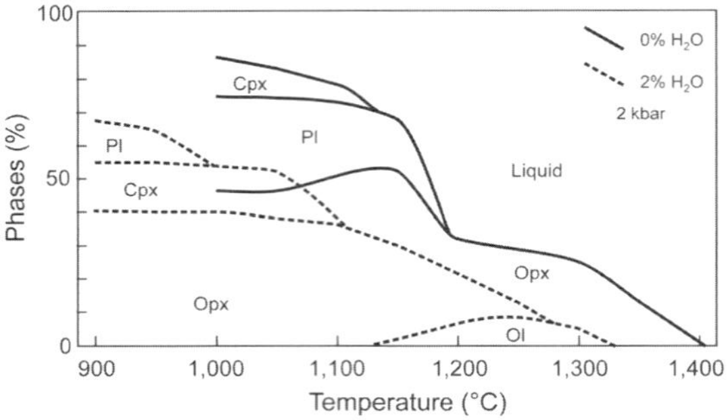

Figure 9.1 The effect of H_2O on altering the crystallization behaviour of a Bushveld Lower Zone magma composition. Results of a MELTS (Ghiorso et al., 1994; Ghiorso and Sack, 1995), calculation showing the equilibrium crystallization behaviour of a dry (solid black lines) and with 2% H_2O added (dashed black lines), both at 2 kbar pressure. Dry magma liquid composition is proposed B1 liquid from Wilson (2012; Table 3, analysis 4,with Cr_2O_3 ignored as MELTS does not allow for Cr in pyroxene). Note the delayed crystallization of plagioclase and the presence of olivine in the wet liquid as compared with the dry liquid.

Almeev et al. (2012) have a similar expression for plagioclase-saturated liquidus suppression, also at 200 MPa:

$$\Delta T(^\circ C) = 77.0(C_{H_2O})^{0.71} \tag{9.3}$$

Furthermore, the addition of volatiles, but particularly H_2O, to a basaltic liquid can affect phase boundary relationships such that the liquid line of descent of a 'wet' magma can be markedly different from that of a 'dry' magma. These changes manifest themselves as shifts in the location of cotectic and peritectic boundaries on phase diagrams. More recently, these changes can now by calculated using programs such as MELTS (Ghiorso et al., 1994; Ghiorso and Sack, 1995). For example, the effect of H_2O on the crystallization sequence of a Bushveld parent liquid composition is illustrated in Figure 9.1. For a 'dry' magma with 0.0 wt.% H_2O, the equilibrium crystallization sequence is orthopyroxene → plagioclase → clinopyroxene. For a magma with 2 wt.% H_2O the crystallization order is olivine → orthopyroxene → clinopyroxene → plagioclase.

The effect of the addition of pure H_2O on a basaltic magma can be summarized as follows: (1) Olivine is stabilized at the expense of pyroxene (e.g., Kushiro et al., 1968; Kushiro, 1975). (2) Plagioclase crystallizes compositions that are more anorthitic than in the dry system and plagioclase stability is depressed relative to the mafic silicate minerals (e.g., Holloway and Burnham, 1972; Helz, 1976; Rutherford et al., 1985). However, both igneous and metamorphic fluids are likely to contain significant alkalis and the shift to more anorthitic plagioclase would be suppressed. (3) Spinel stability is enhanced relative to the silicate phases (Ford et al., 1972). An example of the later is presented in Chapter 12, where a model for the formation of chromitites involves the expansion of the chromite field at the expense of olivine and pyroxene by hydration of a magma initially multiply saturated in pyroxene and plagioclase.

The reverse of these effects can occur as the result of volatile loss to degassing. For an example, it has been estimated that as much as 30% of the crystallization of Kilauea lavas is the result of isothermal degassing (Lipman et al., 1985). Chapter 6 summarized studies that have suggested that magma chambers as deep as 10 km can lose a significant amount of their original volatile content by vertical mixing with shallow (P < lithostatic) magma chambers or even long-lived lava lakes. The loss of volatile to degassing in layered intrusion is consistent with the overall loss of Cl in the more evolved rocks seen in most intrusions described in Chapter 6.

Changes in an observed crystallization sequence that may have been caused by cotectic boundary shifts resulting from a gain or loss of volatiles can be difficult to discern. In part, this is because modal changes can also be produced by injections of magmas with different crystallization sequences or by magma mixing of primitive and evolved magmas that otherwise have the same liquid line of descent. Some petrologic/geochemical indicators can include variations in grains size and textural variations, an association with pegmatoids or other replacement features, enhanced compaction, changes in Cl/F ratios of hydrous minerals (especially apatite) and the presence of hydrous melt inclusions. These are summarized at the end of this chapter.

9.3 Volatile Fluxing

The situation becomes more complicated when volatile components are lost or gained from a liquid + crystal mush. For a mush that is degassing and losing its volatile components, the main effect is to hasten solidification. The situation is potentially more complicated for a mush seeing an influx of volatile components. This can happen when vapour lost from crystallizing interstitial liquids near the base of the intrusion move up into hotter, vapour-undersaturated crystal + liquid assembles. In this case, the system will undergo partial melting. In part, the solids act as a buffer to limit the changes that can take place. Nonetheless, modal changes can occur as the system undergoes a hydration-induced, incongruent melting event.

An example of how the solid assemblage can be affected by this process is seen in the transition from a poikilitic harzburgite (olivine and a coarse-grained, poikilitic orthopyroxene) to a granular harzburgite (euhedral to subhedral olivine and orthopyroxene) that occurs within the cyclic units of the Peridotite Zone of the Stillwater Complex. The conventional interpretation is that the magma initially crystallized olivine alone and this olivine partially resorbed by a later, peritectic reaction with intestinal liquid to produce the poikilitic harzburgite. In contrast, the granular harzburgite crystallized from a more evolved magma that was co-saturated in olivine and orthopyroxene without a reaction relationship.

However, whereas most modal layering contacts in the Stillwater are planar, this contact is generally described as irregular and in some places the poikilitic harzburgite forms finger structures into the overlying granular harzburgite (Figure 9.2a). These finger structures suggest that the poikilitic harzburgite is a replacement of the original granular harzburgite.

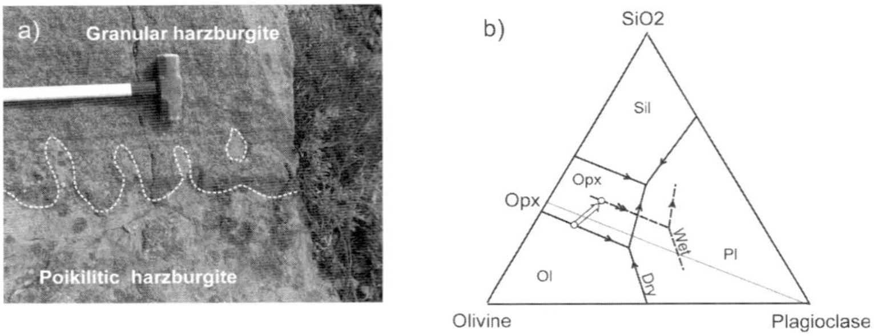

Figure 9.2 Example of the effects of hydration of a crystal-liquid mush. (a) Fingers of poikilitic harzburgite intruding into overlying granular harzburgite in the Peridotite Zone of the Stillwater Complex. From Boudreau (2016), with permission. (b) Schematic diagram illustrating the effect of H_2O on shifting phase boundaries in the system olivine-plagioclase-SiO_2. Phase boundaries in the H_2O-poor system as shown as solid dark lines labelled 'Dry', and those for the H_2O-enriched system are shown dashed and labelled 'Wet'. Expansion of the olivine field by the addition of H_2O also changes the olivine-opx boundary from a cotectic (olivine + opx co-crystallizing) to a peritectic (olivine reaction) boundary.

Boudreau (2016b) suggested that this replacement reaction was driven by an increase in H_2O as underlying crystal mush degassed and migrated upward into hotter olivine + opx + liquid assembles that were not yet vapour saturated. A shown schematically in Figure 9.2b, addition of H_2O will lead to the expansion of the olivine phase field. This shift of phase boundaries can cause the olivine + orthopyroxene boundary to shift from a cotectic boundary to a peritectic (reaction) boundary, resulting in the recrystallization of the granular harzburgite to form the poikilitic harzburgite.

9.4 Constitutional Zone Refining

McBirney (1987) discussed how the crystallization of interstitial liquid in a solidification zone can lead to the build-up of incompatible elements, some of which can have strong fluxing effects on the mineral assemblage. In regards to H_2O, the crystallization of anhydrous minerals leads to steep chemical gradients which leads to diffusive transport to regions where the H_2O concentration is not as high. The high rate of diffusion of H_2O, coupled with the fact that H_2O diffusivity increases as the concentration of H_2O increases, implies that H_2O will always be moving down its chemical potential gradient to regions of less advanced crystallization. This movement can be aided by compaction, compositional convection in the interstitial liquid or vapour saturation.

The enrichment in H_2O in the interstitial liquid above what would have occurred without the diffusion, compaction, etc. has several results. (1) The migration of H_2O down its chemical potential gradient into hotter rocks lowers the liquidus and solidus of the anhydrous solids and results in some of the solid being resorbed (Figure 9.3). The partial melt generated will dilute the introduced components. (2) The fluxing effect shifts

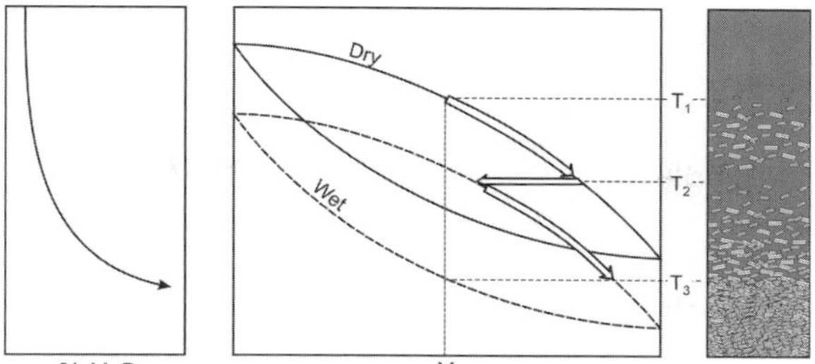

Figure 9.3 Cartoon illustrating the constitutional zone refining model of McBirney (1987). Liquidus and solidus relations in a crystal mush are assumed to crystallize like a simple solid solution. Crystallization of anhydrous solids leads to a build-up of H_2O in the interstitial liquid. Owing to either steep compositional gradients, compaction or eventual saturation in a fluid phase, an H_2O-enriched liquid moves upward into hotter mush where it lowers the nominally 'dry' solid solution loop to the dashed 'wet' curves. The result is the partial melting of the original assemblage as it re-establishes local equilibrium between solid and the hydrous liquid. Redrawn after McBirney (1987), with permission.

phase boundaries such that some minerals, typically the ferromagnesian minerals, are resorbed less that other minerals, typically the sialic minerals. (3) The increase in partial melt increases the porosity/permeability of the mush and enhances compaction, further enhancing the upward transport of H_2O into hotter rocks. (4) Mixing of the partial melt generated by the hydrous fluxing with drier interstitial liquid or with liquid above the crystal mush will lead to a blended liquid enriched in those components that were preferentially melted, again typically the sialic minerals. Owing to the rapid diffusion of H_2O, this blended liquid will lose is H_2O and preferentially crystallize minerals such as plagioclase. (5) The gain or loss of H_2O can affect both the nucleation and growth rate of minerals as the liquidus temperature changes. (6) The zone of flux melting as it moves through a solidifying mush is similar to conventional zone refining as used in industrial refining processes. The general effect is to scavenge excluded elements and enhance their fractionation (McBirney, 1987). As illustrated in Figure 9.4, the process can lead to modal variations as the volatile-rich liquid favours mafic minerals over felsic minerals.

An example of a stratigraphic section in which the crystallization was affected by a constitutional zone refining process is seen in the macrorhythmic units in the upper most part of Gabbronorite Zone II (GNII) of the Stillwater Complex described by Crescenti (1984). The units are located immediately below Olivine-Bearing subzone II of the Stillwater Complex (Figure 9.5a), and overlie a section where modal layering is rare and the rocks are instead uniform in appearance. The macrorhythmic units are defined by an upward increase in modal plagioclase relative to pyroxene. The lower units are defined a sharp planar basal contact and a marked increase in colour index, while the units towards the top grade from mafic to felsic. The lower mafic parts of each unit contain abundant, small pyroxene crystals while the more felsic upper layers contain fewer but larger pyroxene crystals. As the pyroxene crystals grow in size, their habit changes from acicular

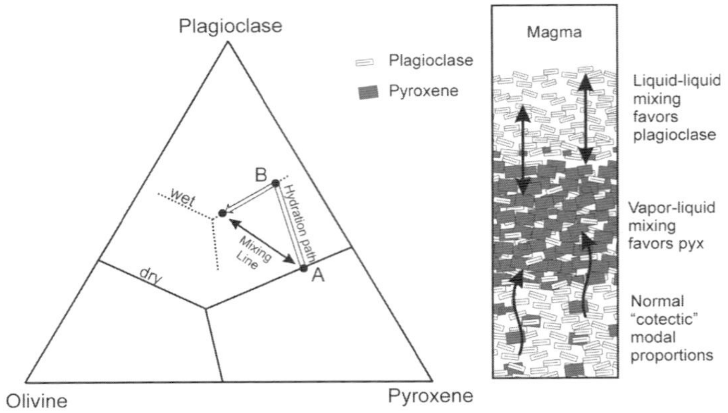

Figure 9.4 Schematic model for the formation of mineral variations and modal layering by a constitutional zone refining process. Vapour migrating from a solidifying mush with cotectic mineral proportions of plagioclase and pyroxene (point 'A' on the phase diagram Figure, left) leads to hydration of interstitial liquid and a shift of the phase boundaries to the dashed 'wet' lines. The liquid adjusts to the phase boundary shift by resorption of plagioclase, leading to a solid assemblage preferentially enriched in pyroxene (point 'B'), which can now continue to crystallize down to the ternary eutectic. Any mixing of the hydrous, plagioclase-enriched liquid with the cotectic liquid in the chamber can lead to a liquid saturated in plagioclase alone. Redrawn after McBirney (1987, with permission).

and subhedral to more equant and poikilitic (Figure 9.5b). Pyroxene size and crystal frequency vary over several orders of magnitude while plagioclase size and crystal frequency are much less variable.

Crescenti (1984) noted that the maximum orthopyroxene crystal frequency may be stratigraphically offset from maximum clinopyroxene crystal frequency, and suggested that an oscillatory nucleation mechanism was involved in the formation of these units. Because the macrorhythmic units formed in the centre of the intrusion where the cooling rate was extremely slow, she discounted the possibility that the nucleation was due to temperature fluctuations and variable degrees of undercooling. Instead, it was suggested that fluctuations in the supersaturation levels of the three minerals was the result of metastable crystallization and constitutional supercooling was driving the oscillatory nucleation mechanism.

The section is topped by the discordant 'pillow troctolite' (Hess, 1960) that defines the thin Olivine-Bearing Zone II. As described in the next chapter, these troctolites are most likely the result of the infiltration of silica-poor fluids that resulted in the replacement of pyroxene by olivine. Thus, the upper section of the Lower Banded series records the build-up of volatiles, initially affecting the nucleation and growth of the various minerals to produce increasingly pronounced modal layering and then a volumetrically abundant fluid that lead to reaction and replacement of the original assemblage. A similar change from massive norite and gabbronorite to sharply modally layered rocks occurs just below Olivine-Bearing Zone I, host zone of the JM Reef. The modal layering in this instance is associated with fluid inclusion-bearing pegmatoids, and is again consistent with the modal variations being the result of a constitutional supercooling/zone refining mechanism.

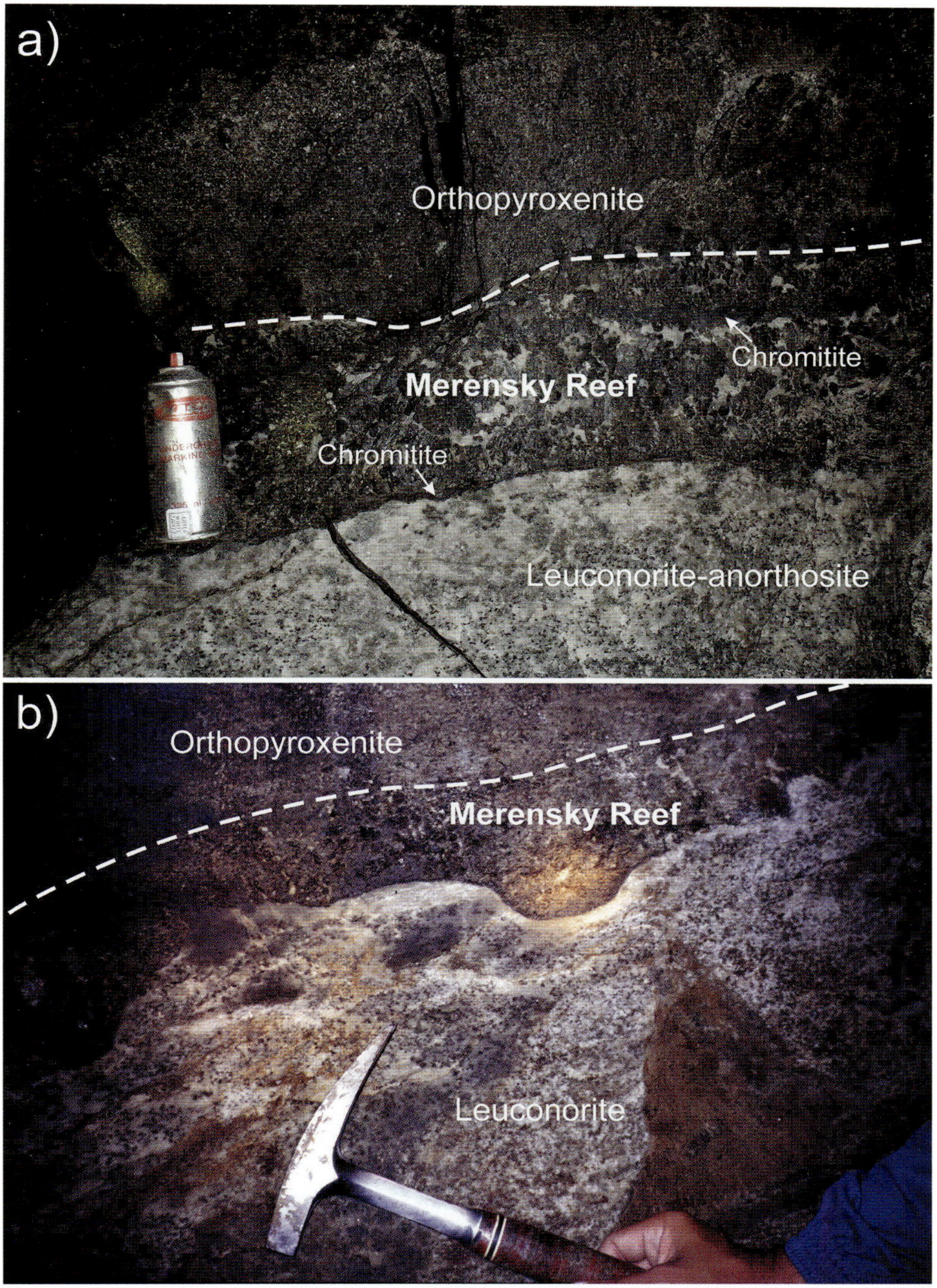

Figure 2.5 The Merensky Reef, Bushveld Complex. (a) Example of normal Merensky Reef, Thembelani Mine. Note coarser grain size compared of the reef rocks with both the underlying plagioclase-rich rocks and especially the overlying orthopyroxenite. Photo curtesy of Rais Latypov, University of the Witwatersrand. (b) Example of dimpling or 'teats' where Merensky Reef protrudes into underlying leuconorite-anorthosite, Western Platinum Mine.

Figure 2.9 Example of the texturally and modally mixed nature of the olivine-bearing rocks of Olivine-Bearing Zone I, host zone of the J-M Reef of the Stillwater Complex. Photo by author.

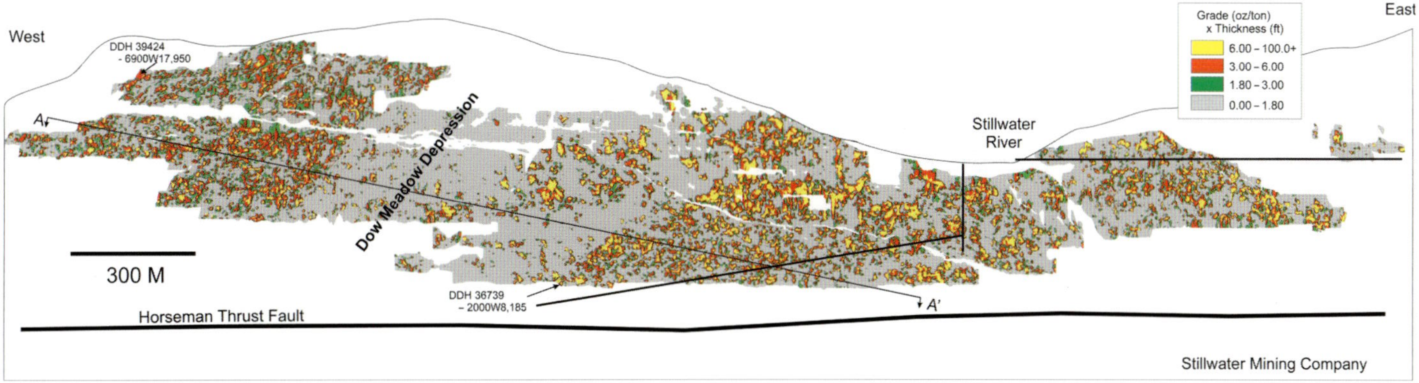

Figure 2.10 Long section of *grade x thickness* along the J-M Reef illustrating the non-uniform nature of mineralization along the mineralized surface. Note especially the lack of mineralization in the region of the Dow Meadow Depression (Figure 2.11).

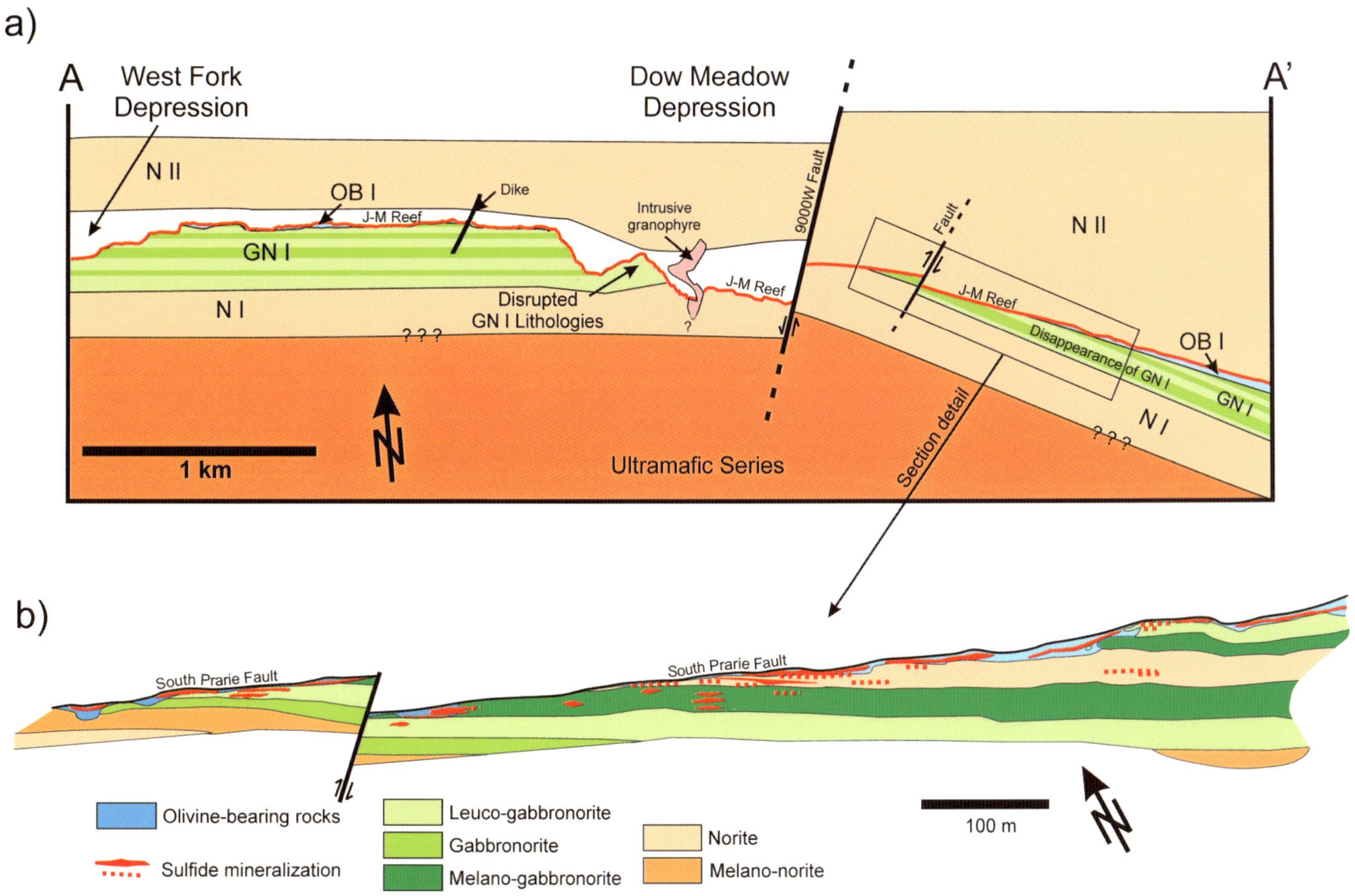

Figure 2.11 Regional downcutting of the J-M Reef into the underlying footwall rocks at the 'Dow Meadow Depression' illustrating stratigraphic variability of the J-M Reef. (a) Regional section constructed along the line of section A-A' shown in Figure 2.12. The reef cuts down or replaces original footwall lithologies. From Boudreau et al. (in review), with permission. (b) Detail of the reef development into the underlying GN-I and N-I Zones showing the presence of olivine-bearing rocks along the unconformity and the irregular distribution of sulfide mineralization that defines the J-M Reef. Redrawn after Zientek et al. (2002), with permission from the Canadian Institute of Mining, Metallurgy and Petroleum.

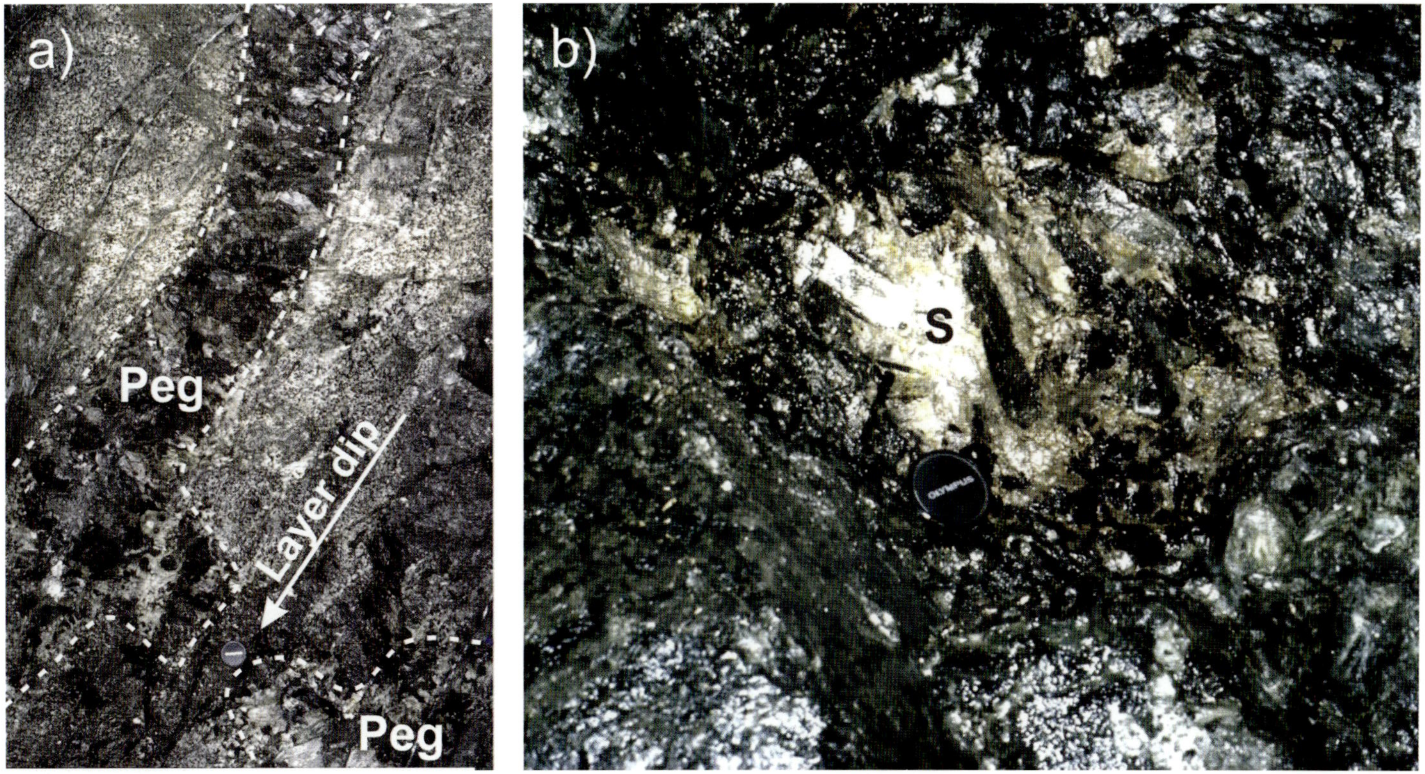

Figure 2.12 Pegmatoids associated with the J-M Reef, Stillwater Complex. (a) Crudely layer-parallel to discordant/crosscutting pegmatoids developed in gabbronorite from Olivine-Bearing Zone I. Arrow shows local dip of layering and also points to lens cap for scale. (b) Local core of massive sulfide in pegmatoid from the same region as in (a). Cast of surrounding pegmatoidal minerals are preserved in the massive sulfide. From Boudreau (2016), with the permission of the Mineralogical Society of Great Britain & Ireland.

Figure 2.14 The Picket Pin PGE Zone of the Stillwater Complex. (a) Discordant, pipe-like body of disseminated sulfide located approximately 150 In below the top of the Upper Anorthosite of the Middle Banded Series. Stratigraphic 'up' is towards top and slightly into the plane of the photograph. The pipe has separated along a joint which runs diagonally through its long axis; this allows one to see that the sulfide-rich pod has a roughly circular section in plan view. From Boudreau and McCallum (1986), with permission. (b) Photo of the typical weathered podiform occurrence of the disseminated PGE-sulfide mineralization of the Picket Pin Pt-Pd deposit that occurs in the upper part of the Upper Anorthosite Zone at the top of the Middle Banded Series. Note hammer (at arrow) for scale. Photo by author.

Figure 8.13 Examples of discordant troctolites from the Middle Banded series of the Stillwater Complex. (a) Discordant troctolite bodies in a boulder from Olivine-Bearing Zone III. The boulder is also cut by a large plagioclase + quartz + hornblende vein, which are otherwise rare in the complex. (b) Discordant troctolite in OB IV just below the Upper Anorthosite at the top of the Middle Banded series. Stratigraphic 'up' is towards the top of the photo. (c) Example of the 'egg' texture in an olivine gabbro described by Hess (1960), with a transition to discordant troctolite in OB IV. F) Detail of 'egg' texture; large, black olivine defines the 'yoke' in the centre-bottom of the surrounding plagioclase-rich 'egg white', both hosted by a medium-grained gabbro. From Boudreau (2016), with the permission of the Mineralogical Society of Great Britain & Ireland.

Figure 11.6 Examples of textural and mineral variation at the top of the Bronzitite Zone, Stillwater Complex, attributed to effects of compaction on interstitial liquid migration and entrapment at the boundary between the Bronzitite Zone and the overlying Banded series. (a) Increase in interstitial plagioclase (white) and large oikocrystic clinopyroxene (dark green) in orthopyroxenite (orthopyroxene, brown). Stratigraphic 'up' towards top of photo. (b) Detail of (a). Hammer head is covered in part by the photo label. (c) Plagioclases-rich orthopyroxenite layer attributed to tensional increase in porosity during interstitial liquid migration. Width of picture is ~20 cm, stratigraphic 'up' towards upper right.

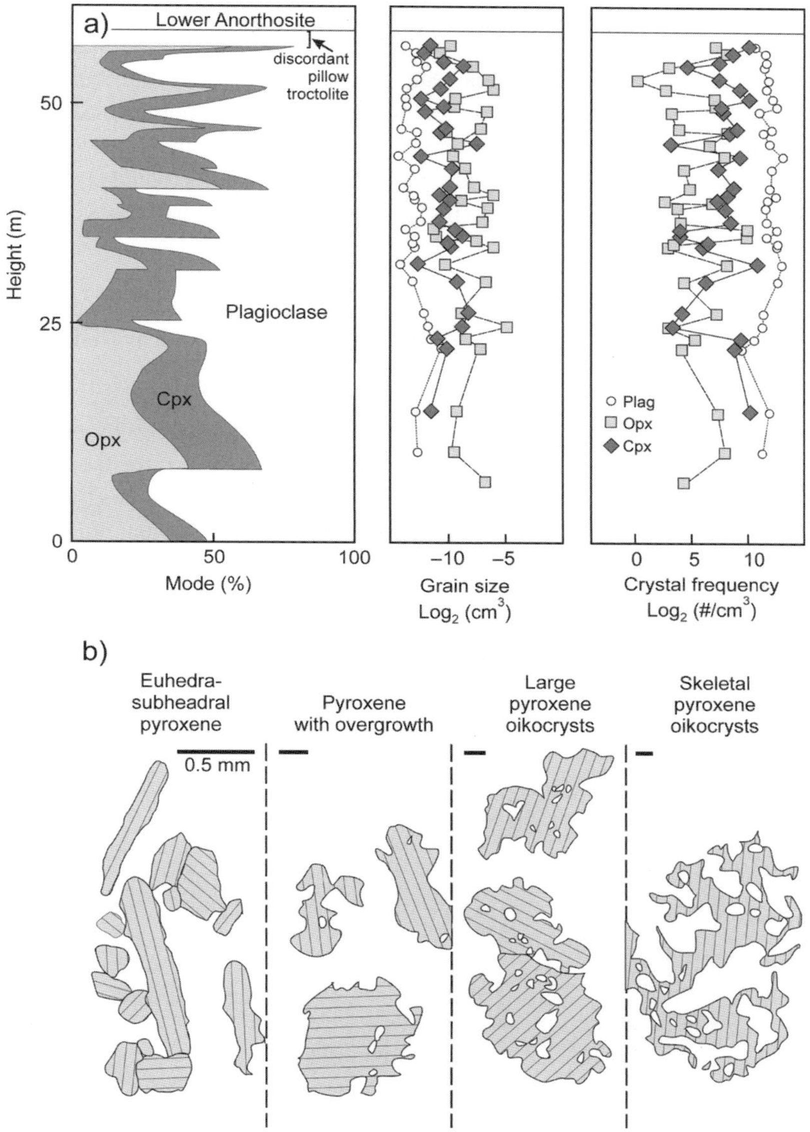

Figure 9.5 A section of the Stillwater Complex interpreted as showing the effects of constitution zone refining. Section is from the top of Gabbronorite Zone II of the Lower Banded series, just below Olivine-Bearing Zone II and the thick Lower Anorthosite Zone of the Middle Banded series. (a) Modal proportions, grain size and crystal frequency of the primary phases plagioclase, orthopyroxene and clinopyroxene. The location of the discordant 'pillow troctolite' of Hess (1960) and shown in Figure 8.20 occurs just below the Lower anorthosite. and Grain size and volumetric crystal frequency are defined as $(0.01M/(N/A))^{1.5}$ and $(N/A)^{1.5}/(0.01M)^{0.5}$, respectively, where N = the number of crystals counted in area A and M = mode (vol. %) of the mineral. (b) Examples of the textural variations in pyroxene. Redrawn after Crescenti (1984), with permission.

It is suggested that a similar volatile-driven segregation of mafic from felsic minerals produced the modal layering associated with the Platinova Reef of the Skaergaard intrusion. As noted in Chapter 2, mineralization occurs in a series of stacked layers of decreasing size, with mineralization preferentially occurring in mafic silicate and oxide-rich layers. It is proposed that the addition of a mineralizing fluid to vapour-undersaturated silicate interstitial liquid both introduced the mineralization and favoured crystallization of mafic over the plagioclase minerals. This process repeated several times as the vapour saturation front moved upward.

9.5 Examples of Partial Melting and Metasomatic Effects in Layered Intrusions

9.5.1 The J-M Reef, Stillwater Complex

A simple phase diagram description of incongruent hydration melting by the addition of H_2O is illustrated by the isothermal phase relations in the system forsterite-silica-water (Figure 9.6). Consider an assemblage of enstatite and a liquid of composition L_1 that has not yet solidified to the point where it would separate a fluid. The initial bulk composition of the assemblage will lie on a tieline connecting enstatite and the original liquid composition, shown here as point B. Any migrating H_2O-rich fluid that encountered this mush would dissolve in the silicate liquid. This addition of H_2O will cause the bulk composition to move upward along the dotted 'fluid-mixing line' towards the H_2O apex. Because the

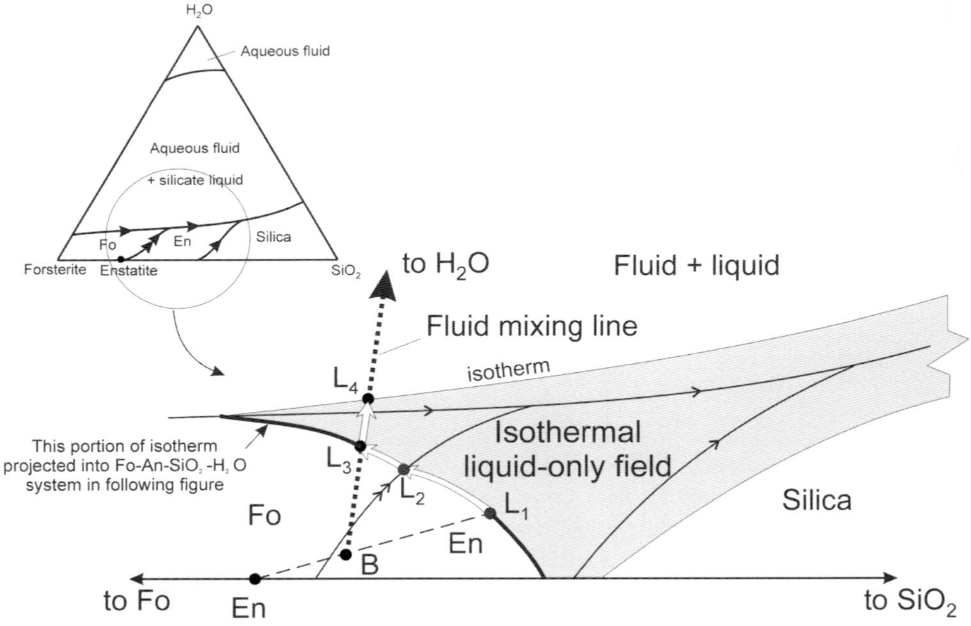

Figure 9.6 Detail of the Forsterite-Silica-H_2O system, illustrating the incongruent melting of pyroxene by isothermal hydration. See Text for details. Redrawn after Boudreau (1988), by permission.

bulk solid composition initially remains fixed (i.e., pyroxene is the only solid present), a change in bulk composition must result in a change in liquid composition. Thus, the isothermal addition of water to bulk composition B causes the liquid composition to change (by melting of enstatite) along the isotherm on the liquidus surface, as shown by the open path. Once the liquid composition reaches the forsterite boundary at L_2, however, the liquid composition is held constant while the enstatite melts incongruently to forsterite, the reaction again being driven by the bulk increase in water. Once all the enstatite is converted to forsterite, the liquid is free to move into the forsterite phase field. With continued influx of fluid, the forsterite itself will melt. At the point, the liquid composition reaches the dotted fluid-mixing line at L_3, all the solid will have melted, and the liquid can move into the liquid-only field until it reaches fluid-saturation at point L_4. Proportions of liquid, enstatite and forsterite at any point can be determined by use of the lever rule for the changing bulk composition. At low pressures, only a few weight percent H_2O is required to completely melt the original liquid + enstatite assemblage and saturate the liquid: a small amount of fluid can produce many times its weight in partial melt.

The extension of the isothermal reaction in the forsterite-silica-water system shown in Figure 9.6 can be extended into the anorthite-forsterite-silica-H_2O system, as shown schematically in Figure 9.7. This hydration-mixing model was proposed by Boudreau (1999a) to explain the features of the JM Reef of the Stillwater Complex. The initial assemblage consist of a liquid (L_1) in equilibrium with an anorthite + enstatite solid assemblage (S_1), the bulk solid + liquid composition of which is at B_1. Addition of H_2O to this bulk composition leads to progressive melting, with the partial melt following the open path L_1–L_2–L_3 along the isothermal surface (shown shaded, with the heavy black lines being the intersection of the

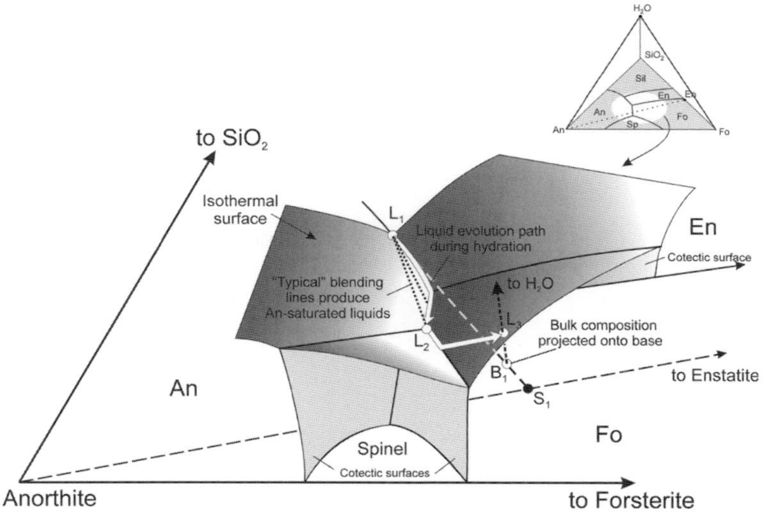

Figure 9.7 Detail of the four-component forsterite-anorthite-silica-H_2O system, illustrating the isothermal melting path by hydration of a partially molten norite (plagioclase + enstatite + liquid). See text for details. Redrawn after Boudreau (1999a), with permission.

various cotectic and peritectic surfaces with this isothermal surface). The system is completely molten when the evolving liquid reached the hydrous bulk composition at L_3. During partial melting, the solid assemblage changes by incongruent melting from anorthite + enstatite → anorthite + forsterite → forsterite + spinel → forsterite. In this case, the restite (residual solid assemblage) that is produced by progressive hydration of a partially molten norite is a forsterite ± spinel-bearing assemblage.

It is also noted that mixing of the partial melts with the original liquid can produce blended liquids that fall within the plagioclase field. Mixing of any of the partial melts generated along the path from L_1 to L_2 with the initial melt at L_1 can produce hybrid liquids that fall in the anorthite field, as shown by the dotted 'liquid blending' lines. Depending on the proportions of the endmembers, this hybrid liquid can have the crystallization sequence (anorthite) → (anorthite + olivine) or (anorthite) → (anorthite + enstatite). Hence, above the hydration melting zone one might expect anorthosite, troctolite and norite to represent crystallization products of this hybrid liquid, as is observed.

Boudreau (1999, 2016b) suggested that the anorthosites that make up the upper part of OB I was a result of this mixing of the hydration melts formed by incongruent melting of

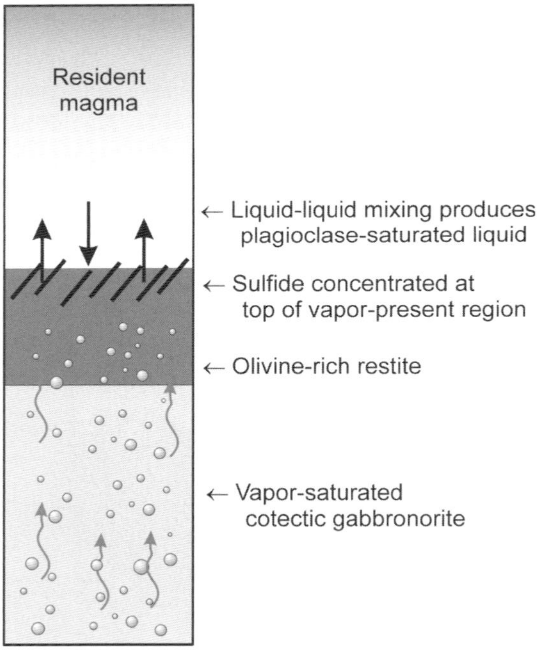

Figure 9.8 Cartoon of hydration melting of a gabbronorite protolith. Vapour migrates upward into hotter parts of the crystal pile until it reaches a vapour-undersaturated mush near the top of the crystal pile, where it redissolves and induces partial melting to produce an olivine restite by incongruent melting. Mixing of the liquid produced by partial melting with the resident magma in the chamber results in a hybrid liquid saturated in plagioclase. Sulfides are preferentially precipitated at the transition between vapour-dominant to silicate liquid-dominant regions. After Boudreau (2016b), with the permission of the Mineralogical Society of Great Britain and Ireland.

the gabbroic rocks beneath it with the resident magma in the chamber (Figure 9.8). This mixing zone also represents the change from vapour-saturated, fluid-dominated rocks in the lower part of OB I to silicate liquid dominant, vapour-undersaturated conditions in the Anorthosite subzone of OB I. PGE-sulfides are preferentially precipitated at the top of the vapour-saturated region. This magma mixing model is similar to those proposed by Irvine et al. (1983) and Barnes and Naldrett (1986) but in this case the liquids are not from different parental liquids but are generated from events involving the crystallization of a single magma.

9.5.2 The Merensky Reef, Bushveld Complex

An idealized section through normal (non-potholed) Merensky reef from the Rustenburg section of the western Bushveld Complex is shown in Figure 9.9. In this area, the Merensky reef consists of a pegmatoidal feldspathic pyroxenite (melanorite) typically 15–30 cm thick. The base of the pegmatoid is separated from the underlying norite by an undulating contact consisting of a thin (~1 cm) chromitite above a thin (1–3 cm) anorthosite seam. A similar chromitite may be present at the top, separating the pegmatoidal pyroxenite from a medium-grained pyroxenite; elsewhere it may be expressed as a zone of disseminated chromite or may be absent altogether. On a regional scale, these, and the immediately surrounding rocks, show considerable variability in thickness, modal mineralogy, the degree of pegmatoid development, and the distribution of sulfur and PGE grade across the section. For example, the lower norite may be more leucocratic to

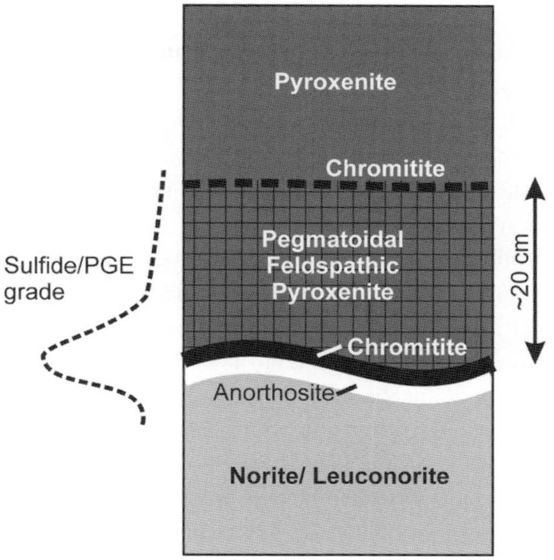

Figure 9.9 Idealized section of the Merensky Reef from the Rustenburg area, western Bushveld Complex. Redrawn after Nicholson and Mathez (1991), with permission.

anorthositic in composition, and the pegmatoid itself may be absent (Nicholson and Mathez, 1991; Viljoen, 1999; Cawthorn and Boerst, 2006). PGE-sulfide distribution is usually described as being 'top-loaded'; that is, it mostly occurs with or just below the upper chromitite seam (Viljoen, 1999; Cawthorn and Boerst, 2006). However, this is not ubiquitous, and published trends across the reef suggest the PGE are concentrated at the bottom chromitite seam and the PGE and the base-metal sulfides (BMS) may decouple from each other (Barnes and Maier, 2002b; Godel et al., 2006).

An example of how the Merensky sequence can develop by local mass transport was presented by Nicholson and Mathez (1991). In their model, the assemblages initially consisted of a partially molten pyroxenite overlying a leuconorite. To this initially vapour-undersaturated assemblage is added a volatile component that degassed from and migrated through fractures in the mostly solidified underlying rocks and which induced remelting and recrystallization. The observed mineral sequences resulted from the eventual cooling and inward crystallization of this remelted zone. In effect, the Merensky pegmatoid was modelled as a 20 cm thick layered intrusion.

Boudreau (2008) investigated the consequences of the model of Nicholson and Mathez (1991) using the numerical modelling program IRIDIUM, which links diffusive and advective mass and heat transport with a phase equilibration routine based on the MELTS program (Ghiorso et al., 1994; Ghiorso and Sack 1995). The initial system modelled by Boudreau (2008) consists of a simple stratigraphic succession of a partially molten leuconorite overlain by a partially molten pyroxenite, both initially at 1,190 °C and 2 kbar. A S-O-H volatile fluid is then added to the lower 20 cm of the pyroxenite. The system is then allowed to evolve under conditions of chemical diffusion in the liquid, and a snapshot of the system after 5.2 years of simulated time is shown in Figure 9.10.

The addition of the volatile components at the start of the simulation initially results in a modest increase in the amount of melt in the pyroxenite. After 5.2 years the following is observed to have occurred: (1) Chemical diffusion across the leuconorite-pyroxenite boundary leads to more extensive melting at and below this boundary, with preferential loss of pyroxene from the underlying leuconorite to produce a 'pure' anorthosite layer. (2) Just above this, sulfide and chromite preferentially precipitate at the lower boundary, leading to concentration of the PGE at this boundary. These results mimic actual mineral and compositional profiles across the Merensky pegmatoid and illustrate that a long-term diffusion process can effectively produce mineralogical and compositional layering not present in the original assemblage. The observation by Nicholson and Mathez (1991) that norite blocks embedded in the pegmatoid are mantled by plagioclase and chromite rinds, similar to that seen at the base of the pegmatoid, are consistent with the results presented here: the rinds form by diffusion-reaction between the two contrasting assemblages.

These model results demonstrate that the first order features of the Merensky pegmatoid can develop over time as a result of chemical diffusion between a hydrated pyroxenite and the leuconorite. In particular, initially homogeneous assemblages can become compositionally and mineralogically segregated over time, particularly for the minor minerals, and

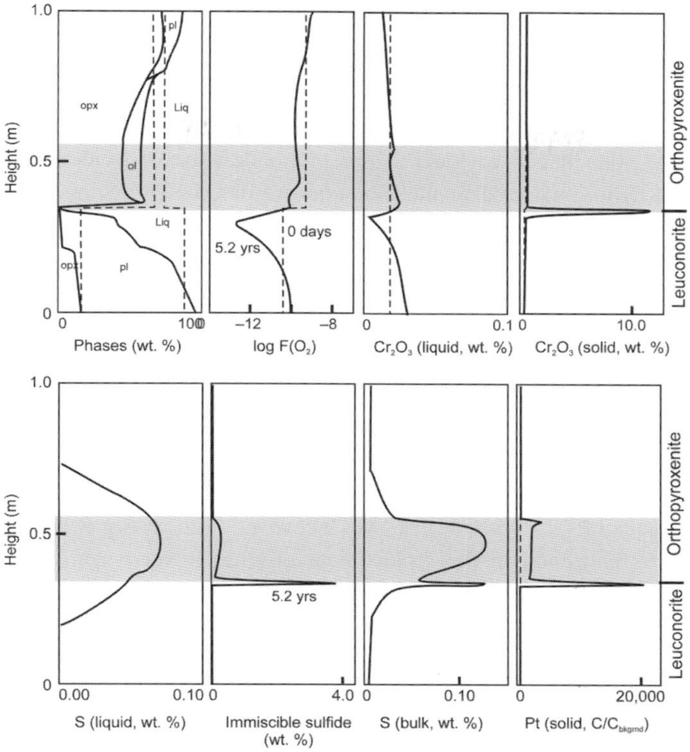

Figure 9.10 Results of IRIDIUM model calculations after addition of 2.0 wt.% volatile fluid and 1,000 × background concentration of Pt to the lower 20 cm of the Merensky orthopyroxenite, after evolving over 5.2 years. Grey area is the originally hydrated region at the base of the orthopyroxenite, dashed lines show initial profiles at the start of the simulation, solid lines are after 5.2 years. Minor phases are not shown in the graph of phase mass wt.%. Note the concentration of sulfide, chromite and Pt that develops at the boundary between the leuconorite and the pyroxenite. From Boudreau (2008), with permission.

this layering need not be indicative of crystal settling or magma mixing in the conventional sense. In the words of Nicholson and Mathez (1991), the Merensky reef 'is igneous, but not cumulate'.

In the model of Boudreau (2008) no compaction or crystal settling was allowed so that the effects of chemical diffusion and the localize behaviour of melting are more clearly displayed. One can imagine, however, that the more extensive melting seen at and below the pyroxenite/leuconorite contact would lead to deformation of the boundary to produce the characteristic irregular contact. Compaction of the partial melt produced by melting of the plagioclase-rich leuconorite into the pyroxenite can lead to the crystallization of more plagioclase within the pyroxenite on cooling. Loss of the secondary hydration melt would also help to preserve the minor olivine commonly found in the pyroxenite.

9.6 Summary of the Evidence for a Secondary Origin for Olivine or Chromite

There is not one diagnostic chemical or textural features that might be used to definitely say if olivine or chromite formed by primary precipitation form a silicate liquid or by a secondary process (incongruent hydration melting or metasomatic replacement of pyroxene). As in law, one needs to look at the preponderance of evidence for or against orthomagmatic, hydromagmatic or metasomatic origin. The following are suggested for evidence for a secondary origin for these two minerals.

Hydrous melt, molten salt and fluid inclusion evidence. The presence of polyphase inclusions in olivine and chromite as described in Chapter 7 that can be confidently interpreted to have crystallized from trapped hydrous silicate liquids, molten salts or fluid inclusions are the most direct evidence that the minerals grew from a hydrous liquid or in the presence of a fluid.

Partially resorbed and reverse zoning in plagioclase. Rounded plagioclase grains and reverse zoning in plagioclase are commonly observed where a troctolite (olivine + plagioclase) replaces a (grabbo)norite protolith (plagioclase $\pm$ orthopyroxene $\pm$ clinopyroxene) (e.g., Meurer et al., 1997; Boudreau, 1999a). These are particularly well-preserved where the plagioclase is entrained in the replacement olivine. Reverse zoning is an expected consequence of flux melting (Figure 9.3), particularly is the partial melt is lost after it is formed, but it can also occur during the infiltration of silica-poor fluids (see Chapter 12).

Relative abundance of hydrous minerals and textural considerations. Hydrous minerals can be more common in the secondary rocks, particularly if formed by a flux melting event. This need not be universal, as the replacement/melting rations may occur at temperatures above the stability of common hydrous minerals. In regard to textures, replacement or remelting can destroy an original igneous lamination, but may enhance it elsewhere by allowing more efficient compaction of the residual solids. Coarser textures or varitextured rock would be expected for mineral growth from a volatile-rich silicate liquid. Partial melts can become segregated, to produce rocks of mixed lrock types (Figures 2.8 and 8.13).

Geochemical/mineral anomalies. Remelting of pyroxene can free up the Cr component in pyroxene to precipitate as chromite (Chapter 12), but it can also liberate Ti as well. In the J-M Reef, this appears to have led to the precipitation of relatively abundant ilmenite rather than a mixed Fe-Ti oxide as the interstitial liquid that was not lost from the mush began to crystallize (Boudreau, 1988). Boudreau (2016b) also suggested that the high TiO_2 concentrations of Stillwater chromite would not be expected if the chromite grew from a Ti-poor siliceous high magnesian basalt (*SHMB*) Stillwater parent magma, but would be if the Ti were inherited from pyroxene during incongruent melting.

Halogen variations. Invariably, one finds higher Cl concentrations in apatite in secondary olivine-bearing rocks than in the host (gabbro)norites and pyroxenites.

Discordant or irregular contacts. Remelting or replacement of a protolith that was itself showing strong modal layering can produce planar replacements, but need not be. Discordant or laterally abrupt changes in rock type are common (e.g., pothole-like features).

Apophyses of remelted rock may be found intruding into the surrounding host rock (Boudreau, 1988) (Figure 9.2).

Evidence for a build-up of volatiles beneath the zone of partial melting or metasomatic replacement. This can be obvious, such as an observed increase in the abundance of pegmatoids in the rocks immediately below the secondary olivine-bearing rocks. It would also be evident in the change from crystallization of thick sections of homogeneous rock to an increase in frequency of modal layering, suggesting that the build-up of volatiles is beginning to affect the crystallization behaviour of the magma.

Lack of stratigraphic compositional offsets to more primitive compositions. The rocks above and below the olivine-rich rocks of the J-M Reef show a continuous fractionation trend without any obvious step-back in, for example, plagioclase compositions. This characteristic can be problematic, because melting can lead to more Mg-rich composition (assume one is just adding H_2O and not what might be carried in with the fluid, such as Fe). Also, it is possible to have a true compositional stratigraphic break to more primitive compositions, but is may have been that the discontinuity was not initially accompanied by olivine or chromite. Instead, these minerals formed later as H_2O came into the stratigraphic break at a later time and induced partial melting. The final rock shows a compositional return to a more primitive composition, but this was not the cause for the presence of olivine or chromite.

Magma mixing considerations. A more generalized consideration is the difficulty of a thin olivine-saturated magma to keep its chemical composition without mixing and homogenizing as it enters the chamber and moves many 10s–100s of km.

Consistent mineral sequences. Changes in the mineral assemblage should be consistent with chemical potential variations, phase diagram considerations, and locally transport of chemical components owing to chemical gradients between the rocks undergoing reaction and the surrounding host rock. This one can also be problematic, as direct precipitation from a magma of changing composition could form a similar sequence of rocks.

A Ni problem? The concentration of Ni in olivine might be indicative of precipitation from a magma, but Ni can also be freed up by incongruent melting or incongruent dissolution of pyroxene, or if the melting is happening near the top of the pile such that Ni could be added from the resident magma. Generally for Ni and Mn in olivine, if not partitioned into olivine from the magma, their concentration in olivine would broadly be appropriate for the stratigraphic position such that the Ni/Mn ratio would be low in the lower parts of the intrusion where the silicates would be expected to be relatively Ni-rich, Mn-poor. Similarly, the Ni/Mn ratio should be lower for olivine formed from the incongruent melting of more evolved rocks that would be expected to have lower Ni and higher Mn. This is observed in the Stillwater Complex (Raedeke, 1982), where the Ni/Mn ratio of Banded Series olivine is much less that is the olivine of the Ultramafic Series, suggesting that these elements are largely controlled by bulk rock abundances.

10

Chromatographic Effects

10.1 Introduction

In a number of layered intrusions that have PGE-rich sulfide zones, there is a distinct and characteristic stratigraphic separation of the PGE, base metals, gold and sulfur that have been termed *offsets* (Barnes, 1993). Intrusions where this is observed include the Skaergaard intrusion and Kap Edvard Holm complex of East Greenland (Bird et al., 1991, 1995; Andersen et al., 1998), the Great Dyke of Zimbabwe (Prendergast and Wilson, 1989) and the Munni Munni complex of Western Australia (Barnes, 1993). The stratigraphic ore element distribution is characterized by a lower, typically S-poor, Pt- and Pd-enriched zone overlain by one or more zones enriched in the base metals, sulfur and gold (Figure 10.1). The offset separations can range from a few tens of centimetres to several tens of metres. The Pt and Pd peak concentrations can occur well below the peak in sulfide abundance and in some instances the PGE peaks are associated with trivial amounts of sulfide. Other changes associated with these offsets may include changes in the nature of the sulfide minerals that mimics the S-poor nature of the stratigraphically lowermost mineralization. For example, the Great Dyke Main Sulfide zone is characterized by a lower troilite-bearing assemblage whereas the overlying assemblages contain pyrrhotite $\pm$ pyrite, indicative of a marked increase in sulfur fugacity as a function of height (Oberthür et al., 1998).

Orthomagmatic explanations for the offsets typically envision that the PGE are largely incorporated in the first sulfide to precipitate, with sulfide saturation perhaps initiated by a magma mixing event. This can be further modified by the mixing of sulfide equilibrating with different parcels of magma with different silicate liquid/sulfide mass ratios (the R factor discussed in Chapter 1) (e.g., Prendergast and Keays, 1989; Barnes, 1993), or even kinetic effects involving different diffusivities between the sulfide liquid and the silicate liquid (Mungall, 2002).

10.2 Offsets as Chromatographic Separations

In contrast to the orthomagmatic interpretations, Meurer et al. (1998), Boudreau and Meurer (1999) and Boudreau (2004) suggested that degassing crystal mushed can act as chromatographic columns to produce the observed PGE, base metal, and gold offsets.

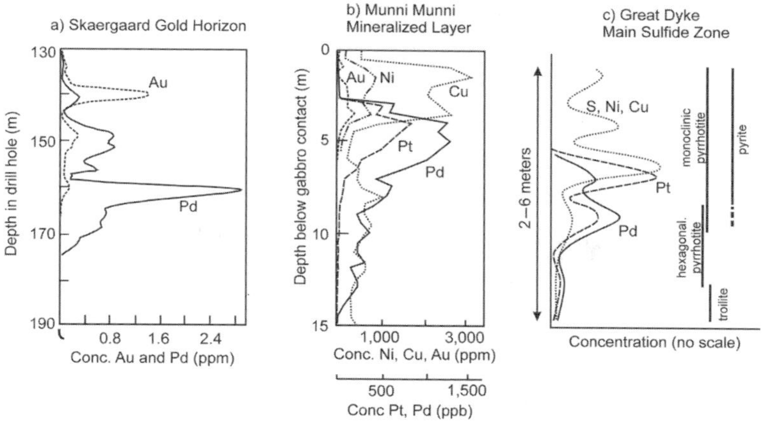

Figure 10.1 Examples of PGE, base metal and gold offsets in three intrusions. (a) Skaergaard intrusion gold horizon. (after Bird et al., 1991) (b) Munni Munni complex, mineralized layer at the top of the Ultramafic zone. (after Barnes 1993) (c) Generalized offset trends and variations in iron sulfides in the Great Dyke, Main Sulfide zone. (after Oberthür, 2011, with permission)

This can also affect the isotopic character of the rock as well as the ore element distribution if the infiltrant is isotopically different from the host rock (see Chapter 13). The simplest models of chromatographic theory envision a solid matrix of a fixed composition with fixed bulk partition coefficients for element partitioning between matrix and the infiltrating phase. In more advanced treatments, the infiltrating fluid can react with the solid assemblage such that the mineral assemblage of the matrix can change, which affects the bulk partitioning between the matrix and the infiltrant. In the case of a degassing crystal pile, the presence or absence of sulfide will strongly control element partitioning, and for which the sulfide itself can be precipitated or redissolved during vapour migration.

The general features of an infiltration-reaction model have been described by Godard et al. (1995). Although Godard et al. were interested in silicate liquid infiltration, the general form of the mass balance equations are indifferent to nature of the infiltrating medium (volatile fluid or silicate liquid). In the following, the infiltrating phase is considered a volatile fluid.

For an example, consider the case in which a volatile fluid, undersaturated with respect to a sulfide phase, migrates through a sulfide-bearing solid + liquid matrix assemblage. As the fluid moves through the matrix, sulfide is redissolved and a reaction boundary is produced between sulfide-bearing rocks and those that are sulfide-depleted (Figure 10.2). One can define a *reaction zone* as the zone in which reaction between the fluid and host solid + liquid matrix leads to a change in the mineral assemblage. Where the reaction zone is of minimal thickness, one may instead speak of a *reaction front*. In the case of interest the reaction front is a *sulfide-out front* defined by the loss of sulfide in the crystal pile; i.e., sulfide is absent below the reaction zone and present above it. Similarly, a

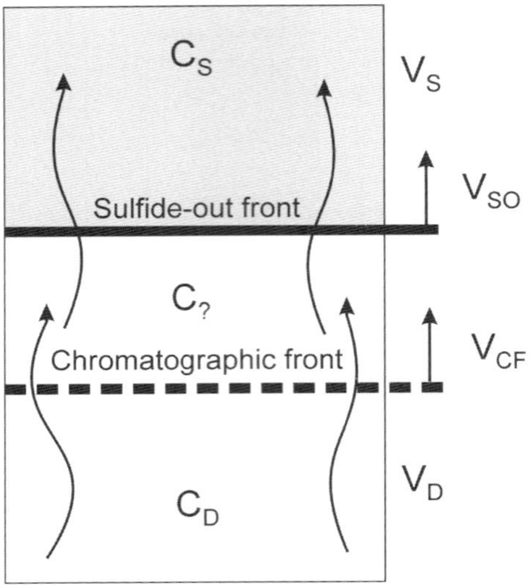

Figure 10.2 Cartoon illustrating the case of metals + sulfur being carried by an interstitial vapour in a crystal + liquid + vapour mush. The mush pile also contains a sulfide-out front, above which the assemblage contains a sulfide phase and below is a depleted zone in which sulfide is absent. The chromatographic front can occur above, below or on both sides of the sulfide-out front, as illustrated in the next Figure. Abbreviations: C_S = initial fluid concentration is equilibrium with a sulfide-bearing assemble, C_D = concentration in infiltrating fluid in equilibrium with a sulfide-depleted assemblage, $C_?$ = region of anomalous fluid concentration, V_S = velocity of element in sulfide-saturated region, V_D = velocity of element in sulfide-depleted region, V_{SO} = velocity of sulfide-out front, V_{CF} = velocity of chromatographic front.

chromatographic or *chemical front* occurs where there is a sharp change in the concentration of a trace element. The mass balance equation for a trace element partitioning between the fluid and the solid + liquid matrix at any point is:

$$\left(\frac{\rho_{maxrix}}{\rho_{fluid}}(1-\Phi)\overline{D}_i+\Phi\right)\frac{\partial C^i_{Fluid}}{\partial t}+\Phi V_{fluid}\frac{\partial C^i_{fluid}}{\partial z}-\boldsymbol{D}\frac{\partial \Phi}{\partial z}\frac{\partial C^i_{fluid}}{\partial z} = \frac{\rho_{maxrix}}{\rho_{fluid}}C^i_{fluid}\left((\Phi-1)\frac{\partial \overline{D}_i}{\partial t}+(\overline{D}_i-1)\frac{\partial \Phi}{\partial t}\right) \quad (10.1)$$

(Symbols are as defined in Table 10.1) The three terms on the left describes the change in the trace element concentration in the fluid and matrix and the change in volume of these phases with time, the advective transport, and diffusive transport of the element in the fluid, respectively. The right-hand side describes the change with time in the trace-element concentration that occur because of mineral reactions that change both the bulk distribution coefficient, $\frac{\partial \overline{D}_i}{\partial t}$, and the porosity, $\frac{\partial \Phi}{\partial t}$.

If one neglects diffusion in the fluid phase and assumes that the thickness of the reaction zone is also negligible, then this equation simplifies to:

Table 10.1 *List of symbols*

C^i_{fluid}	Fluid concentration of a trace element, *i*
C_S, C_D, $C_?$	Fluid concentration of trace element in equilibrium with the original sulfide-bearing assemblage, the sulfide-depleted assemblage, or uncertain, respectively
$\boldsymbol{D}$	Mass diffusion coefficient
$\bar{D}_i$	Bulk matrix/fluid distribution coefficient for element, *i*: $\overline{D}_i = \sum_{j=1}^{n} k_d^{ij} X_j$
ΔH_{cryst}	Heat of crystallization
Q_i	Mass of element *i* at reaction front
k_d^{ij}	Phase j/fluid partition coefficient for a trace element, *i*
t	Time
T	Temperature
V_{fluid}	Velocity of fluid
V_S, V_D	Velocity of chromatographic front for a trace element in a sulfide-bearing and sulfide-depleted matrix, respectively
V_{SO}, V_{CR}	Velocity of the sulfide-out and chromatographic reaction fronts, respectively
X_j	Wt. fraction mineral j in matrix
z	Height in the crystal pile
ρ_{fluid}, ρmatrix	Density of fluid and bulk matrix, respectively
κ	Thermal diffusivity
Φ	Fluid weight fraction
Ψ	Solid matrix weight fraction

$$\frac{\partial C^i_{fluid}}{\partial t} + V_i \frac{\partial C^i_{fluid}}{\partial z} = 0 \tag{10.2}$$

Where the velocity of a trace element is given by:

$$V^i = \frac{V_{fluid}}{\left(1 + \theta \overline{D}_i\right)} \tag{10.3}$$

and

$$\theta = \frac{(1 - \Phi)}{\Phi} \frac{\rho_{matrix}}{\rho_{fluid}} \tag{10.4}$$

This applies to all regions except at the sharp sulfide-out reaction front, whose position over time is defined as $z = V_{SO}\, t$. Note that the velocity that a trace element moves through any part of the system is inversely proportional to the bulk matrix/vapour partition coefficient; an element that strongly partitions into the solid will move slower than one that only modestly prefers the solid + liquid matrix.

If one considers the sulfide-absent and sulfide-bearing zones to each be homogeneous in their properties, then the diffusion coefficient $\boldsymbol{D}$, V_{fluid} and Φ all have constant values in

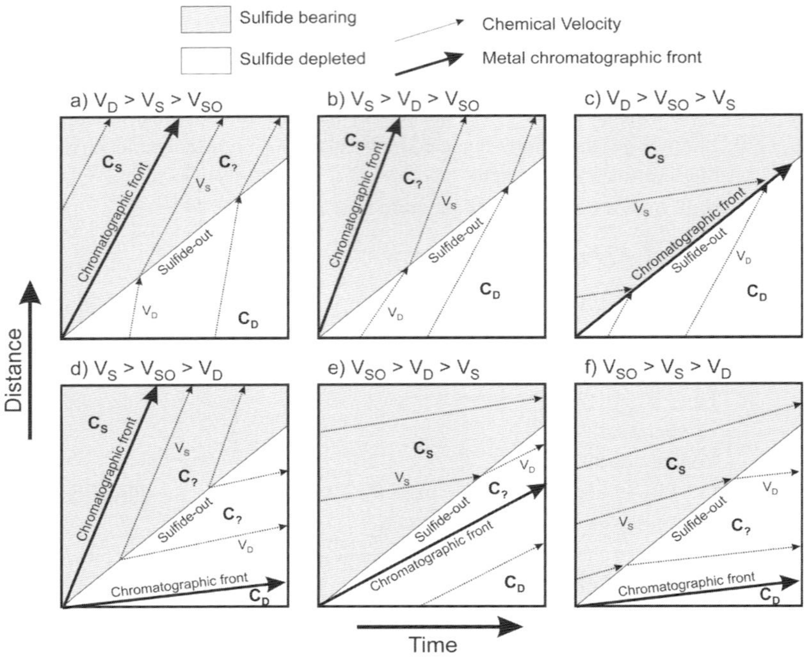

Figure 10.3 Evolution regions of constant concentration in the fluid as a function of distance and time. The rock has two modal domains divided by a 'sulfide-out' reaction front that coverts a sulfide-bearing assemblage (grey region) to a sulfide-depleted assemblage (white). Abbreviations as for the previous Figure. The velocity of a chromatographic front is shown by the heavy solid lines. The position of chromatographic fronts depends on the chemical velocities of the element in the sulfide-bearing assemblage (V_S), in the sulfide-depleted assemblage (V_D), and the velocity of the sulfide-out reaction front (V_{SO}), the latter defining the migrating boundary between the sulfide-bearing and sulfide-depleted regions. The steeper slope of the chemical velocities imply a higher velocity. See text for additional discussion. Modified after Godard (1995), with permission.

the homogenous regions on either side of the reaction front. In a plot in the plane (z, t), curves of constant concentration plot as straight lines, the slope of which is the velocity of the chromatographic front for that element in that zone. A given set of trace elements with different bulk partition coefficients, $\overline{D_i}$, will move at different velocities through the sulfide-depleted and sulfide-bearing zones, V_D and V_S, as defined by Equations 10.3 and 10.4. High values of the chemical front velocity V_D or V_S are associated with small bulk partition coefficients, all other factors being equal.

The time-distance behaviour of a trace ore element with different chromatographic velocities as a sulfide-bearing assemblage loses its sulfide to an infiltrating fluid at a sharp sulfide-out reaction front that is also moving with time is illustrated in Figure 10.3 (Godard et al., 1995). The slopes of the lines in this Figure are proportional to the chromatographic and reaction front velocities. These are shown relative to the velocity of the sulfide-out reaction front that is defined by the boundary between the sulfide-bearing region (grey) and the sulfide-depleted region (white). The metal concentration in the fluid initially in equilibrium with the sulfide-bearing assemble is C_S, whereas that of the infiltrating fluid equilibrated with a sulfide-depleted matrix is denoted C_D.

Using Figure 10.3a as an example, for which $V_D > V_S > V_{SO}$ (e.g., the velocity of the reaction front is less than the chromatographic velocity in the sulfide-bearing region which is itself less than the velocity of the chromatographic front in the sulfide-depleted region). As the infiltrating fluid moves through the system, sulfide is lost and the sulfide-out front moves downstream with time. In this case, the chromatographic fronts move faster that the sulfide front such that the chromatographic front is in the sulfide-bearing matrix. Between this chromatographic front and the sulfide-out front the infiltrating liquid encounters a different matrix assemblage and it is in this region that anomalous fluid concentrations, $C_?$, can occur. In most of the cases shown in Figure 10.3 (cases a, b, e and f), trace element anomalies occur in the regions bounded by the sulfide-out reaction front and the chromatographic front. However, for the case for which $V_S < V_{SO} < V_D$, the chromatographic front and the sulfide-out front are coincident (Figure 10.3c). Similarly where $V_D < V_{SO} < V_S$ then the anomalous concentration is contained by two chromatographic fronts on either side of the sulfide-out reaction front (Figure 10.3d).

Consider the case where V_S and V_D are velocities of chromatographic fronts for some element '*I*' in the sulfide-bearing and sulfide-depleted sides of the reaction front, respectively. In the instance where $V_S < V_{SO} < V_D$ the element is concentrated at the reaction front (Godard et al., 1995) and the instantaneous concentrations profile is given by:

$$\frac{\Delta Q_i}{\Delta t} = C_D\left(V_D - V_{SO}\right)\left(\Phi' + \left(1 - \Phi'\right)\bar{D}_i'\right) + C_S\left(V_S - V_{SO}\right)\left(\Phi + (1 - \Phi)\bar{D}_i\right) \quad (10.5)$$

Where $\frac{\Delta Q_i}{\Delta t}$ is the quantity of the trace element accumulated per unit time in the reaction front, the primed quantities represent values for the sulfide-depleted region, and C_S and C_D are the initial concentration of the trace element in the fluid in equilibrium with the original sulfide-bearing assemblage and in the sulfide-undersaturated infiltrating fluid, respectively. Note that the term $(V_S - V_{SO})$ is negative. Dividing by the velocity of the reaction front gives the increase in the quantity of the trace element as a function of the position of the reaction front:

$$\frac{\Delta Q_i}{\Delta t} = \frac{\Delta Q_i}{\Delta t} \Big/ V_{SO} \quad (10.6)$$

This simply states that the metal content at the reaction front will increase as the sulfide-out front advances with time. Equations 10.5 and 10.6 further state that the sulfide-out front contains all the trace element lost from the upstream side of the front. This is essentially the case modelled by Boudreau and McCallum (1992, case 1) for Pd concentrating at a sulfide dissolution front.

These theoretical considerations suggest how the observed ore metal offsets in layered intrusions can develop in an infiltration-reaction environment. In the sulfide-free regions below the sulfide-in front, metal transport is relatively rapid and they are easily carried to the reaction front. In the sulfide-bearing region above the front, however, the distribution of metals largely reflects their transport velocity. The stratigraphic distribution of the offset peaks would suggest that the relative transport velocity of Pt and Pd is less than that of Cu and Au. Metal partitioning between silicate liquid, sulfide and silicate minerals is a good first approximation as to how strongly the metals will be held by the matrix in both regions. Elements such as Pt and Pd are strongly partitioned into sulfide whereas the base metals are less so. Measured sulfide liquid/silicate liquid partition coefficients in both experimental and natural systems are on the order of 10^{4-5} for Pt and Pd, whereas, Cu and Au have partition coefficients 1–2 orders of magnitude smaller (*c.f.*, summary of Barnes and Maier, 1999). None of these elements are significantly partitioned into common silicate minerals. Hence, one would expect Pd, Pt, Cu and Au to have relatively high chemical velocities in the sulfide-free region compared to the sulfide-poor region. Within the sulfide-bearing region, Cu will have a relatively higher transport velocity than Pd and Pt but perhaps be similar to that for Au. This is consistent with the observation that Cu peaks are above PGE peaks but that Cu and Au peaks are broadly coincident,

10.3 Numeric Models of Ore Element Transport during the Degassing of a Compacting and Solidifying Crystal Pile

The above theoretical description, and especially its simplifications, is strictly true only when the material being infiltrated is uniform on either side of the reaction front and for an infiltrating fluid with a constant composition. This situation is not commonly encountered in layered intrusions because fluid evolution and sulfide saturation is a function of both the crystallization and degassing of the interstitial liquid. Thus, both the matrix and the exsolved fluid compositions are functions of time and the location in the crystal pile. Modal changes resulting from crystallization or compaction will result in changes in bulk distribution coefficients independent of any infiltration reactions.

To more realistically illustrate the effects of fluid transport of the PGE and base metals requires a numerical model to evaluate these pertinent variables as a function of time and space. Boudreau and Meurer (1999) and Boudreau (2004) included a simple heat flow and solidification model that included fluid separation and migration with the numerical compaction model of Shirley (1986). The inclusion of compaction is largely to model the amount of interstitial liquid available for degassing as a function of depth in the crystal pile. The compaction equation at described in Chapter 11 and are not describe here.

For heat transport and crystallization, the compacting crystal pile is assumed to be cooling thorough the base. The evolution of temperature with time is governed by two parameters: a) thermal diffusion and b) thermal changes involving the latent heat of crystallization. A simplifying assumption is made that the heat capacities of the liquid and solid phases are equal. The small heat capacity contribution of the fluid phase is also ignored. With these simplifications, the evolution of temperature can be expressed by a numerical analogue to the following one-dimensional heat transport-reaction equation:

$$\frac{\partial T}{\partial t} = \kappa \frac{\partial^2 T}{\partial z^2} + \frac{\Delta H_{cryst}}{\bar{\rho} c_p} \frac{\partial \Psi}{\partial t} \tag{10.7}$$

where $\frac{\partial \Psi}{\partial t}$ is the rate of change with time of the mass of crystalline material, per unit volume.

Heat loss from the crystal + liquid assemblage causes crystallization. The change in the amount of crystalline material can be expressed as a function of the gain or loss of heat, q, with time:

$$\frac{\partial \Psi}{\partial t} = \frac{\partial \Psi}{\partial q} \frac{\partial \mathrm{q}}{\partial t} \tag{10.8}$$

where $\frac{\partial \Psi}{\partial q}$ is the fraction crystallized per unit heat loss, and $\frac{\partial \mathrm{q}}{\partial t}$ is the heat transfer resulting from thermal diffusion. In the following examples, it is assumed that crystallization is a simple function of heat loss such that

$$\frac{\partial \Psi}{\partial q} = \frac{1}{\Delta H_{Cryst}} \tag{10.9}$$

In these models, the silicate liquid is assumed to be fluid-saturated when it water concentration reaches a pre-set vapour-saturation limit. Any fluid exsolved immediately migrates upward until it encounters silicate liquid that is not yet fluid-saturated. Furthermore, the mass of fluid that is evolved in one level is assumed to equilibrate with each overlying zone until it encounters fluid-undersaturated interstitial liquids. On encountering fluid-undersaturated liquid, the sulfur, metal and water content of the fluid is added to the liquid in that zone. Sulfide can be either precipitated or resorbed, depending on the local equilibrium situation.

10.4 Example Model Results

An example of the evolution of vapour and ore element profiles in a growing and solidifying crystal pile is illustrated in Figure 10.4 (After Boudreau, 2004, his *Case 2*). Crystallization of the magma lead to the formation of a crystal-liquid mush growing up from the bottom of the magma chamber. For this case, the initial magma is sulfide-undersaturated. The profiles are shown in a growing pile at time steps of 737, 1,526 and 2,621 years: the solid curves in the Figure 10.4a define to mush zone, the top of which is assumed to initially have 60% liquid and decreasing with depth in the crystal pile to 0% at the bottom of the mush by the combined effects of compaction and solidification. In these time intervals the magma in the chamber above the top of the mush column has not yet reached sulfide saturation. Instead, it is the crystallization of the interstitial liquid (owing to heat loss out the bottom of the crystal pile) and percolation of S-bearing fluid exsolved

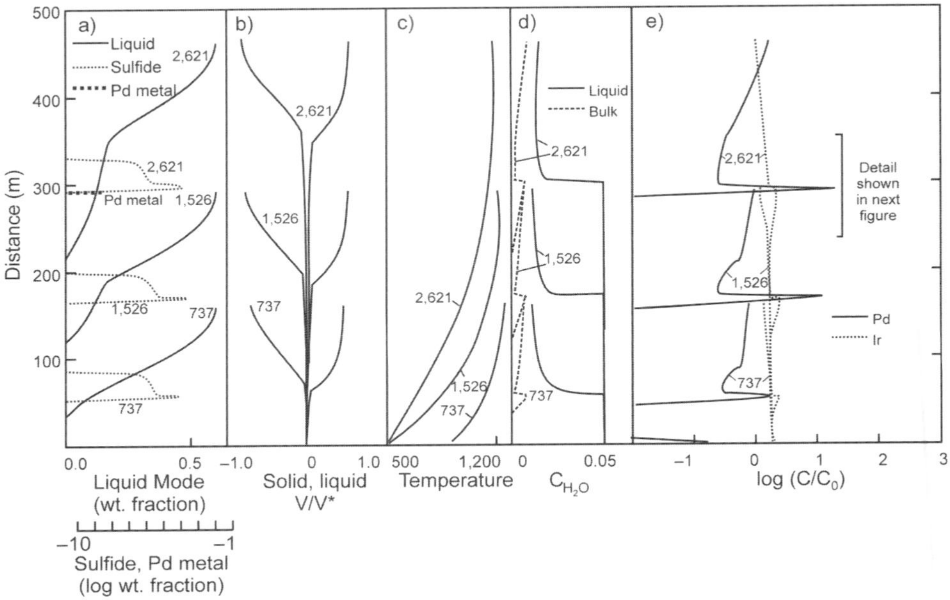

Figure 10.4 Example of numerical modelling of metal and sulfur migration in a compacting, solidifying and degassing crystal-liquid mush. Profiles are shown in a growing pile at time steps of 737, 1,526 and 2,621 years. (a) Wt. fraction liquid (solid line, linear scale), sulfide (fine dashed line, log scale) and Pd metal (heavy dashed, labelled 'Pd metal', log scale). The top of the crystal pile at each time step is located at the top of the solid liquid mode fraction line, above which the liquid fraction is 1.0 but not shown. (b) Velocities of solid matrix and interstitial liquid velocities, scaled to the characteristic velocity (Eqtn. 11.8). (c) Temperature profile. (d) Concentration of H_2O in liquid, $C^{H_2O}_{liquid}$, (solid line) and bulk H_2O content, $C^{H_2O}_{bulk}$, (dashed line). (e) Bulk element concentration normalized to initial concentrations of Pd (solid line) and Ir (dashed line) in the magma. A zone of increasing modal sulfide and Pd/Ir enrichment moves upward over time, and immediately below which eventually develops a zone of Pd metal as shown in the 2,621 year profile. See text for additional discussion. After Boudreau (2004), with permission.

from the interstitial liquid can both lead to precipitation of sulfide in the mush. A detail of the sulfide mode and ore element profile at 2,621 years is shown in Figure 10.5.

The solidification of the interstitial liquid in the lower part of the crystal pile leads to an increase in the water content of the interstitial liquid until it becomes fluid-saturated. The exsolved fluid migrates upward, carrying S and ore elements, until it encounters a zone of fluid-undersaturated liquid, at which point it redissolves into that liquid. The addition of S to the fluid-undersaturated interstitial liquid leads to the precipitation of sulfide, in abundances higher than would have precipitated from the interstitial liquid alone. If frozen in, this sulfide-enriched front might be interpreted as a *sulfide-in* boundary, the point at which the main magma became sulfide saturated, but in reality represents an ephemeral *sulfide-enriched front*. The ore elements also become concentrated at this sulfide-enriched

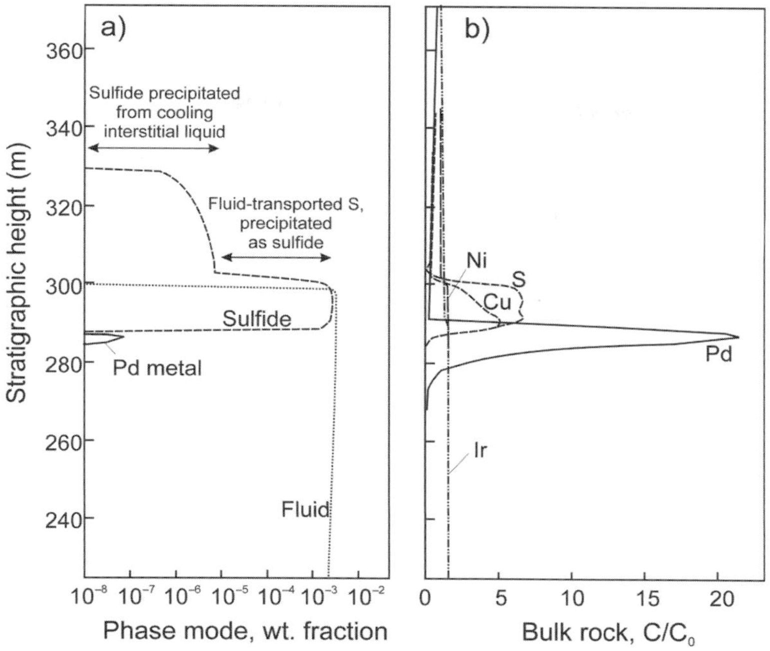

Figure 10.5. Detail of the run at 2,621 years. (a) Wt. fraction of fluid, sulfide and Pd metal as a function of stratigraphic height. (b) Bulk-rock concentration of the elements S, Pd, Ir, Cu and Ni, all normalized to initial magma composition. In (a), the peak in sulfide mode labelled 'Fluid-transported S, precipitated as sulfide' is that portion of the sulfide that is the result of fluid transport of sulfur, and it migrates to higher stratigraphic levels with time. The portion of the sulfide mode curve labelled 'Sulfide precipitated from cooling interstitial liquid' is the sulfide that results from the cooling and crystallizing interstitial silicate liquid. It also migrates upward with time. Note also that a Pd metal zone develops stratigraphically below the sulfide-enriched zone. After Boudreau (2004), with permission.

front. This continues until the interstitial liquid becomes vapour saturated, after which time the sulfide redissolves in the vapour. Over time this results in the upward migration of a sulfide-enriched front and the separation of the more soluble ore elements from the less soluble. The migrating sulfide-in front also produces characteristic offsets in the maximum concentrations of the various elements (Figure 10.5).

Pd is strongly concentrated at the sulfide-enriched front, and becomes more concentrated as the front moves upward over time. Eventually, the Pd concentration in the sulfide can become so great that a Pd metal zone develops just below the sulfide-enriched front as the sulfide redissolves, as shown at 2,621 years in Figure 10.5. This Pd metal zone moves up with the sulfide-enriched front. Finally, in the model Pd and Ir have the same solubility in terms of fluid/liquid partitioning, but Ir (and Ni) is assumed to have some partitioning into the non-sulfide solid matrix. Because of this, degassing of a small mass of fluid does not strongly change their bulk concentration. The result is that fluid transport of Ir is minor, and the bulk-rock Pd:Ir and Pd:Ni ratios at the metal front increase with time. (More recent experiments on Ir partitioning behaviour described in Chapter 4 suggests that this is an over-simplification in the model.)

Other model results have suggested that the various metal fronts can collapse to form a uniform mineralized zone. This can happen, for example, when the sulfide-enriched front moves fast enough to reach the top of the mush column and the overlying magma absorbs all vapour migrating out of the crystal pile (Boudreau and Meurer, 1999).

These model illustrative of a process, and under current state of the science they can only approximate what has occurred in any given intrusion. This is because it also involves a rather simple set of assumptions, including complete equilibration of the system at all times and constant partition coefficients. As discussed by Boudreau and Meurer (1999) there are a number of complications that can occur in a real intrusion. For example, in the model the vapour is assumed to move up instantaneously from its source of exsolving interstitial liquid until it reaches a fluid-undersaturated zone, equilibrating along the way and not changing its composition or metal carrying behaviour, whereas the halogen geochemistry of layered intrusions discussed in Chapter 6 suggests that important metal complexing agents such as Cl can vary widely.

However, these model calculations do demonstrate several critical points. First, they show how PGE-enriched sulfide zones in layered intrusions can be explained as the result of migrating sulfide-enriched fronts accompanying degassing of interstitial liquids and need not be the direct result of sulfide saturation of the main magma body. They also illustrate that high PGE/S ratios can be produced by a chromatographic mechanism without the need to call on high silicate liquid/sulfide liquid mass ratios (R-values; Campbell et al., 1983) of conventional magmatic sulfide-saturation models. For those deposits that show ore element offsets, the stratigraphic order is consistent with known or reasonably inferred partitioning of the ore elements between silicate liquid, sulfide and silicate minerals; the slower transport velocity of Pt and Pd is primarily a consequence of their high affinity for sulfide. Also, the separating fluid need not, in a general sense, 'transport' significant PGE except immediately at the sulfide resorption front. Fluid migrating below or above the sulfide resorption front may contain very little of the PGE in solution. Below the front, the PGE have already been lost and thus cannot supply significant PGE to the fluid, whereas above the front the metals are mostly held in the sulfide phase. Finally, the degree of PGE enrichment is a function of the length the chromatographic front travels. The longer the front travels, the higher are the resulting PGE concentrations. This explains why high-grade PGE zones typically do not occur along the lower contact of layered intrusions and instead require some thickness of cumulates beneath them. The exception is those cases where the ore component moves laterally up-dip. An example would be the gold enrichments seen at the upturned margins of the Platinova Reef of the Skaergaard intrusion noted in Chapter 2.

11

Compaction-Driven Stratigraphic Traps and the Formation of Great Dyke-Type Deposits

11.1 Introduction

The hydromagmatic model does not in most cases predict where in the stratigraphic succession a PGE concentration may occur. This follows from, among other controls, on variations in the size of an intrusion and its initial volatile content. However, similarities in stratigraphic changes associated with the major PGE zones of the Great Dyke of Zimbabwe, the Munni Munni Complex of Australia, and other intrusions point to a role in compaction in the development of stratigraphic traps to migrating ore-bearing fluids at and below the ultramafic–mafic transitions in these intrusions. These deposit that develop at the ultramafic–mafic boundaries are here termed *Great Dyke-Type PGE deposits*.

11.2 The Great Dyke

The Great Dyke is an early Proterozoic intrusion formed in a rifted cratonic environment of Archean granites and greenstone belts (Worst, 1960; Wilson, 1982, 1996). It is 520 km long and up to 11 km wide and forms a distinct geologic feature, cutting across much of Zimbabwe (Figure 11.1a). U–Pb zircon studies have reported ages of 2,579±7 Ma (Armstrong and Wilson, 2000) and 2,575±0.7 Ma (Oberthür et al., 2002).

The geology of the Great Dyke has been summarized by a number of workers, including Wilson (1982), Podmore and Wilson (1987), Prendergast (1988) and Wilson and Prendergast (1989). The exposed portion consists of a line of roughly north-trending set of contiguous, narrow layered mafic-ultramafic complexes (Figure 11.1b). The existence of some stratigraphic differences has led to the subdivision of the original magma chamber into two major North and South Chambers plus a smaller Mvuradona Chamber at the extreme north end (Podmore and Wilson, 1987; Wilson and Prendergast, 1989). The North Chamber has been further subdivided into the Musengezi, Darwendale and Sebakwe Subchambers, and the South Chamber into the Selukwe and Wedza Subchambers, respectively. In the Darwendale Subchamber the rocks are comprised of a lower Ultramafic sequence about 2,000 m thick and an upper Mafic sequence about 1,000 m thick (Figure 11.1c). The Ultramafic sequence consists of fourteen cyclic units, each ideally made up of a basal chromitite overlain by dunite and/or harzburgite, and then by pyroxenite. The chromitite

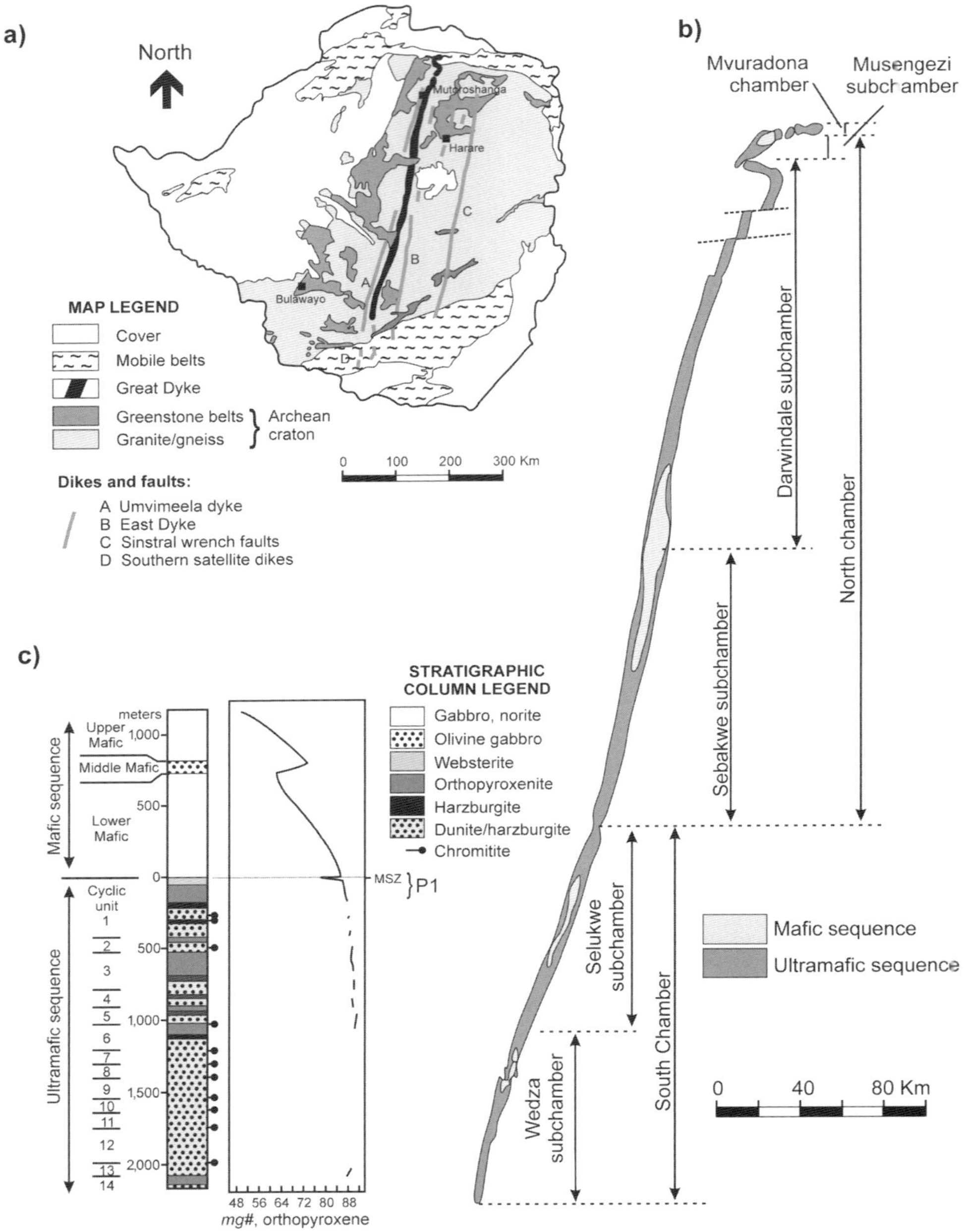

Figure 11.1 Geology of Zimbabwe and the Great Dyke. (a) Generalized geologic map of Zimbabwe, showing location of the Great Dyke and associated smaller dyke and faults. (b) Major subchambers of the Great Dyke. (c) Lithostratigraphic section of the Darwendale Subchamber, showing the location of the Main Sulfide Zone (MSZ). Figures 11.1a and 11.1b after Wilson, (1982); Figure 11.1c after Wilson (1996).

and/or the pyroxenite layer are not present in all units. The overlying Mafic sequence is defined by the presence of modally abundant plagioclase and is composed predominantly of norite and gabbronorite. The flat-lying layered sequence along the axis of the Great Dyke gradually thins and steepens towards the east and west margins to merge with a Marginal Border Group against the walls (Wilson and Prendergast, 1989).

At the top of the uppermost cyclic unit (Cyclic unit 1) at the top of the Ultramafic sequence is a prominent pyroxenite unit denoted as the P1 Pyroxenite and consisting of a bronzitite several tens of metres thick plus a relatively thin websterite immediately beneath the mafic/ultramafic contact. Several sulfide zones are present within the P1 pyroxenite including a PGE-rich zone termed either the Main sulfide zone (MSZ) (e.g., Prendergast, 1988; Prendergast and Keays, 1989) or S1 (e.g., Wilson et al., 1989) situated at the top of the orthopyroxenite. The MSZ is situated at or a few metres below the ultramafic–mafic sequence contact to just below the base of a websterite layer that forms the uppermost layer of the ultramafic sequence. The websterite is characterized by an abrupt decrease in the Mg# of the pyroxene and an increase later crystallizing minerals (Figure 11.1c); these changes are more pronounced along the margins than in the central axis (Figure 11.2). The websterite layer varies in thickness from <10 m along the western margin to >35 m along the central axis and >20 m along the eastern margin of the Darwindale subchamber (Wilson, 1992). Locally, the layering on either side of the ultramafic–mafic contact is

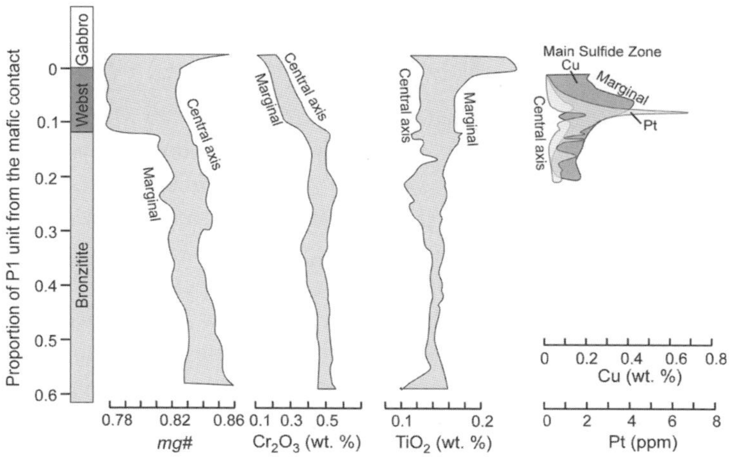

Figure 11.2 Summary of stratigraphic pyroxene composition and ore element trends in the P1 unit of the Great Dyke, Zimbabwe. From the left, *mg#* (molar Mg/(Mg + Fe) ratio), Cr_2O_3 and TiO_2 concentrations of pyroxene, and bulk-rock variations in Pd and Cu in the Main Sulfide Zone. All stratigraphic sections have been normalized to the thickness of the PI layer in the axis, and are represented as the relative proportional distance below the ultramafic–mafic contact. Analysis from the central axis of the intrusion tend to cluster along the side of the shaded regions marked 'Central axis,' and those from the margins cluster along the side marked 'Marginal.' Note that the higher-grade rocks and more evolved pyroxene compositions occur within the websterite and towards the margins. Compiled from data from Wilson (1992) and Wilson and Prendergast (2001).

disturbed, showing soft sediment-like folding and layer disruptions. A stratigraphically lower sulfide zone (LSZ) occurs 10–60 m below the MSZ and is of lower PGE grade. Both sulfide zones are laterally continuous and occur at approximately the same relative stratigraphic position in all subchambers.

The average grade of PGE mineralization in the MSZ of the Great Dyke ranges from 1 to 3 ppm of both Pt and Pd over an average thickness of ~4 m. Ore grade is not uniform across the intrusion, however, with grades being higher along the western margins than along the central axis (Wilson and Prendergast, 2001) (Figure 11.2). The host rocks to both the LSZ and MSZ are orthopyroxenites that are composed of >80 wt.% orthopyroxene with interstitial plagioclase, clinopyroxene, phlogopite and quartz. Both zones are composed of variable amounts (0.2–5 wt.%) of disseminated sulfides (pyrrhotite + chalcopyrite + pentlandite ± pyrite) are closely associated with interstitial silicates. The PGE-rich zone MSZ is also marked by the presence of Pt and Pd sulfides, arsenides, tellurides and antimonides. The footwall and hanging wall to the LSZ and MSZ are also orthopyroxenites, but have lower sulfide contents (<0.2 wt.%) than the LSZ and MSZ.

11.3 The Munni Munni Complex

The Munni Munni Complex is a well-preserved PGE-bearing Archaean (2,850 Ma) layered mafic-ultramafic complex occurring near the margin of the Pilbara Craton. Details of the regional geology and the geology, petrography and geochemistry of this complex and its mineralized zones can be found in Donaldson (1974), Hoatson and Keays (1989), Hoatson (1991), Barnes et al. (1992) and Hoatson et al. (1992). In brief, the complex intrudes older granites generally dated between 3,500 and 2,800 Ma. The Munni Munni Complex covers an area of roughly 9 × 25 km, with a minimum thickness of 5.5 km. The complex has been described as a boat-shaped layered intrusion, divided into a lower Ultramafic series 'keel' up to 1,850 m thick overlain by a Gabbroic series with an exposed thickness of 2,500 m. It has a smaller ultramafic 'keel' that expands from the Cadgerina Dyke, interpreted to be the feeder to the complex (Figure 11.3).

The Ultramafic series is layered sequence of about 20 cycles of olivine- and clinopyroxene-bearing rocks. Orthopyroxene and chromite only become significant in the 20–200 m thick Porphyritic Webserite Zone (PWZ) at the top of the Ultramafic series, the thicker portions of which occur along the margins of the complex. The parent magmas are believed to have been siliceous high-Mg (9–12 wt.% MgO) basalts of Al-depleted ($Al_2O_3/TiO_2 \approx 11$) Barberton-type komatiitic affinity (Hoatson et al., 1992).

The Gabbroic Series includes a lower interval of inverted pigeonite-bearing gabbro and gabbronorite and an upper zone of interlayered magnetite bearing anorthositic gabbro and gabbronorite, with granophyres at the top and a series of quartz monzonites and quartz syenites in the south-east of the complex.

As for the Great Dyke, PGE mineralization is associated with an interval of copper-nickel sulfide accumulation generally found 0–15 m below the Ultramafic-Gabbro series contact and extending to a few m above the contact, hosted by a PWZ of the Websterite

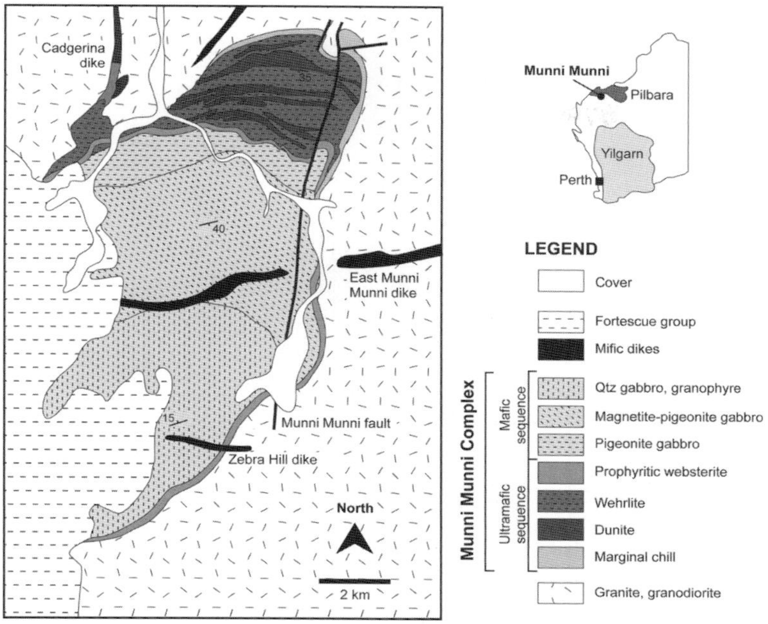

Figure 11.3 Simplified geological map of the Munni Munni layered intrusive complex. After Barnes et al. (1990), with permission.

zone. The porphyritic websterite is composed of 0–15% olivine, 10–30% porphyritic orthopyroxene, and 60–80% augite with up to 20% interstitial plagioclase. The PWZ marks a pronounced discontinuity in mineral composition trends within the intrusion, between uniform relatively high-Mg# and Cr-rich pyroxenes in the main part of the Ultramafic Series, to gabbroic rocks with a monotonic decrease in Mg# with stratigraphic height. The Cr_2O_3 content of augite decreases from about 0.2% to less than 0.05% over less than 4 m near the top of the PWZ, where it is accompanied by a sharp increase in bulk-rock Zr. It contains 1–2% chalcopyrite and pyrrhotite with trace pentlandite (Figure 11.4).

In detail, Barnes et al. (1992) note that mineralization occurs in two distinct styles. The first is a relatively PGE-poor component, characterized by coincidence peaks in Cu, Ni, S and PGE abundances. The second is a variable composition component, strongly PGE-enriched at the base and depleted at the top, characterized by systematic offsets between the peak PGE and Cu abundances. The 'offset' sulfides occur above the 'coincident' sulfides, and the offset sulfides broadly correspond to the sudden drop in Cr_2O_3 content of clinopyroxene within the PWZ. The PGE-enriched layer is generally 2–5 m thick, averaging 2.5 m thick, and occurs 0–15 m below the gabbro-ultramafic contact. PGE grades in the PGE-enriched section average 1.0 g/t Pt, 1.6 g/t Pd, 0.3 g/t Au, 0.3% Cu, 0.2% Ni. PGE grades are not uniform across the intrusion, however, and are distinctly higher above the footwall high separating the Cadgerina Dyke from the Ultramafic series. In this region the offset sulfide style is absent and mineralization is of the coincident style (Figure 11.5c).

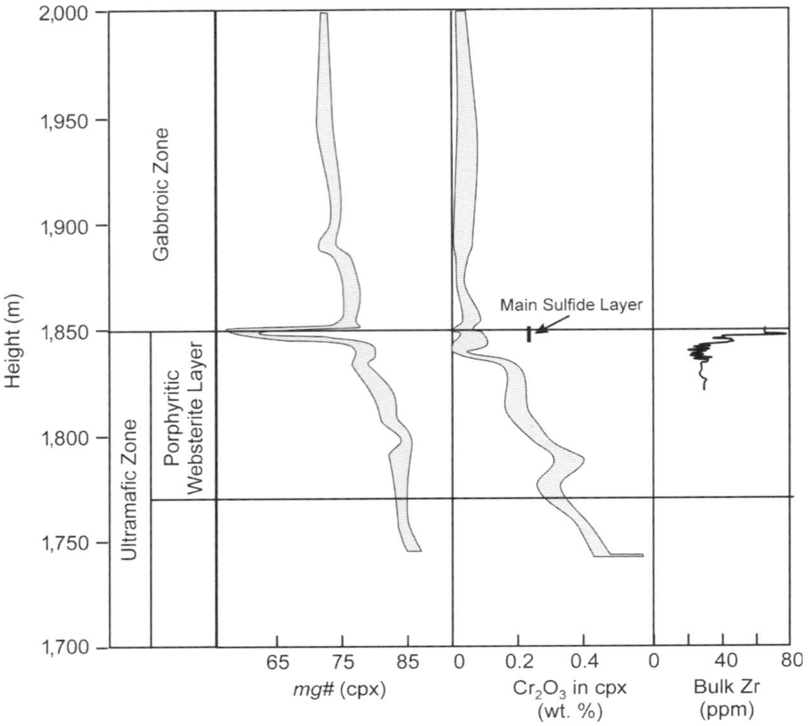

Figure 11.4 Stratigraphic variation in the *mg#* (Mg/(Mg + Fe) molar ratio) and Cr_2O_3 of clinopyroxene and bulk-rock Zr concentrations across the ultramafic-mafic transition in the Munni Munni Complex. Compiled from data of Barnes and Hoatson (1994), with permission.

11.4 Other PGE Occurrences at Ultramafic–Mafic Transitions

Like the Great Dyke and Munni Munni intrusions, the upper part of the Bronzitite zone at the top of the Ultramafic series in the Stillwater Complex is relatively enriched in interstitial minerals (clinopyroxene and plagioclase; Figure 11.6). A region of discontinuous mineralization occurs near the top of the Bronzitite zone in what is known as the Janet 50 claim block. It extends over an area of 30 m by 50 m and consists of a pegmatoidal orthopyroxenite with disseminated sulfide minerals and local concentrations of graphite (Volborth and Housley, 1984; Mcllveen, 1996). Five analyses of samples from surface exposures average 0.38 wt.% sulfur, 0.09 wt.% copper, 0.27 wt.% nickel, <2 ppb iridium, 5 ppb ruthenium, 6 ppb rhodium, 692 ppb platinum, 1,278 ppb palladium and 130 ppb gold (Zientek et al., 2002).

The Jimberlana intrusion also shows an increase in incompatible elements at the top of the ultramafic section (Campbell, 1987) similar to that seen in the Great Dyke and Munni Munni intrusions, but does not contain mineralization.

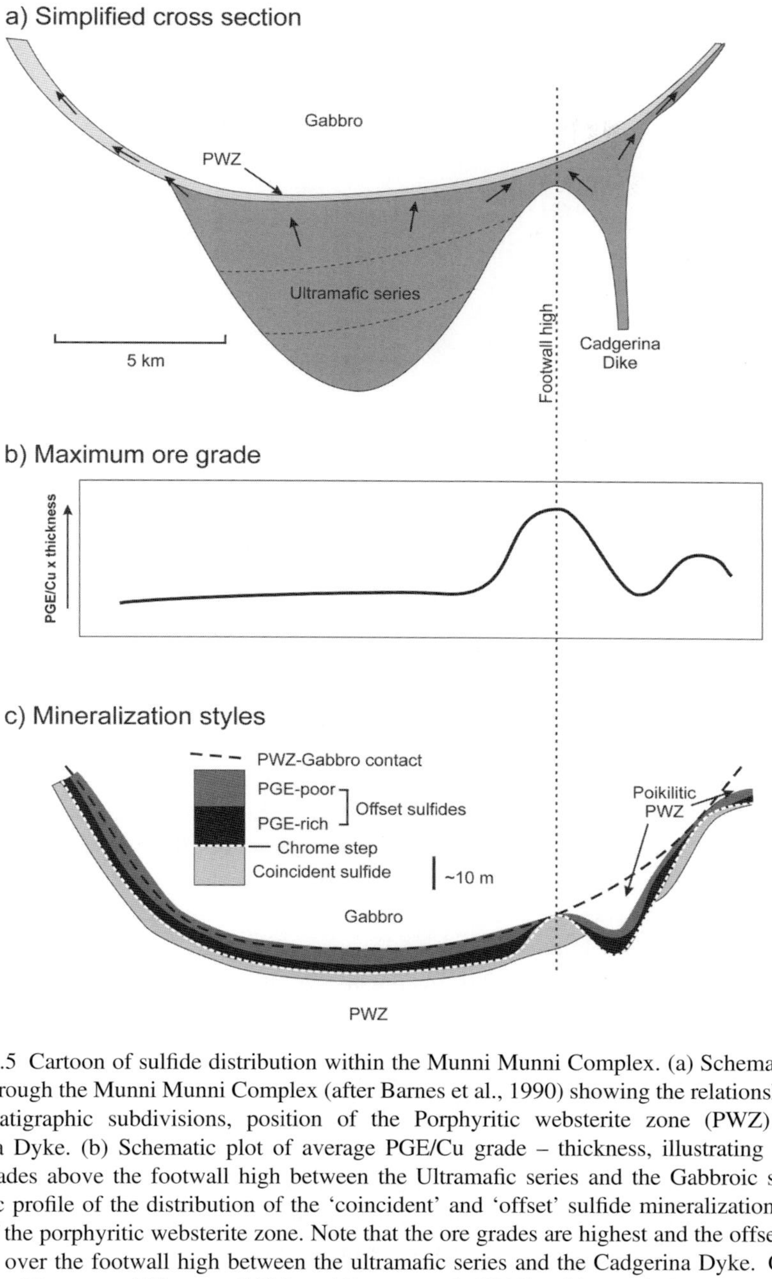

Figure 11.5 Cartoon of sulfide distribution within the Munni Munni Complex. (a) Schematic cross-section through the Munni Munni Complex (after Barnes et al., 1990) showing the relationship of the major stratigraphic subdivisions, position of the Porphyritic websterite zone (PWZ) and the Cadgerina Dyke. (b) Schematic plot of average PGE/Cu grade – thickness, illustrating generally higher grades above the footwall high between the Ultramafic series and the Gabbroic series. (c) Schematic profile of the distribution of the 'coincident' and 'offset' sulfide mineralization styles at the top of the porphyritic websterite zone. Note that the ore grades are highest and the offset sulfides disappear over the footwall high between the ultramafic series and the Cadgerina Dyke. Compiled from data of Barnes and Hoatson (1994) and Barnes et al. (1992), with permission.

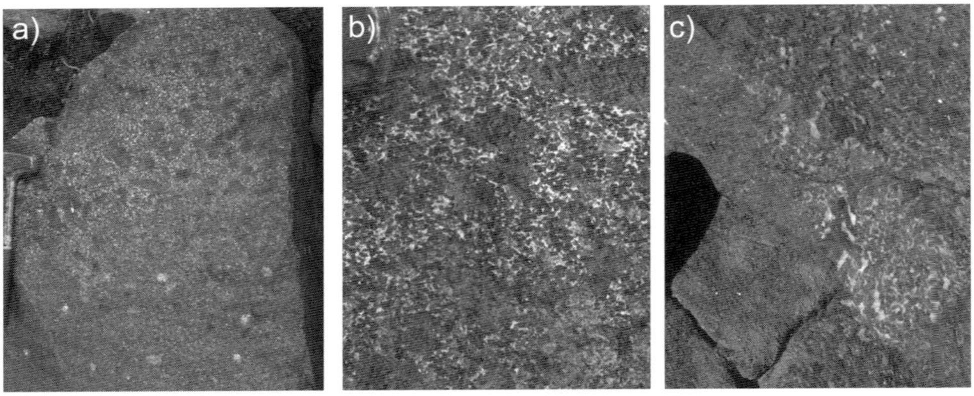

Figure 11.6 Examples of textural and mineral variation at the top of the Bronzitite Zone, Stillwater Complex, attributed to effects of compaction on interstitial liquid migration and entrapment at the boundary between the Bronzitite Zone and the overlying Banded series. (a) Increase in interstitial plagioclase (white) and large oikocrystic clinopyroxene (larger dark grains) in orthopyroxenite. Stratigraphic 'up' towards top of photo. (b) Detail of (a). Hammer head is covered in part by the photo label. (c) Plagioclases-rich orthopyroxenite layer attributed to tensional increase in porosity during interstitial liquid migration. Width of picture is ~20 cm, stratigraphic 'up' towards upper right. A black-and-white version of this figure appears in some formats. For the colour version, please refer to the plate section.

11.5 Summary and Discussion

The Great Dyke and the Munni Munni deposits have a number of common features. (1) Each is characterized by a lower sequence of ultramafic rocks consisting mainly of pyroxene ± olivine overlain by mafic rocks containing 50–60% plagioclase. The ultramafic–mafic transition is sharp (commonly on the scale of a few grain diameters) and results in a marked density change in the rock and has consequences for compaction as described below. Although the eventual precipitation of plagioclase along the liquid line of descent is an expected consequence of fractional crystallization of an initially plagioclase-undersaturated magma, a number of authors have suggested that the incoming mafic rocks instead represents the injection of a cooler, more evolved magma (e.g., Barnes and Hoatson, 1992). This is based, in part, on shifts to more Fe-rich compositions and sharp decreases in Cr content of pyroxene. (2) The PGE-sulfide mineralization begins to show up a few metres to a few tens of metres below the ultramafic–mafic contact. In addition, there are typically significant stratigraphic 'offsets' between the maximum concentrations of the PGE, base metals and sulfur as discussed in Chapter 10. (3) Occurring just below the ultramafic-mafic boundary are rocks that have higher incompatible trace-element concentrations, greater modal proportions of interstitial and oikocrystic minerals and more evolved mineral compositions. These features have been interpreted to reflect a higher porosity and higher proportion of interstitial liquid in these rocks than in the immediately surrounding rocks. (4) Ore grades tend to be higher along the up-dip portions of each intrusion. In the Great Dyke, this occurs along the margins, whereas in the Munni Munni is also occurs where layering dips away from the footwall high between the ultramafic series and the Cadgerina feeder dyke.

A number of investigators have suggested that these zones are the result of a complex interplay of locally extreme fractionation, magma mixing and sulfide saturation with variable 'R-values' (e.g., Hoatson and Keays, 1989; Barnes, 1993). For example, Barnes et al. (1992) have suggested that the coincident sulfides in the Munni Munni are interpreted to have formed by batch equilibration of sulfide liquid accompanying a magma mixing event that produced the PWZ. The offset sulfides they suggest formed subsequently by fractional segregation and gravity settling of sulfide liquid droplets from a stagnant hybrid melt layer dominated by the gabbroic component. They suggest that it was the mixing of a more evolved and cooler liquid with the resident magma was the driver for sulfide precipitation.

Barnes et al. (1992) also noted that the maximum ore grades do not correlate with the observed thickness of the underlying crystal pile and that the stratigraphically lowest coincident sulfides have lower PGE concentrations that do the overlying offset sulfides. Thus, they conclude that sulfides in the PWZ is inconsistent with derivation of PGE from vapour coming up from the crystal pile. This objections to the hydromagmatic transport assumes that the current distribution of mineralized rocks did not change after its initial precipitation of an ore-bearing fluid. Ore components could have been lost, in part, to later, metal-poor fluids moving through after the formation of the ore zone. The objection also does not account for the possibility of layer-parallel flow. It is noted that the highest grades occur where the dip of layering would focus fluids towards either the margin of the intrusion or to the region just above the footwall high between the UMS and the flaring of the Cadgerina Dyke (Figure 11.5). Similar layer-parallel flow away from synclinal fold hinges and towards anticlinal fold hinges is observed in metamorphic rocks (e.g., Skelton et al., 1995; Skelton, 1996). As shown by Barnes et al. (1992), the U-shaped layering in the UMS is asymmetric, with most of UMS dipping away from the footwall high. Thus, it is not unreasonable that higher grades are observed where there is flow focusing by dipping layers. This would also agree with the higher-grade rocks occurring at the margins of the Great Dyke, where the inwardly dipping layering would again focus fluid flow to the margins.

The main PGE-sulfide zones also typically show significant shifts in the *mg#* of the mafic minerals and higher incompatible trace-element concentrations, as shown in Figures 11.1c, 11.2 and 11.4. The sections are characterized by a sudden drop to lower *mg#* as the ore zone is encountered, although it is more pronounced along the margins of the Great Dyke than in the central axis. The crystallization sequence is otherwise as expected for a magma that has been crystallizing a thick section of mafic minerals to eventually become saturated in plagioclase.

It is commonly accepted that the *mg#* shift is due to the trapped liquid effect (Chapter 2). For a given amount of trapped liquid, the shift to lower *mg#* and higher incompatible element concentrations in the mafic sections would be an expected consequence of a higher proportion of interstitial liquid. Barnes and Hoatson (1992) attribute this to the more rapid crystallization accompanying the mixing with a cooler magma. In contrast, Meurer and Boudreau (1996) suggested that the high porosity zones arise spontaneously in compacting systems where a sharp density break occurs, such as across ultramafic–mafic boundary. In regards to the latter, there is abundant evidence for the role of compaction during crystallization of layered intrusions. The development of mineral foliations without significant lineations is one line of evidence for a pure shear expected by compaction (e.g., Meurer and Boudreau, 1998; Boorman et al., 2004). For example, Wilson (1992) shows that the long c-axis of orthopyroxene from the P1 unit preferentially lie in the plane of layering. Also, the downwarping of layering in the central sections of the Great Dyke and the draping of the layered rocks over the footwall high in the Munni Munni is an expected consequence of more extensive compaction in the slower cooled crystal mushes away from the cold margins or in regions of thinner sections of crystal mush.

Meurer and Boudreau (1996) note that both the Great Dyke and the Munni Munni PGE deposits occur in conjunction with a characteristic stratigraphic sequence that is indicative of compaction-driven porosity changes that can develop at the top of the ultramafic sections in layered intrusions. The model they proposed envisions that the compaction of a density-stratified crystal pile leads to an *increase* in porosity and evolved interstitial liquid towards the top of the ultramafic section. As the interstitial liquid below this zone degasses, ore components begin to migrate upward. When the ore front reaches the top of the ultramafic section it encounters a zone that both contains more liquid and is more iron-rich. This high porosity zone can become a trap to the ore front by physically retarding the upward movement of the vapour as it contains more silicate liquid which takes longer to solidify to vapour saturation. In addition, the vapour can precipitate the PGE sulfides because of the Fe-rich bulk composition. The physics and chemical consequences of compaction are summarized further below.

11.6 Modelling Compaction

The physics describing gravity-driven compaction of partially molten rocks have been defined in a number of papers (e.g., McKenzie, 1984, 2011; Ribe, 1985; Scott and Stevenson, 1986). These have been applied compaction of igneous rocks in general (Shirley 1986) and in an increasing number of igneous systems in particular (e.g., the Palisades Sill, Shirley, 1987; The Bushveld Complex, Boorman et al., 2004; Holyoke Flood basalt flow, Boudreau and Philpotts, 2002; the Skaergaard intrusion, Tegner et al., 2009; McKenzie, 2011).

Most models assume that the density of the crystal assemblage is constant with height. While this simplification seems appropriate when considering the cause of melt extraction from a region of partially molten mantle (McKenzie, 1984; Scott and Stevenson, 1984), it is not appropriate for modelling layered intrusions. For example, many layered intrusions as well as the oceanic crust, are composed of a lower sequence of ultramafic rocks consisting primarily of olivine and pyroxene overlain by a sequence of mafic rocks that contain 50–60% plagioclase, along with one or more mafic minerals. The transition from ultramafic to mafic rocks can be sharp (often on the scale of a few grain diameters) and results in an abrupt change in rock density from approximately 3.3 g/cm^3 to approximately 3.0 g/cm^3. For an interstitial liquid with a density of 2.7 g/cm^3, the ultramafic rocks have a density contrast with the liquid that is twice that of the mafic rocks ($\Delta\rho = 0.6$ g/cm^3 vs 0.3 g/cm^3, respectively).

Table 11.1 *Description of Variables*

Symbol	Meaning	Units	Typical Values
f	Liquid fraction	none	variable
g	Gravity	m s^{-2}	9.8
K_0	Permeability constant	none	1.5×10^{-3}
r	Crystal radius	m	0.0003
t	Time	s	variable
v_s	Velocity of solids	m s^{-1}	variable
v_l	Velocity of liquid	m s^{-1}	variable
z	Height	m	variable
μ	Liquid viscosity	Pa s	10
$(\xi^s + 4/3\eta^s)$	Bulk and shear viscosity of solids	Pa s	5 x 10^{14}
ρ_s	Density of solids	kg m^{-3}	3000 – 3300
ρ_l	Density of liquid	kg m^{-3}	2700

11.6.1 Compaction of a Crystal Pile

Compaction of a growing crystal-liquid mush has been modelled by Shirley (1986) using the equations derived by McKenzie (1984) and applied to a system with a change in the solid density by Meurer and Boudreau (1996). The following two momentum and continuity equations are used to describe the physical process of compaction (symbols, units and typical values are summarized in Table 11.1):

$$\left(\xi^s + \frac{4}{3}\eta^s\right)\left((1-f)\frac{\partial^2 v_s}{\partial z^2} - \frac{\partial f}{\partial z}\frac{\partial v_s}{\partial z}\right) = (1-f)\,(\rho_s - \rho_l)g - \frac{\mu f}{K_f}\,(v_l - v_s) \quad (11.1)$$

$$\frac{\partial f}{\partial t} = -\frac{\partial (f\, v_s)}{\partial z} \quad (11.2)$$

The left side of equation 11.1 describes the reaction of the matrix of crystals to the forces acting on the crystal pile. The first term on the right side is the downward force due to gravity, which is itself a function of the density contrast between liquid and solid. The second term accounts for the upward force produced by the viscous interactions of the liquid moving up and the solids moving down. When there is no net stress on the solids, the left side of equation 11.1 is equal to zero and the viscous interactions between the liquid and solid is equal to the force due to gravity. Under these conditions, the weight of the solid is exactly supported by the viscous interaction between the upward flowing liquid and downward moving solids and the solids are said to be *fluidized.* The net velocity difference between the velocity of the liquid and solids that achieves this balance is called the fluidization velocity. The ratio of the fluidization velocity to the net velocity difference between the solid and the liquids is called the *fluidization ratio* and provides a means of understanding the behaviour of the system (McKenzie, 1984; Shirley, 1986, 1987).

Equation 11.2 expresses the constraint that the space in the crystal pile must be occupied by either the solid or the liquid and assures the continuity of the liquid throughout the crystal pile.

The shear and bulk viscosities of the matrix (ξ^s and η^s, respectively on the left-hand side of Equation 11.1) do not vary greatly within a crystal pile for liquid fractions less than 0.8 (McKenzie, 1984). Expression for the permeability of the mush (K_f) can take a number of forms, and Shirley (1986) assumed it to have a simple functional relationship to liquid fraction:

$$K_f \;=\; K_0 r^2\, g(f) \tag{11.3}$$

Where

$$g(f) = \frac{f^3}{(1-f)^2} \tag{11.4}$$

Characteristic time, compaction length scale, compaction velocities and fluidization ratio are given by Shirley (1986):

$$z = r\left(\frac{K_0\left(\xi^s + \frac{4}{3}\eta^s\right)}{\mu}\right)^{1/2} z' \tag{11.5}$$

$$t = \left(\frac{\mu\left(\xi^s + \frac{4}{3}\eta^s\right)}{K_0}\right)^{1/2} \frac{1}{rg(\rho_s - \rho_l)}\, t' \tag{11.6}$$

$$v_s = \frac{K_0(\rho_s - \rho_l)g}{\mu}\, v_s' \tag{11.7}$$

$$\textit{Fluidization ratio} = \frac{v_s'}{(1-f)g(f)} \tag{11.8}$$

In Equations 11.5–11.8, the primed variables (z', t' and v') are dimensionless. Substitution of Equations 11.4, 11.6–11.8 into Equation 11.1 along with the boundary constraints that there is no compaction at the top of the pile and that solids cannot move below the base of the pile leads to a dimensionless form for the momentum equation:

$$(1-f)\frac{\partial^2 v_s'}{\partial z'^2} - \frac{\partial f}{\partial z'}\frac{\partial v_s'}{\partial z'} = (1-f) + \frac{v'}{g(f)} \tag{11.9}$$

Shirley (1986) modelled how the porosity will change in a crystal pile growing by the addition of crystals at the top of the pile at a constant rate that is undergoing simultaneous compaction. An example for the slow accumulation of crystal is shown in Figure 11.7. For the case of uniform conditions, most of the actual physical compaction occurs at the very top of the mush column. Beneath this is a zone where the system is perfectly fluidized and the fluidization ratio (Equation 11.5) is equal to 1.0. In this region, liquid moves up and solids move down, but the velocities of each are nearly constant such that at any point the solids moving into the zone are match by solid moving out such that the proportion of

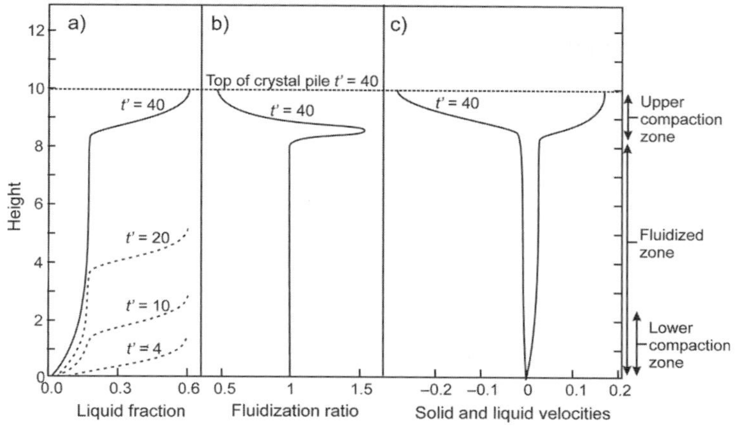

Figure 11.7 Normal compaction behaviour for slowly deposited crystals. All quantities are in dimensionless units and can be dimensionally scaled using Equations 11.4–11.6. Shown are (a) Liquid fraction, (b) the relative velocity of the solid divided by the fluidization velocity and (c) the matrix and liquid velocities. Crystals accumulate on an impermeable floor with an initial liquid fraction of 0.6 at a constant rate of 0.5. Profiles of various quantities in the cumulate pile are plotted at dimensionless time $t = 40$ except for the liquid fraction which shows the profile at the earlier times of 4, 10 and 20 as well. Note that the system is essentially at the fluidization velocity everywhere except in the upper compaction zone (and in a zone at the base of the crystal pile which is too thin to be seen here). Although the lower part of the pile is fluidized, owing to the solid floor the liquid fraction decreases near the base of the crystal pile to produce the lower compaction zone. Compiled from Shirley (1986), with permission.

solids to liquid stays constant. The exception is near the base, where the loss of liquid again allows for the solid fraction to increase.

The situation in the upper compaction zone is more complicated. In the upper part of the upper compaction zone, permeability is high and the solids are not fully supported by the upward moving liquid. The result is that the solids exert a compressive force on the underlying solids. For the solids in the lower part of the upper compaction zone, there is not enough liquid coming out from the underlying fluidized zone to fully support the solids moving down, and they are also bearing the weight of the overlying crystals. The net effect is that they move downward and compact at a velocity that is high relative to the fluidization velocity, as seen in the peak in the fluidization ratio in this region.

The simple cases modelled by Shirley have been extended to more realistic situations in several studies. Shirley (1987) used a numerical model that included compaction, thermal and compositional processes, and crystallization to model differentiation of the Palisades Sill, New Jersey. Boudreau and Phillpotts (2002) and Boudreau and Simon (2007) showed how crystal mush systems cooling from both top and bottom can lead to tensional openings to explain features such as segregation sheets in the Holyoke flood basalt flow. Boorman et al. (2004) included crystallization and heat loss with compaction to explain features of the Lower Zone–Lower Critical Zone transition in the Bushveld Complex, and Tegner et al. (2009) and McKenzie (2011) examined differentiation and compaction in the Skaergaard.

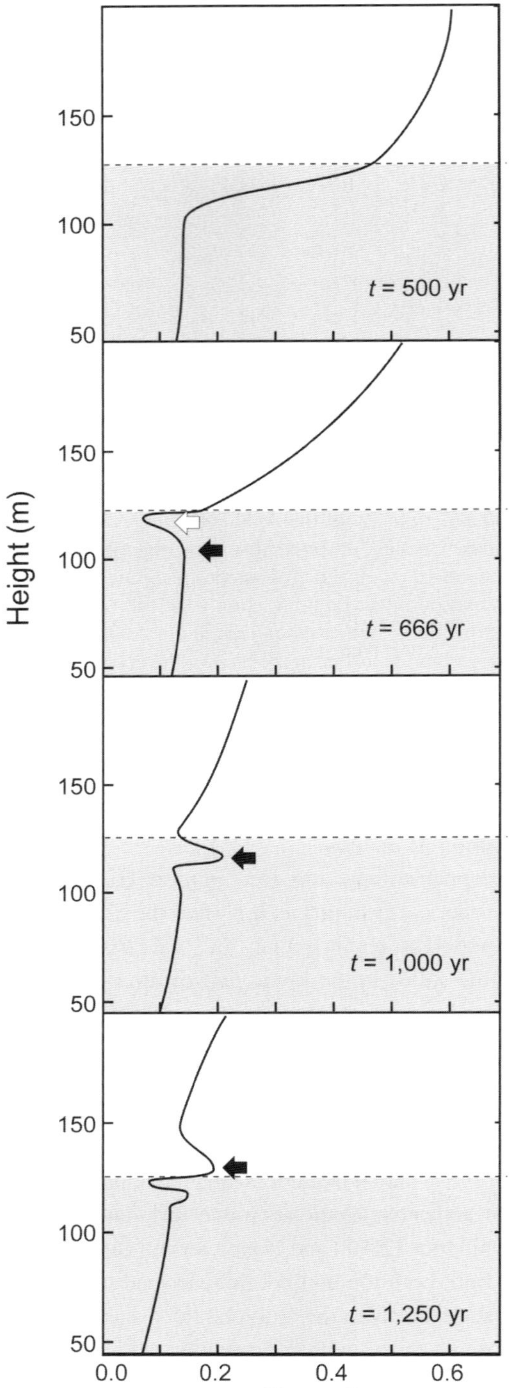

Figure 11.8 (*cont.*)

Of particular interest here is that of Meurer and Boudreau (1992), who modelled the evolution of porosity during compaction when there is a sharp density break such as occurs across an ultramafic–mafic transition as occurs in Great Dyke and Munni Munni intrusions. They noted that the transition from the ultramafic to mafic assemblage typically involves a change in density from approximately 3.3 g/cm^3 to 3.0 g/cm^3. Compared with a typical liquid density of 2.7 g/cm^3, the liquid–solid density contrast in the ultramafic section is about twice that of the overlying mafic section. Other parameters being equal, this means that the Ultramafic Zone will compact faster than the mafic section. The results of numerical simulations show that sharp density changes can produce an uneven distribution of interstitial liquid at the top of the ultramafic section as shown in the time sequence of plots in Figure 11.9. In brief, the presence of a density change at the contact between the Ultramafic and Mafic Zones has the following time-dependent effects:

1. Compaction slows once the system reaches plagioclase saturation at the base of the mafic zone, owing to the lower density contrast between the liquid and solid (Figure 11.8a).

2. As a consequence of the slow compaction in the Mafic Zone, permeability remains high at the base of the Mafic Zone and the upper part of Ultramafic Zone is able to compact due to the ease by which liquid can move up and into the overlying Mafic Zone. Owing to loss of liquid from the top of the Ultramafic Zone, however, permeability is reduced and liquid below this level has a much harder time moving through this low-permeability 'lid' (Figure 11.8b). The effect is similar to what happens in a French coffee press. Pushing down on the plunger removes the coffee grounds from the upper part of the coffee pot but causes a build-up of coffee grounds beneath the plunger. The net effect of this build-up of coffee grounds is to clog up the filter pores and make it harder to push down the coffee plunger.

4. The relatively impermeable lid separates the column into two different compaction regimes. Beneath the lid, crystals continue to compact and move down from the lid while liquid moves up. The net effect is an increase of interstitial liquid beneath the low-permeability lid (Figure 11.8c).

5. Eventually the ponded liquid below the low-porosity region causes the mush to dilate, increasing the permeability and allows the liquid to migrate upward. The net result is a porosity 'wave' that can move upward with time (Figure 11.9c and d).

Figure 11.8 Profiles of the developing porosity profile as a function of height in a growing and compacting pile with a sharp density break between the ultramafic (shaded) and mafic (unshaded) solid assemblages. Profiles are shown at 500, 666, 1,000 and 1,250 years. The density transition originally occurred at 150 m above the base; compaction causes it to move down by about 25 m after 1,000 years. Note that rapid compaction at the top of the ultramafic sequence initially produces a low-porosity 'lid' at the top of the ultramafic sequence after 666 years (marked with the unfilled arrow), which then induces a zone of high porosity to develop beneath it at 1,000 years (marked with the filled arrow). This high porosity zone migrates over time into the base of the mafic sequence as a moving porosity wave. At 1,250 years, a second high porosity zone begins to develop about 15 m beneath the original high porosity peak. Redrawn after Meurer and Boudreau (1996), with permission.

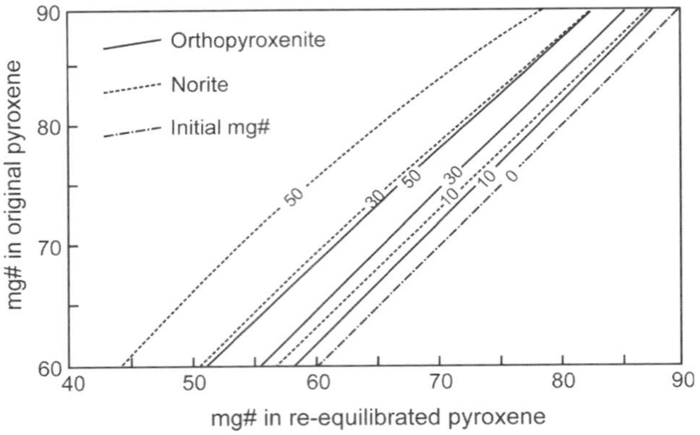

Figure 11.9 Calculated curves for final bulk composition of re-equilibrated orthopyroxene (horizontal axis) versus initial composition of orthopyroxene (vertical axis) for simple two-component crystal-liquid systems. Lines are for different values of the different wt.% trapped liquid as indicated on each curve for both an orthopyroxenite (solid lines) and a norite (dashed lines). The line at 0 wt.% trapped liquid (dot-dashed line) is the same as the initial *mg#*. Note that the trapped liquid shift effect is more extreme for the norite than for the orthopyroxenite for a given initial amount of trapped liquid. Redrawn after data of Barnes (1986), with permission.

Although it seems counter-intuitive that compaction can lead to dilation of the mush and an increase in interstitial liquid, one should realize that the interstitial liquid can effectively transfer the lithostatic pressure from deeper in the crystal pile to higher levels. In this regard, the behaviour of the interstitial liquid has parallels with the formation of magmatic sills. In the latter, the magma is driven upward by the lithostatic load imposed in the upper mantle and lower crust on the magma. As this magma moves upward into the upper crustal environment, the pressure can exceed the downward weight of the overlying rocks and the liquid can lift the overlying rocks and form a sill.

11.7 Compaction-Induced Porosity Changes and the Trapped Liquid Shift Effect

As described in Chapter 2, the presence of some residual liquid in the crystal mush at the point the mush begins to solidify will affect the composition of the pre-existing minerals, causing them to have a more evolved composition than they started with. This is referred to as the *trapped liquid shift effect*. As conventionally modelled, the mush is taken to be a simple a two-component system composed of 'cumulus' minerals and the liquid that crystallized these minerals. Knowing the initial liquid one can find the proportion of solids and liquid and presumable the amount of trapped liquid shift effect by a method discussed below. If the porosity is reduced by compaction, it is implicitly assumed that the compaction occurred isothermally and at the same temperature as the mush originally formed, before any solidification of the interstitial liquid could change the composition of the liquid or the solid. The corollary of this is that one can theoretically use the bulk composition to

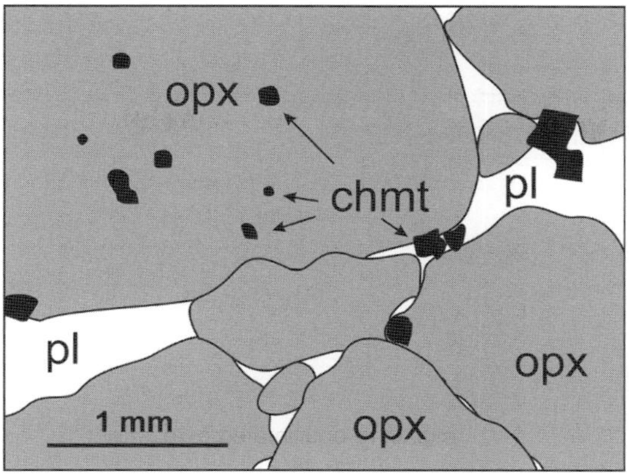

Figure 11.10 Sketch of photomicrograph of the UG2 orthopyroxenite. Abbreviations: chmt = chromite, opx = orthopyroxene, pl = plagioclase. Redrawn after Mondal and Mathez (2007), with permission.

find the composition of the original liquid by simply removing the 'cumulus' component from the bulk-rock composition (e.g., Bédard, 1994)

The trapped liquid shift effect is most effective with olivine and pyroxene for which diffusion is relatively rapid, resulting in these mafic minerals to have a lower *mg#* than they initially started with. The magnitude of the effect is a function of the amount of liquid present as the rocks solidify, as illustrated in Figure 11.10. The shift is less effective in plagioclase, where slow diffusion is more likely to produce normal zoning in individual grains rather than re-equilibration of the whole grain in the absence of extensive recrystallization. But even for plagioclase, zoning would be expected to be more extensive in rocks that crystallized a large amount of interstitial liquid.

There are two main methods by which the amount of trapped liquid can be estimated. The first is to use the modal abundance of late crystallizing interstitial and oikocrystic minerals, with the assumption that they grew entirely from the residual trapped liquid. For example, the thick anorthosite units of the Middle Banded series of the Stillwater Complex the modal abundance of zoned interstitial and oikocrystic pyroxene is variable on the outcrop scale, with regions containing up to 20% pyroxene (Haskin and Salpas, 1992). Ignoring minor phases and assuming that plagioclase and pyroxene grew at a 60:40 ratio from the interstitial liquid, one can infer an initial porosity 50% in the pyroxene-rich regions.

The second method uses the abundance of a strongly incompatible element such as P or Zr along with an assumption of their concentration in the liquid that initially precipitated the solid assemblage. For example, if the magma that initially crystallized a rock had 1,000 ppm P, and the final solidified rock now contains 50 ppm P, one would deduce that the mush contained 5% trapped liquid.

However, as pointed out be Meurer and Meurer (2006), estimates of trapped liquid using an element such a P is at best just monitoring the amount of liquid present at the time

Table 11.2 *Percent trapped liquid in a Bushveld UG2 orthopyroxenite, as calculated by five different methods*

Method	Whole rock[1]	B1 sill[2]	Est. trapped liquid %
P_2O_5 (wt.%)	0.01	0.09	11%
Zr (ppm)	14	98	14%
Al_2O_3 (wt.%)	4.87 (3.61)[3]	13.06	42% (27.6%)[3]
Na_2O (wt.%)	0.65	1.89	34%
Plag (norm%)[4]	16.3 (12.0)[3]	——	27% (20%)[3]

Notes: (1) Element abundances are taken from the average whole rock UG2 orthopyroxenite from the Critical Zone of the Bushveld Complex (Eales et al., 1994, analysis 12). (2) Average B1 Sill composition from Maier et al. (2000), assumed parent liquid from the Lower and Critical zones. Fractionation of this liquid has been ignored for this example, but would result in even lower estimates of trapped liquid for P and Zr concentrations than shown. (3) Number in parenthesis corrects for ~1.5 wt.% Al_2O_3 contained in pyroxene. (4) Estimate of trapped liquid based on normative percent of plagioclase, assuming the residual pore liquid crystallized 60% plagioclase, 40% orthopyroxene and other minerals and that original mode % plagioclase was the same as the normative %.

the interstitial liquid became saturated in apatite. If one uses elements that saturate earlier, or uses elements that are to some degree compatible in the original minerals, then in a system that is simultaneously undergoing compaction and cooling these different methods can give different estimates of trapped liquid. Those elements that crystallize earlier, when the amount of interstitial liquid is high, will have a higher abundance in the final rock than will those elements that only precipitate a solid phase late, after much of the liquid has been lost to compaction. Indeed, estimates in those rocks that have used both the modal and incompatible element methods invariably do not agree. Again, using the Stillwater anorthosites as an example, estimates of the trapped liquid using La are typically less than 5% (Salpas et al., 1983) as compared to the 50% estimated from the pyroxene mode noted above.

An example of the problem is shown in Figure 11.10 and Table 11.2 for a Bushveld UG2 orthopyroxenite. In this orthopyroxenite, the crystallization of orthopyroxene alone is assumed to have initially produced an orthopyroxene + liquid mush. One can make the reasonable assumption that the crystallization order from the interstitial liquid is orthopyroxene → plagioclase → apatite, zircon, with the onset of apatite and zircon crystallization occurring well after the start of crystallization of interstitial plagioclase. If the plagioclase, apatite and zircon all grew from an initially fixed amount of trapped liquid, then the amount of trapped liquid deduced from the bulk-rock abundances of Al and Na (now contained mainly in plagioclase, although some Al can be in the orthopyroxene), P (contained mainly in apatite), and Zr (contained in zircon) should all be similar. This is not generally observed. Instead, estimates from plagioclase suggest it grew from a volume of interstitial liquid 2–3 time larger than did the apatite and zircon. In contrast, for plagioclase, the amount of trapped liquid based on an element such as Na or Al also agrees with the amount of trapped liquid estimated the modal abundance of plagioclase; all three estimates agree within about ±10% (Table 11.2).

The conclusion is that compaction (or some other mechanism of liquid advection) caused at least part of an evolved interstitial liquid fraction to be lost prior to apatite and zircon saturation. In this case, if the original orthopyroxene + liquid assemblage started at 60% porosity (a typical value for loosely packed crystals; e.g., Shirley, 1986) then plagioclase began to crystallize at ~30% porosity, equal to a loss of 50% of the original porosity. Apatite and zircon crystallized after a further porosity reduction down to about 10–15%. These numbers are approximate as the system is undergoing porosity reduction as the interstitial minerals grew.

The effects of concurrent crystallization and compaction is most clearly seen in the marginal rocks to many intrusions. Here, the mineral compositions may show what appear to be 'reverse fractionation trends,' in that the mafic minerals become more Mg-rich with height away from the base. This apparent reverse from an expected normal fractionation trend can be explained by the higher liquid fraction near the base, as rapid cooling and solidification prevents the loss of interstitial liquid to compaction (e.g., Boudreau and Philpotts, 2002). Away from the base, cooling is slower and compaction is more effective at causing loss of interstitial liquid.

With the discussion of the development of high porosity zones described above, one can envision three different ways the composition of the final solid assemblage may be modified during concurrent compaction and crystallization. These are illustrated in the plots shown in Figure 11.11a–c. The plot of Figure 11.11a shows the conventional interpretation in which the bulk composition does not change over the course of crystallization and the *mg#* shift of the final solid is a simple function of the original amount of trapped liquid. In Figure 11.11b, compaction leads to loss of interstitial liquid causing the bulk composition to become more Mg-rich with time and to lose its incompatible element component. Using the observed incompatible element concentration would be to underestimate the amount of trapped liquid and thus underestimate the magnitude of the *mg#* shift as compared to Figure 11.11a. This is essentially the case described in Table 11.2. For Figure 11.11c, dilation of the crystal mush by the mechanisms discussed previously, and/or the migration of partly crystallized, evolved liquid, can both result in the bulk composition to become more Fe-rich and enriched in incompatible elements in general. The net effect is that one would overestimate the trapped liquid shift effect. In addition, if the interstitial liquid has lost its compatible elements prior to entering the inflating region (e.g., lost Cr to additional crystallization of pyroxene from the interstitial liquid), then the original liquidus minerals would have lower compatible element concentrations than they originally started with. One could envision more complex scenarios. For example, the mush could initially begin to dilate but later compact (as suggested by Figure 11.8 in which the high porosity zone moves upward over time) while cooling and crystallizing at the same time.

Barnes (1986) has argued that there is very little migration of residual liquid in most cases owing to the lack of observed chromatographic effects. The classic example is that of

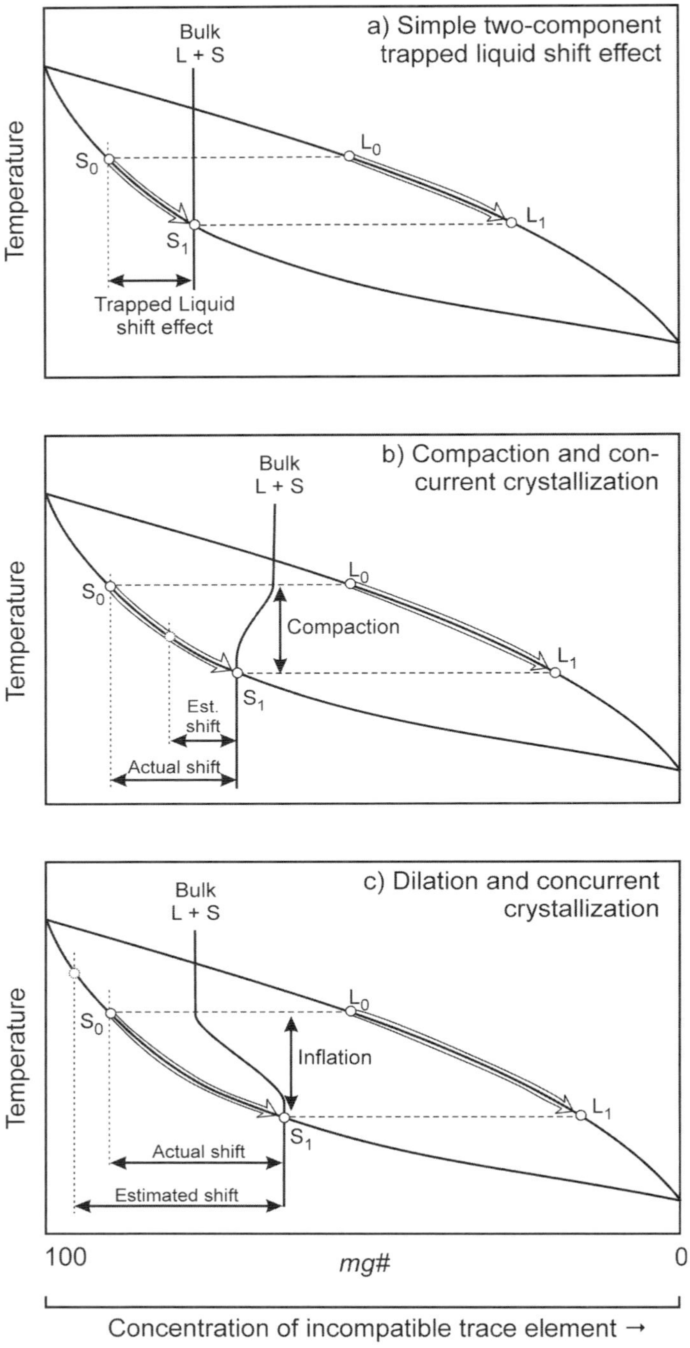

Figure 11.11 (*cont.*)

Irvine (1980), in which a stratigraphic break in the Muskox Intrusions was the result of the influx and crystallization of a more primitive magma on top of a partly crystallized and more evolved mush. Compaction of the sequence drives liquid from the more evolved section into more primitive crystal assemblage, leading to compositional breaks to be offset upward from the modal breaks. Barnes suggested that this is rarely seen, and concluded that liquid migration was minor at best in most intrusions. More liberally, the lack of observed offsets would imply that most compaction-driven liquid migration would occur soon after the mush was formed. However, the analysis here would suggest that stratigraphic compositional breaks themselves can be caused by sharp stratigraphic variations in the amount of interstitial liquid that is initially in equilibrium with the solid assemblages. Indeed, in some cases the dilation of the mush can result in the formation of segregation sheets that are entirely liquid (Boudreau and Philpotts, 2002). It is only as the interstitial liquid crystallizes that the sharp compositional breaks develop.

11.8 High Porosity Zones as Mineralizing Traps

The high porosity zones predicted to be located near the ultramafic–mafic transition may have consequences in addition to causing unusual trace-element concentrations.

A cartoon of how the profile can develop is shown in Figure 11.12. Figure 11.12a shows the initial sequence as a fractionating magma goes from crystallizing pyroxene alone (to form an ultramafic sequence of composed of opx + liquid), to crystallizing opx + plagioclase (to form an overlying mafic sequence composed of opx + plagioclase + liquid). Compaction in the crystal pile is driven by density differences between the bulk solid assemblage and the interstitial liquid. Because the mafic section compacts slower than the ultramafic section, the base of the mafic section takes longer to reduce the porosity with time. The result is that the very top of the Ultramafic Zone compacts faster than does the underlying mush owing to the easy that interstitial liquid can move out of the ultramafic section, resulting in a low-porosity 'lid' as shown in Figure 11.12b. Compaction behaviour then begins to evolve differently both above and below the lid.

Figure 11.11 Simple and complicated causes for the 'trapped liquid' *mg#* shift effect. The line labelled 'Bulk L + S' is both the *mg#* of the bulk solid + liquid and the bulk concentration of a perfectly incompatible trace element. (a) Effect of crystallization of interstitial liquid from path L_0 to L_1 results in the solid changing its composition from S_0 to S_1. During crystallization, the bulk composition remains constant. If one knows the concentration of an incompatible element in liquid L_0, one can get a good estimate of the original proportions of liquid and solid and estimate the trapped liquid shift effect. (b) Compaction of the mush reduces the amount of interstitial liquid from L_0 to L_1 as the rock solidifies. By the time the rock solidifies at L_1, the amount of a late crystallizing incompatible element in the bulk composition is lower and the estimated trapped liquid shift effect is also smaller than it actually is. (c) Migration of evolving liquid from L_0 to L_1 during a local inflation event causes the interstitial liquid to become more enriched in an incompatible element than it would have without the concurrent inflation and crystallization. The result is to overestimate the trapped liquid shift effect based on the high incompatible element concentration.

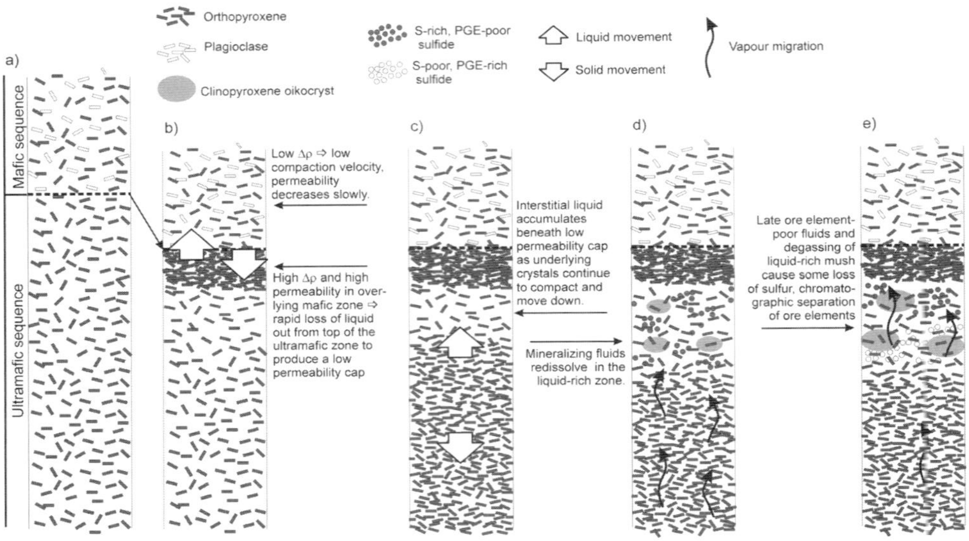

Figure 11.12 Cartoon showing the development of a Great Dyke-type deposit at an ultramafic–mafic transition zone. (a) Initial sequence of pyroxene ± olivine + liquid assemblage of the ultramafic sequence overlain by pyroxene + plagioclase + liquid assemblage of the mafic sequence. (b) The relatively slow compaction of the plagioclase-rich rocks at the base of the mafic zone keeps porosity relatively high. This allows for the rapid loss of liquid from the top of the ultramafic sequence, producing a relatively low-permeability 'lid' at the top of the ultramafic sequence (the 'French coffee press' effect). Beneath this lid, compaction of the ultramafic sequence leads to the formation of a high porosity (liquid-rich) region. (c) The high porosity zone crystallizes relatively abundant interstitial plagioclase and oikocrysts of clinopyroxene. (d) Late migrating mineralizing fluids react with this relatively liquid- and Fe-enriched zone to precipitate PGE sulfides. (e) Continued loss of S from the underlying rocks produces very late, S-poor fluids that resorb some S and base metals to produce compositional offsets of the ore elements.

Beneath the lid, the ultramafic matrix continues to compact downward and liquid moves upward, producing a region rich in interstitial liquid below the lid (Figure 11.12c). As the interstitial liquid deeper in the mush continues to cool and crystallize, it eventually begins to degas and transport S and the ore elements upward (Figure 11.12d). The high porosity region at the top of the ultramafic section can act as a fluid trap in two ways. First, zones enriched in interstitial liquid will have lower solidus temperatures and will take longer to solidify and reach vapour saturation than the underlying rocks. Thus vapour entering the liquid-rich rocks redissolves. The added sulfur drives the interstitial liquid to sulfide saturation. Second, the more iron-rich compositions can lead to reactions between the fluid and the solid assemblage that favours precipitation of sulfide (Boudreau, 1988; Boudreau and McCallum, 1992a). One consequence of crystallizing more interstitial liquid in the upper portions of the ultramafic section is that it creates a discrete compositional layering. The crystallization of interstitial liquid produces more evolved minerals, more abundant interstitial minerals and an increase in the bulk-rock Fe content. The higher Fe content can induce sulfide precipitation as fluids exsolved from magnesian cumulates below react with the more Fe-rich silicates via reactions of the type:

$$\underset{fluid}{H_2S} + \underset{liquid}{FeO} \rightarrow \underset{po}{FeS} + \underset{fluid}{H_2O} \tag{11.10}$$

Because of stratigraphic variability, more than one metal/sulfide zone may develop as the infiltrating fluid equilibrates with different bulk compositions. The presence of several discrete porosity waves just below the ultramafic–mafic boundary also creates the potential for multiple mineralized horizons, such as observed in the Great Dyke. When the upward migrating fluid encounters a porosity wave that is undersaturated, it will redissolve into that liquid until it is saturated and then continue to move upward. Depending upon the porosity and permeability of the overlying mush, sulfur and other ore elements may become trapped in a porosity wave that crystallizes more slowly than the overlying cumulates.

As the system continues to cool, the regions rich in interstitial liquid will themselves become vapour saturated, resulting in the loss of some sulfur and chromatographic separations as the vapour moves upward. This may be aided continued degassing or the underlying mush, these later fluids becoming more and more depleted in sulfur. When these late, S-poor fluids encounter the earlier precipitated sulfides of the developing ore zone, they result in sulfur loss from the lower portions and a chromatographic separation of the more soluble base metals from the less soluble Pt and Pd. Both mechanism will result in a sulfur-poor but PGE-rich zone overlain by a base metal, S-rich sulfide assemblage as shown in Figure 11.12d.

The sequence of events shown in Figure 11.12 does not consider that the high porosity, liquid-rich zone can itself migrate upward with time. In such a case, it may be that the interstitial liquid-rich zone moves upward after sulfide precipitation but before the liquid-rich zone cooled to vapour saturation. In such a case the ore in the lower part of the mineralized zone can preserved its initial sulfur and metal content, whereas the ore assemblage associated with the (now shifter upward) liquid-rich zone experiences some metal redistribution as the interstitial liquid degases. This can explain the 'offset' sulfides overlying the 'coincident' sulfides and the Cr break that separates the two sulfide zones in the Munni Munni: the Cr break represents the base of the liquid-rich zone at the point it began to become frozen-in.

The model proposed by Meurer and Boudreau (1996) explains why the tops of ultramafic sequences are particularly prone to PGE enrichment. This coincidence of these compaction-derived features that develop over a period of time in the crystal pile and associated PGE sulfides would be otherwise be an extremely fortuitous occurrence if the sulfides precipitated by a strictly magmatic mechanism and the compaction profile developed later. Instead, the compaction model does not require anything more than that the magma follow the apparent liquid line of descent in which the magma eventually becomes plagioclase saturated and that ore metals can be carried upward by hydrothermal fluids. Fluid movement results in both the original mineralization and the formation of 'offset' mineralization, the later developing as the high porosity regions degas on further cooling. The association of ore mineralization with compaction-derived porosity changes is perhaps some of the strongest evidence that stratabound zones of PGE mineralization can develop within an evolving and degassing crystal mush.

12
Chromitites

12.1 Introduction

A broad correlation between whole rock Cr and PGE concentrations is seen in a variety of magmatic environments (e.g., Pagé et al., 2012, and numerous references therein). More specifically, the observed enrichment of *IPGE* (Iridium-group PGE; Ru, Ir, Os) in chromite from layered intrusions, ophiolites and alpine-type peridotites have been used to suggest a genetic link between PGE concentrations and Cr-rich spinel (e.g., Peck et al., 1992; Ahmed and Arai, 2002; Buchl et al., 2004). Both empirical and experimental studies have generally shown that the IPGE, but particularly Ru, are preferentially partitioned into chromite whereas, except for Rh, the *PPGE* (Pt-group PGE; Rh, Pt, Pd) are less so (e.g., Capobianco and Drake, 1990; Hart and Ravizza, 1996; Puchtel and Humayen, 2001). Furthermore, this partitioning behaviour is enhanced in more oxidized magmas (e.g., Brenan et al., 2012).

Significantly complicating the correlation, however is that chromite from both plutonic and volcanic rocks can have a variety of mineral and melt inclusions. These can include a variety of platinum-group minerals (PGM), of which minerals of the laurite group, (Ru, Ir, Os)S_2, are the most common, but also include Pt–Fe and Ir metal alloys (*c.f.*, summary of O'Driscoll and González-Jiménez, 2016). The IPGE in particular are extremely insoluble in silicate liquids and hence they have a tendency to form trace minerals such as laurite that are amongst the last to dissolve during partial melting and amongst the first to form during crystallization (Peck et al., 1992; O'Neill et al., 1995; Ertel et al., 1999; Borisov and Palme, 2000; Luguet et al., 2007). These have been termed the *IPGE minerals*, or simply as the *IPGM*. Chromite can also have polyphase inclusions of silicate $\pm$ sulfide and other minerals. The presence of PGM is particularly troublesome as the phases represent a concentration of one or more of the PGE that are in ppt to ppb concentrations in the silicate liquid. Thus the degree to which the association is due to PGE incorporation into the spinel solid solution (and perhaps later exsolved on cooling) or the entrapment of co-precipitating PGE-rich minerals is still an open question.

12.2 Volcanic Chromite

Studies of the PGE-chromite association tend to lump together observations from the volcanic and plutonic environment. However, chromite in the mantle and in plutonic rocks (mainly layered intrusions and ophiolites) formed in a slowly cooling environment where crystal growth by ageing and annealing, prolonged re-equilibration with surrounding silicates (particularly Fe–Mg exchange), and interaction with evolving and migrating interstitial silicate liquid and volatile fluids can complicate interpretation. These problems can nominally be minimized by study of chromite in quickly cooled extrusive rocks, and for this reason, volcanic chromite has been the subject of several recent studies.

Because of the common occurrence of PGE-rich inclusions in chromite (with the possible exception of chromite from komatiite), mineral separate analysis is generally unsuitable to determine the PGE in chromite solid solution. Thus, *in situ* techniques such as laser ablation–induction-coupled plasma mass spectrometry (LA-ICPMS) are required. However, early attempts did not find measureable PGE in most samples. Except for chromite from komatiite and one example of chromite from an oxidized calc-alkaline magma, studies of the PGE concentrations by *in situ* laser ablation do not find any PGE above the detection limits of a few to ~10 ppb except as local PGE-rich inclusions in chromite in magmas ranging from komatiite basalt, ferropicrite, tholeiite basalt and boninite (Fiorentini et al., 2004; Locmelis et al., 2011, 2013; Pagé et al., 2012; Park et al., 2012). In contrast, komatiite chromite can have Ru concentrations to several hundred ppb, 10–20 ppb Ir, 20–30 ppb Os and up to 16 ppb Rh. Chromite from sulfide-bearing komatiite can have distinctly lower Ru contents (<150 ppb; Locmelis et al., 2013). The chromite in oxidized calc-alkaline lavas have concentration of ~100 ppb Ru, and from 35–70 ppb of Ir, Os and Rh and are positively correlated with Fe^{3+}, Ni and V (Park et al., 2012), consistent with experimental and theoretical work that oxidizing conditions favour higher partitioning into chromite (e.g., Brenan et al., 2012). More recent work (e.g., Pagé and Barnes, 2013) have used a newer generation of LA-ICPMS with an order of magnitude better detection limits, but currently there are only a handful of analyses (Figure 12.1a).

These studies imply chromite has a strong preference for Ru over the other PGE, and that Pt and Pd are the most incompatible. The higher concentration of the IPGE in komatiite is largely a function of the higher IPGE concentration of the silicate liquid; estimated chromite/silicate liquid distribution coefficients for the IPGE are otherwise lower for komatiites than for all other magma types (Pagé and Barnes, 2013)

While komatiite chromite can have high IPGE in solid solution, it is otherwise generally free of any type of PGM inclusions (Fiorentini et al., 2004; Pagé et al., 2012). In contrast, chromite from other magma types commonly have one or more inclusion types (Fiorentini et al., 2004; Locmelis et al., 2011, 2013; Pagé et al., 2012; Park et al., 2012). These are generally described as a poorly defined Ir-rich mineral/alloy or a Pt–Fe alloy. For example, the Fiorentini et al. (2004) found about half of 25 analysis of komatiitic basalt and two-thirds of 33 analysis of tholeiite contained Ir-rich inclusions, whereas none were observed in approximately 500 true komatiite analyses. In addition, one-third of the tholeiite and most of the ferropicrite samples also had Pt-rich inclusions. In contrast to layered intrusions where PGE-sulfide minerals are common, particularly those of the laurite group, (Ru, Ir, Os)S_2, the inclusions in the volcanic rocks generally are lacking S (e.g., Proenza et al., 2007; Pagé et al., 2012). The size of inclusions range from ~1 to ~10 μm, the larger Pt–Fe inclusions are big enough to analyse with an electron microprobe. These sizes are similar to

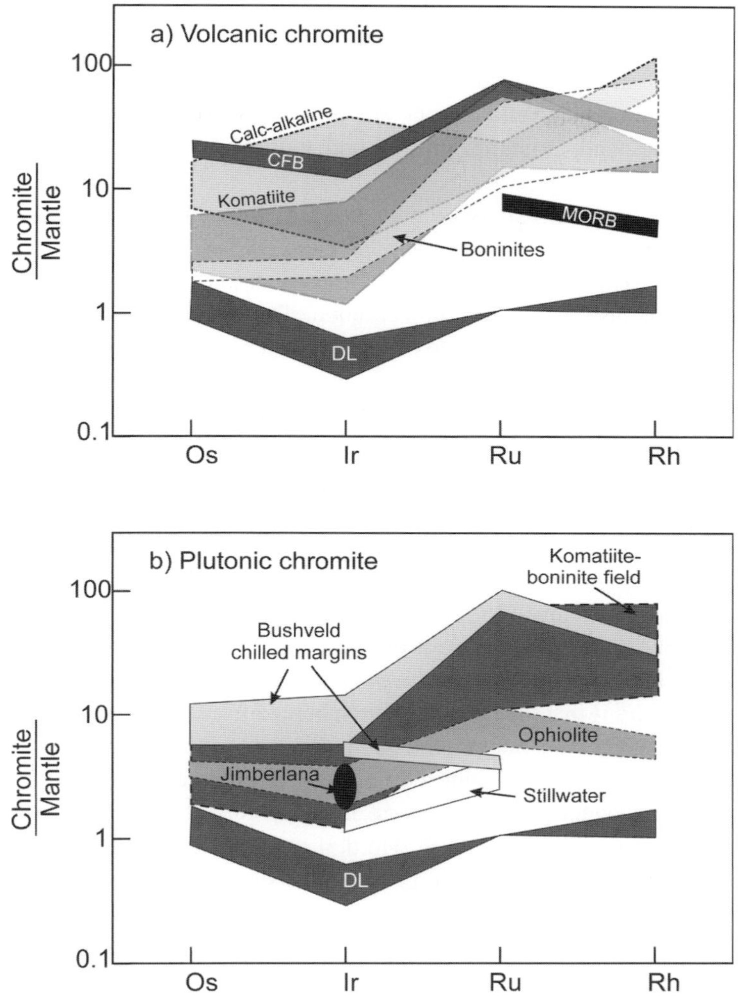

Figure 12.1 Mantle-normalized PGE abundances in chromite from (a) volcanic rocks and (b) layered intrusions and ophiolites. Normalizing values (in ppb) are Os = 3.4; Ir = 3.2; Ru = 5.0; Rh = 0.9. DL = Detection limit. Redrawn after Pagé and Barnes (2016), with permission.

those reported for the UG2 and for Pt–Fe alloy in the pothole regions of the Merensky reef, where they have been interpreted to have formed from PGE-sulfide during late S loss (e.g. Kinloch, 1982).

There have been a number of models to explain the origin of the PGM inclusions in chromite, but their origin is still uncertain. For example, Gijbels et al. (1974) suggested that the PGE-rich inclusions exsolved on cooling from solid solution in the host chromite. This would not appear to be a viable option as they are not seen in chromite from komatiite.

These have among the highest measured IPGE concentrations and cooled from a higher temperature, which would have enhanced diffusion within the chromite. Both of these factors should have strongly favoured the formation of PGE-rich exsolution on cooling in komatiites. In addition, similar inclusions are also seen in olivine (Keays, 1982; Barnes et al., 1985; Barnes and Fiorentini, 2008), and thus the model should work for both minerals.

It also has been suggested that association of IPGM with chromite (and possibly olivine as well) could arise from one of three possible causes. First, chromite could nucleate on pre-existing IPGM (Hiemstra, 1979). Second, the conditions that lead to chromite saturation in the magma also lead to IPGM saturation. Third, the crystallization of chromite itself leads to IPGM saturation at a redox boundary layer next to the growing chromite as demonstrated experimentally (Finnigan et al., 2008). All of these models would suggest that the hot komatiite liquids are far from IPGM saturation and thus the komatiite chromite incorporate the IPGE entirely in solid solution. In contrast, lower temperature magmas are at or near IPGM saturation such that the IPGE go into the co-precipitating alloys rather than into chromite. This can explain both the presence of inclusions in both chromite and olivine and the generally lower PGE concentration of the chromite if the PGE preferentially partition into the inclusion phase. A problem is that, if the magma is saturated in the IPGM and this saturation level does not change, then they should continue to precipitate until the magma becomes saturated in a phase such as sulfide that will deplete the PGE below saturation levels.

Another possibility is that the inclusions are inherited from pre-existing PGM present in the magma prior to chromite saturation. Again, there are a number of possibilities. First, they could be micro-xenocrysts from the mantle (e.g., Luguet et al., 2007; Barnes and Fiorentini, 2008). Second, they could be crystallites that formed during crystallization prior to chromite saturation (e.g., Augé, 1988). Third, they could be residual phases that formed as mantle sulfides lost S as the magma moved to the surface owing to either the negative pressure dependence of the sulfur concentration at sulfide saturation (SCSS) curve or to sulfur loss to degassing during near-surface transport and eruption of the lavas. These models would imply that the lack of inclusions in chromite in komatiite (and perhaps its higher concentration of the IPGE in the komatiite chromite) is due to complete dissolution of these minerals in the hotter komatiite lavas.

In regards to the mantle xenocryst model, Luguet et al. (2007) note that micron- to submicron-sized Ru–Os ± Ir sulfides and Pt–Ir ± Os alloys account for 50–100% PGE budget of refractory depleted harzburgite. These grains are interpreted to have formed as a restite phase during a partial melting event that lead to the complete loss of base-metal sulfides. In contrast, the hypothesis of sulfur loss to near-surface degassing has particular appeal for the Pt–Fe alloy in chromite, owing to the similarity to Pt–Fe alloy formed in the pothole environment where they have been attributed to sulfur loss to late fluids.

It is noted that the experimental work of Finnigan et al. (2008) and Brenan et al. (2012) which suggests that the PGE phases prefer to nucleate in the boundary layer next to growing chromite grains are done in experimental setups that are already saturated in one or more of the PGE, either as Pt loops supporting the experimental charge or as grains of RuS_2 added to the experimental charge. This implies the PGE in these experiments are simply undergoing a re-precipitation to a more favourable environment; it has yet to be shown if chromite grown it a silicate liquid not already saturated in a PGE-bearing phase will also grow PGE-rich minerals at the chromite boundary. A more restricted view of these experimental studies would suggest that xenocrystic PGE phases could be the source of the PGE that reprecipitates at chromite boundaries.

Finally, it has been suggested that the presence of PGE-rich domains in the silicate magma preferentially become incorporated into a growing chromite grains (e.g., Tredoux et al., 1995). This is qualitatively similar to the exotic inclusion interpretation, differing only that the PGE domains are an artefact of the silicate liquid rather than inherited from another phase. It is noted that a 1.0 μm cubic inclusion has on the order of 10^{11} atoms.

12.3 Chromitite in Layered Intrusions

Chromite-rich seams (*chromitites*) within layered intrusions, such as the Great Dyke and the Bushveld and Stillwater Complexes, host the majority of the world's Cr reserves (e.g., von Gruenewaldt et al., 1986; Scoon and Teigler 1994; Oberthür et al., 2002; Zientek et al., 2002; Naldrett et al., 2009). An example of chromitite occurrences in the Stillwater Complex is shown in Figure 12.2. In the Stillwater Complex, the Peridotite Zone of the Ultramafic series is composed of a sequence of cyclic units; a complete cyclic unit averages just over 200 m in thickness and consists of the stratigraphic sequence of poikilitic harzburgite (in which olivine is texturally in a peritectic reaction relationship with orthopyroxene) ± chromite → granular harzburgite (in which olivine and orthopyroxene apparently crystallized together along a cotectic) → orthopyroxenite. This 'ideal' sequence (Figure 12.2a) is not always present and units may be missing from top or bottom. Furthermore, the contact between the granular harzburgite with the underlying poikilitic harzburgite is generally described as irregular with fingers of the poikilitic harzburgite intruding into the overlying granular harzburgite (Figure 12.2b).

Layers composed of both massive and disseminated chromite in dunite and poikilitic harzburgite occurs in the lower portions of many but not all of the cyclic units. The main chromitite seams are traditionally referred to as A (lowermost) through K (uppermost), with the G being the thickest richest in chromite. This labelling can be complicated by the fact that some zones can have multiple thick chromitites, such that the G and H at Mountain View actually have a G, G', H and H' chromitite (Figure 12.2c). Furthermore, these units can be made up of yet smaller interlayered chromitite/dunite-harzburgite units (Figure 12.2d). Modal layering in chromite-bearing rocks ranges from poor to well layered, with the individual chromitite layers occurring on scales from ~metre-thick down to laterally extensive chromitite layers only a grain or two thick. In detail there can be significant textural and modal variations within an individual cycle as well. For example, Page et al. (1972) subdivided the peridotite member of a cyclic unit into nine subunits.

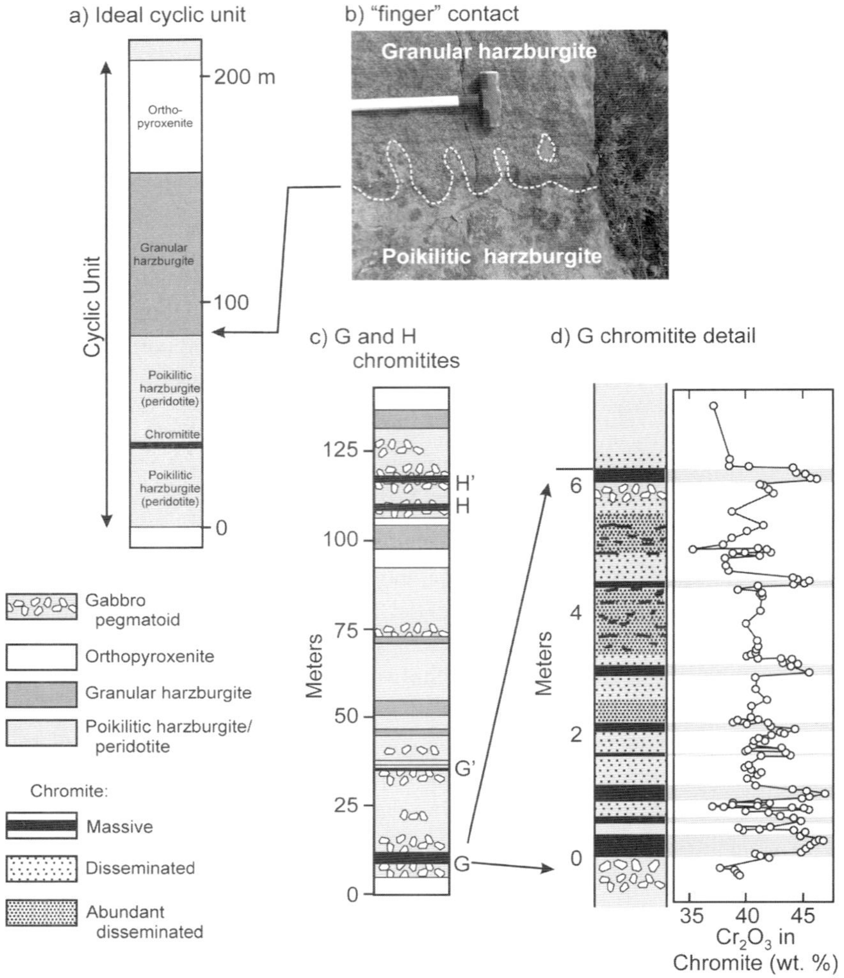

Figure 12.2 Cyclic units and chromitites of the Peridotite Zone of the Stillwater Complex. (a) 'Ideal' Peridotite Zone cyclic unit. Redrawn after Jackson (1961), with permission. (b) Photograph of fingers of poikilitic harzburgite developed in the overlying granular harzburgite. From Boudreau (2015b), with the permission of the Mineralogical Society of Great Britain and Ireland. (c) The G and H chromitite units as exposed at Mountain View. (d) Detail of the G chromitite, note higher Cr_2O_3 in chromite in the massive chromitite layers. Figures (c) and (d) redrawn after Campbell and Murck (1993), with permission.

Chromitites up to a metre or more in thickness occur in many other intrusions as well. These include the UG2 and the spectacular anorthosite-chromitites layers of the Dwars River exposure, the latter discussed in more detail below.

In addition to the massive chromitites noted above, thin chromitite seams can occur throughout the lower parts of the stratigraphy of many layered intrusions. These tend to be

Figure 12.3 Examples of thin chromitite seams. (a) The MG3 chromite seam developed along the irregular contact between the Lower and Upper Critical Zones. Photo by author. (b) Irregularly developed chromite seam in (grey) at the contact between peridotite (green-brown, above) and anorthosite/troctolite (white and speckled, below) in the Rum Intrusion (Scotland). Photo courtesy of Brain O'Driscoll.

on the order of 1 cm thick and commonly occur at the boundary between different lithologic units, where the contacts can be planar to undulating. An example of the MG3 chromitite from the Bushveld Complex is shown in Figure 12.3a, and the chromitite at the base of the Merensky Reef is yet another example. O'Driscoll et al. (2010) document a number of excellent examples from the Rum Layered suite in NW Scotland (Figure 12.3b).

12.4 Chromitite Formation Models

Detailed reviews of orthomagmatic chromitite formation models have been recently summarized by Naldrett et al. (2012) and Maier et al. (2013) and are only briefly reviewed here. If precipitated directly from a magma, these thick chromitite seams require an

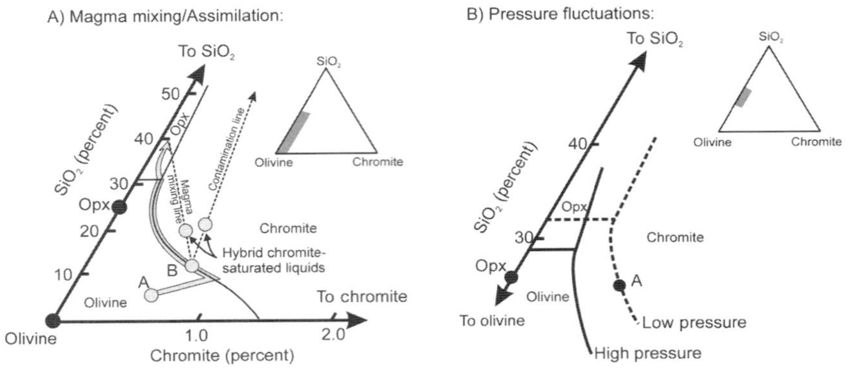

Figure 12.4 Two orthomagmatic models for the formation of chromitite layers (a) Detail of the chromite-olivine-SiO_2 systems showing a possible normal fractionation path (open arrow) for a liquid starting in the olivine phase field at point 'A' (after Irvine, 1977). Also shown are mixing paths that pull an olivine + chromite-saturated liquid (point 'B') into the chromite field. The dotted 'Magma mixing line' is the range of hybrid compositions produced by the mixing of liquid 'B' with more evolved, opx-saturated liquid at the end of the at point 'C'. Alternatively, the 'Contamination line' is the hybrid compositions produced by mixing liquid 'B' with a SiO_2-rich crustal assimilant (e.g., Spandler et al., 2005). (b) Effect of changing pressure on chromite phase field: Liquid on olivine-chromite cotectic at low pressure at point 'A' will become chromite saturated if pressure on the chamber increases. From Boudreau (2016b), with the permission of the Mineralogical Society of Great Britain and Ireland, after the model suggested by Loferski et al. (1990) and Lipin (1993).

extensive period of crystallization of chromite alone, a mineral which otherwise would only crystallize in cotectic proportions of ~1 volume %. A variety of orthomagmatic models have been proposed to cause this shift from cotectic to chromite-only crystallization, two of the orthomagmatic models are summarized in Figure 12.4. The most commonly accepted orthomagmatic model is a magma mixing scheme in which the injection of primitive magma mixes with fractionated magma present in the chamber and drives the hybrid magma into the chromite-only field (dotted mixing line of Figure 12.3a) (e.g., Irvine, 1975, 1977; Campbell and Murck, 1993). A variation on this envisions contamination with a silica-rich composition (perhaps melted roof rock) that moves the liquid into the chromite-only field (dashed mixing line of Figure 12.4a) (e.g., Spandler et al., 2005). Mondal and Mathez (2007) have suggested that the model will not work for the Bushveld chromitite occurring in a Lower Critical Zone pyroxenite and the pyroxenes both above and below the chromitite have the same composition.

Alternatively, it has been suggested that pressure changes in the magma chamber can cause shifts in the chromite cotectic boundary such that an increase in the pressure on the magma moves the cotectic such as to force the magma to crystallize chromite alone (e.g., Loferski et al., 1990; Lipin, 1993) (Figure 12.4b). However, an increase in pressure would also tend to shrink the olivine phase field and would be counter to both the olivine-chromite association and the observed olivine-orthopyroxene peritectic relationship in the

Stillwater Complex. Latypov et al. (2018) suggested just the opposite, proposing that decompression pressure changes as the magma ascended from the mantle can drive a magma to precipitate chromite alone.

More recently, it has been suggested that chromite may not have been precipitated from the magma in the chamber at all. Instead, models have suggest that the Bushveld chromite was carried into the chamber as chromite-bearing magmas that either formed slurries (Mondal and Mathez, 2007; Maier and Barnes, 2008) similar to seafloor density currents or were intruded as sills into the crystal mush (Voordouw et al., 2009; Mukherjee et al., 2017). This moves the origin of the chromite-phyric magmas to a staging chamber or somewhere else in the Bushveld system.

An increase in H_2O and/or other volatiles added to a silicate liquid can also move the chromite cotectic such as to drive the magma to crystallize chromite alone (e.g., Ford et al., 1972; Gaetani et al., 1994; Feig et al., 2010). This is illustrated Figure 12.5a. For a liquid that may have started on the olivine + chromite or orthopyroxene + chromite cotectic, the addition of H_2O expands the chromite phase field such that the liquid will crystallize chromite alone. Boudreau (2016b) suggested that the expansion of both olivine and chromite at the expense of orthopyroxene could explain the change from (dryer) orthopyroxenite and the apparent cotectic crystallization of olivine and orthopyroxene in the granular harzburgite that are generally void of chromitite seams to a poikilitic harzburgite in which olivine is reaction relationship with orthopyroxene that is associated with chromitites of the Ultramafic series of the Stillwater Complex.

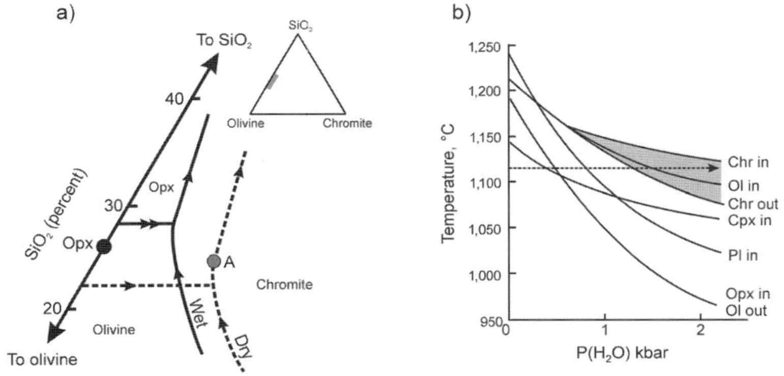

Figure 12.5 Effect of H_2O addition to magma to expand the olivine and chromite phase fields. (a) Schematic of the olivine-chromite-silica system. Liquid on opx-chromite cotectic (point 'A') in a relatively 'dry' magma (dashed lines) become saturated in chromite alone with the shift of phase boundaries after the addition of H_2O (solid lines). Note that expansion of the olivine field changes olivine-opx from a cotectic in the dry system (single arrow cotectic) to a peritectic in the wet system (double arrow cotectic). From Boudreau (2016b), with the permission of the Mineralogical Society of Great Britain and Ireland. b) Experimental phase boundaries showing the stabilization of olivine and chromite with increasing $P(H_2O)$. Region where chromite is stable is shown shaded. Abbreviations: Chr = chromite; Cpx = clinopyroxene; Ol = olivine; Opx = orthopyroxene; Pl = plagioclase. After Mathez and Kinzler (2017), with permission.

Still others have suggested that in influx of hot magma and reaction with the top of the mush column was also important (e.g., O'Driscoll et al., 2010; Mathez and Kinzler, 2017; Pebane and Latypov, 2017; Morse, 2018; Schannor et al., 2018). Both olivine and chromite can be produced by incongruent melting of Cr-bearing pyroxene via a model reaction (e.g., Kushiro, 1975):

$$\underset{\textit{orthopyroxene}}{(\mathrm{Mg,Fe,Cr})_2(\mathrm{Si,Cr})_2\mathrm{O}_6} \rightarrow \underset{\textit{olivine}}{(\mathrm{Mg,Fe})_2\mathrm{SiO}_4} + \underset{\textit{chromite}}{(\mathrm{Mg,Fe})\mathrm{Cr}_2\mathrm{O}4} + \underset{\textit{silicate liquid}}{\text{melt}} \tag{12.1}$$

This melting can be either by direct heating (by injection of a hotter magma as suggested by O'Driscoll et al., 2010) or by hydration melting, the later similar to the production of arc magmas by hydration of mantle source rocks by fluids coming off a subduction lithospheric slab. Hydration melting of a gabbronorite can mush preferentially leaves behind chromite as a residual (restite) mineral (e.g., Brearly and Scarfe, 1986; Armitage, 1992; Webster et al., 1999). This later point is illustrated by Figure 12.5b. As H_2O is added to a partially molten assemblage at 1,020 °C, incongruent melting along the dashed path produces the melting sequence plagioclase + orthopyroxene + clinopyroxene + liquid → orthopyroxene + plagioclase + liquid → olivine + plagioclase + liquid → olivine + liquid → olivine + chromite + liquid → chromite + liquid (Mathez and Kinzler, 2017).

Finally, olivine and chromite can be formed by loss of silica to a silica-undersaturated volatile fluid in a model reaction similar to reaction 12.1 above:

$$\begin{aligned} &\underset{\textit{fluid}}{\text{Si-poor vapour}} + \underset{\textit{orthopyroxene}}{(\mathrm{Mg,Fe,Cr})_2(\mathrm{Si,Cr})_2\mathrm{O}_6} \rightarrow \\ &\quad \underset{\textit{olivine}}{(\mathrm{Mg,Fe})_2\mathrm{SiO}_4} + \underset{\textit{chromite}}{(\mathrm{Mg,Fe})\mathrm{Cr}_2\mathrm{O}_4} + \underset{\textit{fluid}}{\text{Si-rich vapour}} \end{aligned} \tag{12.2}$$

These vapour leaching have been suggest to explain the presence of chromitite in replacement dunite (e.g., McCallum, 1996; see also Chapter 8). The possibility of more extensive replacement are noted further below.

Pagé et al. (2011) note that the Stillwater chromite has high Ti concentrations; this would be unusual for a chromite crystallized from a boninite-like low Ti siliceous high Mg basaltic (SHMB) magma. However, incongruent melting of orthopyroxene would liberate Ti as well as Cr, both of which would then be incorporated into the chromite. Finally, the rock which have both olivine and chromite have chromite far in excess of cotectic proportions, consistent with enrichment by a partial melting model leaving chromite ± olivine as a residual mineral.

There are a number of lines of evidence that fluids have played a role in both the formation of the chromite-rich layers and as both a source and modifier of PGE-sulfide mineralization associated with chromitites. Using the Stillwater chromitites of the Ultramafic series as an example, these include pegmatoids are commonly present below and between the thicker chromitites layers of the Ultramafic Series (Figure 12.2c and d). The pegmatoids are associated with the G chromitite of the Stillwater Complex can contain up to 10% coarse-grained phlogopite and higher than normal abundances of sulfides.

The presence and significance of the unusually Cl-rich nature of the apatite and biotite seen throughout the Ultramafic Zone of the Stillwater Complex as well as the presence of hydrous melt inclusion is chromite and hydrosaline inclusions in olivine (Chapter 7) are consistent with the presence of Cl-rich fluids. In addition, the presence of high temperature carbonate minerals point to a significant CO_2 component to the vapour. The calcite-dolomite geothermometer yields a minimum formation temperature as high as 950°C for the unmixed assemblages (Aird and Boudreau, 2013). Finally, direct evidence for a volatile influence is seen in the polyphase inclusions in chromite containing abundant hydrous minerals interpreted to have crystallized from a trapped silicate liquid described in detail in Chapter 7.

12.5 Chromitites as Metasomatic Features

Models for the thin chromite seams developed between markedly dissimilar rock units as illustrated in Figure 12.3 range from those that envision either the injection of a chromite-saturated magma or a reaction between the existing rocks and by an influx of magma that produces chromite as a result of this reaction (e.g., O'Driscoll et al., 2010; Latypov et al., 2015). However, several features suggest that, in many instances, these thin chromitites may form as Korzhinskii-type metasomatic zones as infiltrating silicate liquids or volatile fluids react and re-equilibrate with a mineral assemblage that is out of equilibrium with the infiltrant (e.g., Korzhinskii, 1959, 1968). Although rarely considered in petrogenetic models, chromite seams can develop within a crystal mush by both orthomagmatic and hydromagmatic mechanisms. For example, O'Driscoll et al. (2010) and Mathez and Kinzler (2017) describe how these seams can be generated reaction zones that can form either in a compositional gradient between hydrous silicate liquid and a crystal mush, or at the melting front between newly injected (hot) magma and the magma chamber floor.

To illustrate how thin chromite seams can be generated within a mush by an orthomagmatic mechanism, consider the irregular chromite seams shown in Figure 12.3 that can developed between orthopyroxenite and a leuconorite-anorthosite. The interstitial liquid in the orthopyroxenite will have its *mg#* buffered by the large volume of orthopyroxene, whereas the *mg#* in the leuconorite will be less so, for reasons discussed in more detail in Chapter 11. Just the opposite can happen for the alkalis, which will be buffered by plagioclase in the leuconorite but less so in the interstitial liquid of the orthopyroxenite prior to plagioclase saturation. Shown in Figure 12.6 are the phase fields at some lower temperature for the orthopyroxenite shown by the dashed lines and that for the leuconorite shown by the solid lines. Thus, even though the interstitial liquid in the orthopyroxenite can become saturated in both orthopyroxene and a late crystallizing plagioclase (shown as point 'A' of Figure 12.6), it is not necessarily in equilibrium with the leuconorite assemblage, for which the bulk assemblage has a lower bulk *mg#* and is in equilibrium with a liquid at point 'B'.

Infiltration of liquid A into the leuconorite assemblage by compaction or advective flow finds itself undersaturated in both plagioclase and orthopyroxene but saturated in chromite

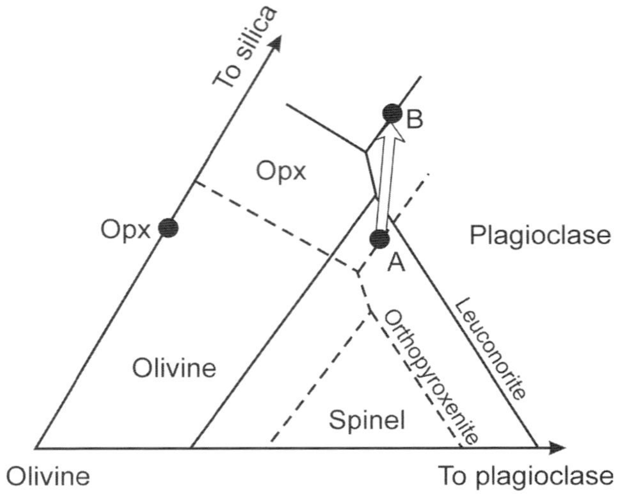

Figure 12.6 A qualitative silicate liquid infiltration metasomatic model for the generation of chromitite seams at contrasting lithologic units. Liquid ‘A’ evolving in a plagioclase-saturated orthopyroxenite (with phase boundaries as shown by the dashed lines) migrates into a leuconorite assemblage that has a lower bulk *mg#* (with phase boundaries as defined by the solid lines) and reacts along the path A–B. See text for discussion.

(spinel), with some of the Cr liberated from the dissolving orthopyroxene. As the liquid continues to react along the path A-B it produce the observed sequence of assemblage: orthopyroxene + plagioclase (= partly crystallized orthopyroxenite at point A), chromite (spinel) alone, plagioclase alone and finally plagioclase + orthopyroxene (=leuconorite at point ‘B’). While some have attributed the undulating nature of the contact to thermal erosion (e.g., Eales et al., 1988), the model presented here would consider the irregular contacts to the consequence of the reaction and its effects on changing porosity leading to flow focusing as discussed in Chapter 4. The lithological sequence is produced as a natural consequence of an evolving and migrating interstitial liquid reacting with contrasting assemblages with different bulk compositions.

Another possibility is that chromite and anorthosite form by leaching of the more soluble component from a norite or gabbronorite protolith. While chromite is a typical refractory igneous mineral and generally considered to be relatively insoluble in aqueous fluids, Watenphul et al. (2014) showed that Cr concentrations in H_2O–HCl solutions can readily exceed 1 wt.% at moderate (400–600 °C) temperatures and upper crustal pressures. Arai and Akizawa (2014) present evidence that chromite with typical igneous compositions can precipitate from C-S-Cl-O fluids in the metasomatic replacement of gabbro by a clinopyroxenite. As noted in Chapter 8, volatile fluids can become out of equilibrium with the host minerals as they move from cooler to hotter rocks. When moving through noritic or gabbroic assemblages the main effect is the preferential loss of SiO_2 followed by loss of MgO and FeO, to produce first troctolite and anorthosite (Meurer et al., 1997). When Cr is

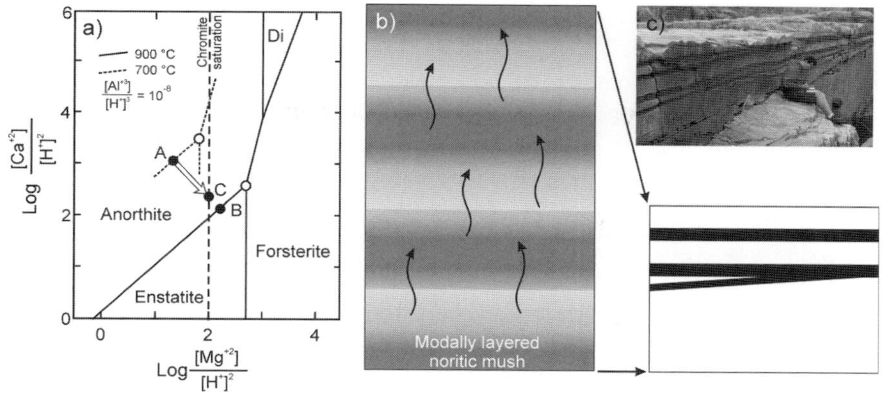

Figure 12.7 Formation of chromite–anorthosite association by fluid leaching of soluble components (e.g., SiO_2). (a) Modified activity-activity diagram from Figure 8.16. Activity of the components are denoted by the square brackets; e.g., $\left[Ca^{+2}\right]^2 = (a_{Ca^{+2}})^2$. Stability fields are calculated based on $[Al^{+3}]/[H^+]$ activity ratio = 10^{-8}, all other phases assumed pure. Phase fields are shown by the solid lines and the chromite saturation boundary by the long dash line, both at 900 °C. The shift of the phase boundaries at the anorthite-enstatite-forsterite triple point at 700 °C are shown by the short dashed lines. (b) Cartoon of the formation of anorthosite-chromite layers as residue resulting from loss of silica and other components from what was initially a modally graded noritic protolith (at left) to produce well-defined chromitite–anorthosite layers (at right). (c) Outcrop of chromitite–anorthosite layering of the UG1 unit exposed at Dwars River, Bushveld Complex. Note layer bifurcations.

significant component of the pyroxene, it can precipitate chromite as a low solubility phase as the pyroxene dissolves.

The stability fields for anorthite, forsterite, enstatite, diopside and chromite are shown on an activity-activity diagram (Figure 12.7a) along with the change in the enstatite-anorthite-forsterite triple point position (open circles) as a function of temperature and a fixed $Al^{+3}/(H^+)^3$ activity ratio in the fluid. Fluids initially in equilibrium with a cooler plagioclase + orthopyroxene (norite) solid assemblage at point 'A' in Figure 12.6a (assumed here to be at 700 °C; phase boundaries shown by the dashed lines), on rising into a hotter but compositionally identical mineral assemblage at 900 °C, will be under-saturated in orthopyroxene (phase boundaries shown by the solid lines). Dissolution of the orthopyroxene along the reaction path A-B leads to chromite saturation at point 'C'. If the reaction is occurring in a modally layered protolith and consideration of surface energy and self-organizing mechanisms that can result in modal segregation over time (similar to the formation of metamorphic banding, e.g., Boudreau and McBirney, 1997; Figure 12.7b). The result is a strongly contrasting chromite-plagioclase segregations (Figure 12.7c). Irregularities in the local amount of material lost can lead to coalescence of layers over time and the formation of bifurcated layers such as are seen in the Dwars River chromitites of the Bushveld Complex (Figure 12.7d).

As noted above, with few exceptions (e.g., Armitage, 1992; Boudreau, 2016; Mathez and Kinzler, 2017) these fluid fluxing and leaching models have not been well-explored in the petrogenetic models for chromitite formation. However, there are several lines of evidence that support them. These plagioclase-chromite associations occur in the Cl-rich portions of the Bushveld Complex, for which associated fluids would also be Cl-rich and could leach significant cations. This would include loss of the alkali elements as well as SiO_2 and result in the residual plagioclase being enriched in the anorthite component by the reaction

$$\underset{\text{-- fluid --}}{H^+ + 2H_2O} + \underset{\text{plagioclase}}{(Ca,Na)(Al,Si)(AlSi_2O_8)} \rightarrow \underset{\text{-- fluid --}}{Na^+ + H_4SiO_4} + \underset{\text{plagioclase}}{CaAl_2Si_2O_8} \tag{12.3}$$

Eales et al. (1990) reports the plagioclase cores from anorthosite associated with the UG1 has the higher An content than does the underlying norite; one might expect the rim values to be even higher. They also note that plagioclase inclusions in orthopyroxene are less anorthositic than plagioclase that is not so armoured. Both of these observations are consistent of loss of the more soluble alkali feldspar components.

Furthermore, if all Cr was originally that incorporated in orthopyroxene, then the amount of orthopyroxenite (containing up to 1.0 wt.% bulk rock Cr_2O_3 as both Cr in orthopyroxene and accessory chromite; e.g. Eales et al., 1994) needed to be destroyed to make chromite containing 50 wt.% Cr_2O_3 (e.g., Teigler and Eales, 1993) is about 50:1 on a weigh for weight basis. Given the higher density of chromite, to make a 10 cm thick chromite seam would require the destruction of about 10 m of orthopyroxenite or roughly twice the thickness of a norite protolith. The total thickness of anorthosite and chromite (~2 m total) is a maximum of about 30 m with some leuconorite beneath this, thus to from chromitite by fluid leaching would require a significant loss of material.

12.6 Hydromagmatic Models for the PGE-Sulfide-Chromitite Association in Layered Intrusions

The chromite-rich layers of the Stillwater (particularly the A and B chromitites; Figure 12.8a) and those of other intrusions, especially the UG2 chromitite of the Bushveld Complex, may contain significant PGE mineralization as well. However, most of these layers contain <100 ppm S, which suggests that collection by a sulfide liquid will not suffice to explain the PGE enrichment unless S was lost later (Barnes et al., 2009). Furthermore, in the case of the Bushveld and Great Dyke, the IPGE and Rh are enriched more than Pt and Pd in most chromite layers. Thus, some other processes must be important in controlling the PGE in these rocks because collection by a sulfide liquid would be expected to enrich the layers in all the PGE.

As for the volcanic chromite studies noted above, most early LA-ICPMS studies of chromite from mantle and plutonic rocks were generally below detection limits, with a few exceptions. More recent work by Pagé and Barnes (2013) using a LA-ICPMS with lower detection limits generally confirms the low IPGE concentrations of Stillwater chromite (Figure 12.1b). In contrast, chromite from the chilled marginal rocks of the Bushveld Complex are broadly similar to those of boninites and komatiites, whereas ophiolites have

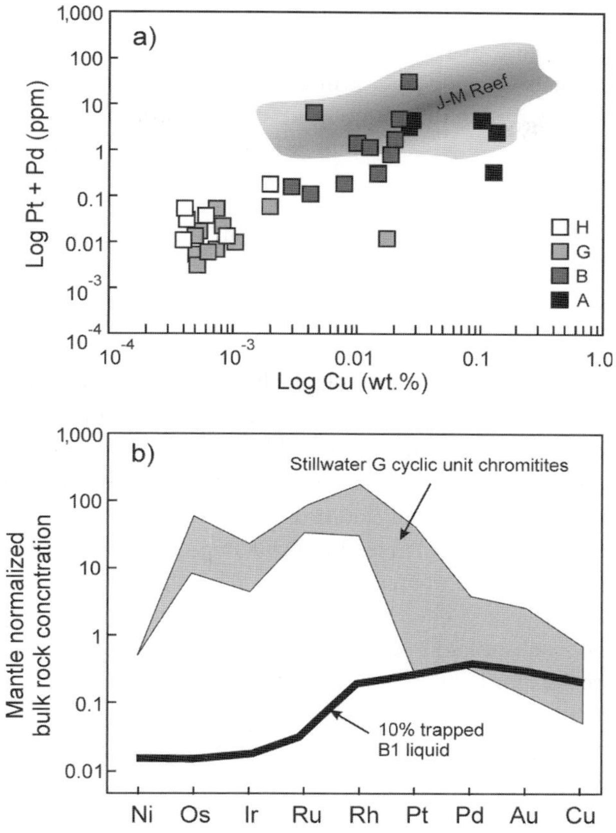

Figure 12.8 PGE concentration in Stillwater chromitites. (a) Bulk rock Pt + Pd vs. Cu concentrations from the A, B, G and H chromitites of the Stillwater Complex, compared with the compositional range of the J-M Reef (shaded, darker region have higher data density). After Zientek et al. (2002), with permission. (b) Mantle-normalized concentration patterns for bulk rock chromitite from the G cyclic unit, Stillwater Complex. The trapped liquid line (solid heavy line) is based on the bulk rock having 10 % trapped liquid with metal concentrations equal to the B-1 parent magma of the Bushveld Complex. Redrawn after Barnes et al. (2015), with permission.

a much flatter mantle-normalized pattern. However, as of yet only a few *in situ* analyses of chromite are available.

Several models have been proposed to account for the IPGE enrichment in the chromite layers. These range from suggestions that the IPGE were at their saturation levels and that laurite co-crystallized with chromite from a nominally sulfide-undersaturated magma; i.e., it is the IPGE that drives saturation in a mineral like laurite and not the saturation of a base-metal rich immiscible sulfide liquid. The presence of laurite would enrich the rocks in Os, Ir and Ru. To explain the lower S but high IPGE of Stillwater chromitites, Barnes et al. (2016) proposed that the IPGE and some of the Rh partitioned into the chromite prior to saturation of the magma in a sulfide liquid. Subsequently, small amounts of base-metal sulfide liquid were added to the chromite layers. During cooling, chromite grains underwent sintering and incorporated some sulfides. As cooling continued Fe and Ni in the sulfide diffused into the chromite and the IPGE and Rh in the chromite diffused into the sulfide, transforming the base-metal sulfide into laurite with remnant sulfides and a few PGM grains.

There are two hydromagmatic models to explain the excess of PPGE in chromitites from layered intrusions that cannot be explained by that fraction of the PGE partitioned or included as minor inclusions in chromite, which are terms here the *strong* and *weak association* models. These are detailed below.

Strong association. In these models, the addition of H_2O-bearing fluid both introduces the PGE-S component and leads to the formation of the chromite as described above. If degassing from the footwall were to stop at this point, one would have an assemblage that was enriched in sulfide and the more soluble PPGE. However, if degassing continues, the chromitite itself eventually becomes vapour saturated and sulfur and the more soluble PPGE can be lost, leaving a residuum enriched in IPGE and Rh (Boudreau and McCallum, 1989). The result is a mantle- normalized PGE abundance pattern either negative-sloped or an inverted 'U' shaped as seen in Figure 12.7b.

The strong hydromagmatic models in which the chromitites and the PGE-S mineralization are both the result of hydrothermal activity have not been extensively called upon outside of the smaller chromite occurrences that occur, for example, at the J-M and Merensky Reefs and in Rum (e.g., Boudreau, 1988, 1999; Mathez and Kinzler, 2017), although Boudreau (2016b) did suggest that it was a viable model for the Stillwater chromitites of the Ultramafic series. More recently, Schannor suggested a strong hydromagmatic model for the UG2 with melting driven by fluids derived by dehydration in the metamorphic aureole beneath the complex.

Weak association. This model envisions that the chromitite is an orthomagmatic accumulation of chromite that represents a marked stratigraphic discontinuity in the crystal pile. Furthermore, the simplest type of discontinuity would be that the chromitite interstitial liquid is far from vapour saturation. This would be the case if the chromite originally precipitated during mixing of relatively dry primitive liquid with the resident magma. Vapour migrating up from below simply redissolves in this drier hybrid interstitial liquid and precipitates PGE sulfides. Partial melting of the original protolith produces the characteristic hydrous melt inclusions. Again, if degassing continues the more soluble PGE can be lost.

13

Isotopic Evidence

13.1 Introduction

The use of both stable and radiogenic isotopes have traditionally been used in layered intrusion studies to infer heterogeneity of mantle source regions, extent of contamination of the parent magmas, and details of magma influx and magma mixing. Early isotopic studies used both bulk rock and mineral separates as proxies for the isotopic composition of their parent magma (e.g., Kruger and Marsh, 1982). Where available, one can more directly sample parent liquids using chilled margin samples or coeval sills and dikes. In this regard, both the Bushveld Complex and the Skaergaard Intrusion are perhaps the most well-studied intrusions based on the stratigraphic coverage and sample density for a number of isotopic systems and these two intrusions are discussed in some detail below. A number of other intrusions show evidence of increased contamination accompanying fractional crystallization. For example, Sorensen and Wilson (1995) and Wilson et al. (1996) observed that Sr-isotopic ratios in the Bjerkreim–Sokndal Intrusion were correlated with the An# of plagioclase and the *mg#* of mafic minerals, and Morse (1983) noted a similar relation in the Kiglapait Intrusion. Palacz and Tait (1985) found an abrupt increase in $Sr_{initial}$ ratios and decrease in ε_{Nd} where plagioclase becomes abundant in the Rum Intrusion. All of these suggest progressive contamination by a crustal component as the magma system evolves.

More recent work, however, has shown extensive evidence for *isotopic disequilibrium* in layered intrusions. This refers to differences in the initial isotopic ratios both within a single crystal (typically core-rim isotopic differences), differences between individual crystals of the same phase (e.g., plagioclase–plagioclase), or more commonly different mineral phases (e.g., plagioclase–pyroxene). Isotopic disequilibrium is increasingly seen to be common in volcanic systems (e.g., Davidson and Tepley, 1997). In volcanic rocks, the most common explanation is contamination of crystal-phyric magmas or inclusions of foreign xenocrysts. However, it is also observed in the mantle where it is attributed to infiltration and incomplete reaction of melts or fluids with a protolith of different isotopic composition. The existence of isotopic disequilibrium in layered intrusions calls into question the use of bulk rock or even mineral separates to infer the isotopic character of the magma from which they crystallized. However, as for the mantle example, it can potentially track migration of isotopically distinct silicate liquids and volatile fluids within the solidifying crystal mush.

13.2 Examples of Isotopic Variability

13.2.1 The Skaergaard Intrusion

The Skaergaard Intrusion is generally considered to have formed by closed system crystallization of a single, homogenous magma. This is based, in part, on the regularity of mineral compositional trends. It was intruded into Archean gneiss overlain by basalts, the latter coeval with the intrusion itself. While the basalts would be isotopically similar to the Skaergaard gabbros, the Skaergaard magma would be sensitive to even small amounts of contamination from the gneiss. Indeed, Stewart and DePaolo (1990) estimated that the assimilation of 2–4 % gneiss could account for the observed variations in Sr and Nd isotopic ratios. McBirney and Creaser (2003) did a more comprehensive study of the $^{87}Sr/^{86}Sr$ and $^{143}Nd/^{144}Nd$ initial isotopic ratios of the Layered Series and found that both isotopes show significant variation on all scales, from intrusion-wide variations down to coexisting minerals in individual rock samples. The intrusion-wide lateral variations are similar to the vertical (stratigraphic) variations (Figure 13.1). Some of the variation can be attributed to contamination along the contacts or from localized contamination in the Layered Series by rare gneissic xenoliths. Otherwise, there did not appear to be evidence of extensive assimilation of the basement rocks.

One of the more interesting findings of McBirney and Creaser (2003) is the isotopic disequilibrium between minerals from apparent replacement bodies in which pods of plagioclase-rich rocks and pyroxene-rich rocks discordantly cut the Layered Series rocks. It was found that the plagioclase and clinopyroxene have distinctive isotopic ratios for both Sr and Nd. Specifically, it was found that the dominant mineral from either the anorthositic

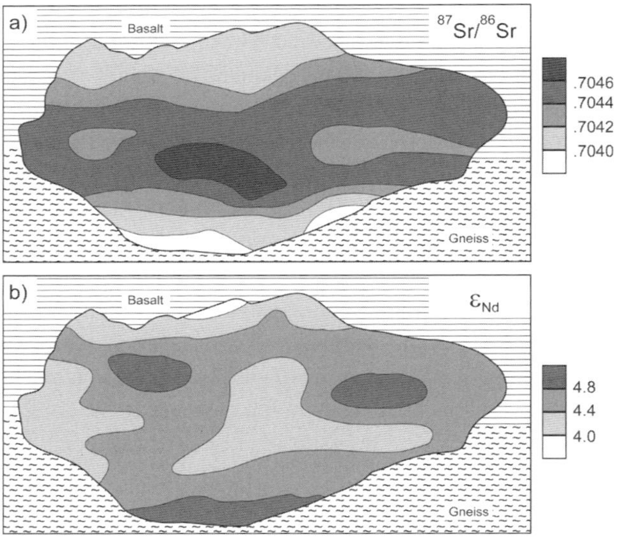

Figure 13.1 Isotopic variations in the Skaergaard Intrusions. Variations of bulk rock (a) initial $^{87}Sr/^{86}Sr$ and (b) ε_{Nd} isotopic ratios. Redrawn after McBirney and Creaser (2003), with permission.

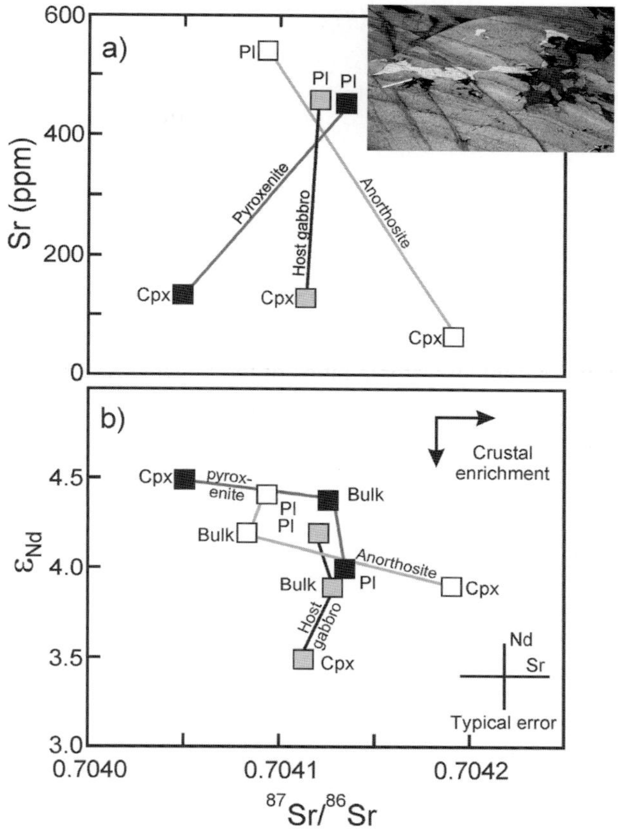

Figure 13.2 Isotopic disequilibrium observed in an anorthosite-pyroxenite metasomatic replacement pods developed in host gabbro in the Skaergaard Intrusion. (a) Sr concentrations vs. $^{87}Sr/^{86}Sr$ for minerals and bulk rock of an anorthositic pod, pyroxenite pod and host gabbro (shown in the inset photograph). Note that the dominant mineral in the anorthosite and the pyroxenite has a lower $^{87}Sr/^{86}Sr$ ratio than that of the host. (b) $^{87}Sr/^{86}Sr$ vs. ε_{Nd}. The dominant mineral in the anorthosite and pyroxenite pods have higher ε_{Nd} values than the rest of the rock. Redrawn after McBirney and Creaser (2003).

pods or the pyroxenite pods was the most 'mantle-like', having a lower initial $^{87}Sr/^{86}Sr$ ratio and a higher ε_{Nd} value (Figure 13.2). They also noted that the same relationships are also seen in modally layered rocks, although the isotopic differences are not as marked. The small-scale inhomogeneity observed in both the modally layered rocks and in the discordant mafic-felsic segregation pods do not seem to be related to low temperature hydrothermal alteration as the variations of Sr and Nd isotopes do not correlate with those of oxygen or hydrogen (Taylor and Forester, 1979; Fehlhaber and Bird, 1991). They concluded that the isotopic disequilibrium between minerals was introduced after the rocks were largely crystalline by whatever metasomatic process caused the mafic-felsic segregations.

13.2.2 Bushveld Complex

Early studies on the Bushveld Complex generally considered it to have crystallized from a magma undergoing a combination of fractional crystallization and replenishment of a single parent magma type. This was inferred from the general modal and mineral compositional changes that show an overall decreasing *mg#* of the mafic phases, decreasing An concentration of plagioclase, and changes in the mineral assemblages that are broadly consistent with a single liquid line of descent. This began to change with the introduction of isotopic studies. This work initially focused on whole rock Sr-isotropic variations through the Complex (e.g., Kruger and Marsh, 1982; Sharpe 1985; Eales et al., 1990a; Lee and Butcher, 1990; Kruger 1994) and later expanded into other isotopic systems (Figure 13.3a). In most instances, the isotopic data suggested that the rocks apparently crystallized from magmas that had progressively larger component of a crustal contaminant in regards to Sr, Pb, Os and Nd (e.g., Hamilton, 1977; Harmer and Sharpe, 1985; Kruger 1994; Harmer et al., 1995; Schoenberg et al., 1998; McCandless et al., 1999; Maier et al., 2000) (Figure 13.3b). Explanations for this component ranged from a contaminated

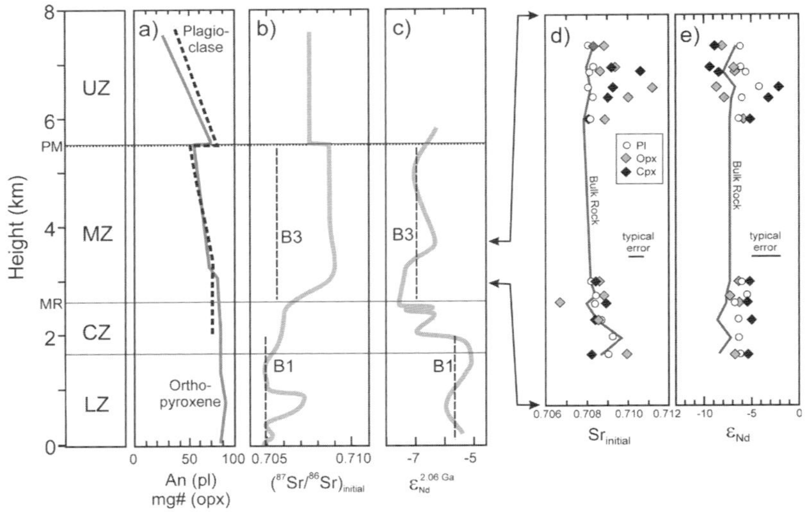

Figure 13.3 Main stratigraphic units of the Bushveld Complex showing stratigraphic variations of (a) Orthopyroxene and plagioclase compositions, (b) General trends of initial Sr isotopic variations, (c) General trends of ε_{Nd} isotopic variations. The isotopic plots in (b) and (c) are generally defined from bulk rock isotopic compositions. The labelled vertical dashed lines in the plots of initial $^{87}Sr/^{86}Sr$ and initial ε_{Nd} isotopic trends in plots (b) and (c) are the isotopic compositions of the B1 and B3 sill/chilled margin rocks thought to be parental to the Lower/Lower Critical zones and Main zone, respectively. The two partial sections at left show isotopic disequilibrium between different mineral types for (d) $Sr_{initial}$ and (f) ε_{Nd} isotopic variation from the Main zone in the Northern Lobe. Abbreviations: LZ, CZ, MZ and UZ = Lower, Critical, Main and Upper zones of the Bushveld Complex, respectively; MR = Merensky Reef; PM = Pyroxenite Marker. Redrawn after Maier et al. (2000) and Roelofse and Ashwal (2012), with permission.

lithospheric mantle to contamination of mantle magmas by both lower and upper crustal rocks. In contrast to these studies showing considerable stratigraphic variation, Hf isotope compositions of zircon reported by Zirakparvar et al. (2014) show little variation throughout the complex, ($\varepsilon_{Hf,\ 2.06\ Ga} = -9.0$ to -6.8). They suggested that these values were inconsistent with extensive amounts of crustal assimilation and instead suggested all the Bushveld magmas were generated by partial melting of a depleted subcontinental lithospheric mantle with an unradiogenic Hf isotopic composition.

Because it involves one of the PGE, Re–Os systematics have been used to infer sources for PGE mineralization in layered intrusions. Schoenberg et al. (1999) found the Bastard Unit initial $^{187}Os/^{188}Os$ to have a uniform value of 0.151, whereas those of the underlying Merensky Reef to be more radiogenic and variable (with values between 0.168 and 0.181). They also noted a change from near chondritic values (~0.120) for the Lower Group chromitites and associated interstitial minerals, which includes sulfides that become more radiogenic in the Middle Group (0.137) and Upper Group (0.150) chromitites. This increase in radiogenic Os is accompanied by an increase in the concentration of Os, attributed to a larger sulfide fraction with height. Similar increase in radiogenic Os were also reported by Hart and Kinloch (1989) and McCandless and Ruiz (1991). Reisberg et al. (2011) found the Os of the Platreef to be similar to slightly more radiogenic that the Merensky Reef, and suggested that the two are genetically related.

The isotopic evidence has challenged classical interpretations of the crystallization sequences. For example Sharpe and Irvine (1983) suggested the appearance of plagioclase as a primocryst at the base of the Upper Critical Zone was the result of the blending of magmas within the plagioclase stability field with plagioclase-undersaturated magmas rather than representing the fractionation of a single magma from orthopyroxene alone to plagioclase + orthopyroxene cotectic crystallization as conventionally suggested. More recent work on isotopic disequilibrium (see below) has even suggested that the different minerals themselves might come from isotopically distinct sources and are not the result of cotectic crystallization.

13.2.2.1 Bushveld Stable Isotopes

Although useful of tracking possible crustal contamination (e.g., Harris et al., 2005) and the role of igneous fluids (e.g., Mathez et al., 1994), the stable isotopes of element such as oxygen, hydrogen and chlorine are typically used to determine the possible influx of metamorphic fluid derived from dehydrated country rock and meteoric fluids. As noted in Chapter 3, crustal rocks can lose 3–6% of their mass to dehydration during metamorphism (Connolly, 2010). A number of stable isotope studies have suggested that very little of this country fluid made it into the Bushveld. For example, it has been suggested that $\delta^{18}O$ in the Platreef, which occurs near the floor of the Northern Lobe of the Bushveld Complex, was mostly magmatic, with perhaps some interaction with calcsilicate footwall rocks resulting in lower $\delta^{18}O$ (Harris and Chaumba, 2001). Measured lithium isotope variations at the base of the Bushveld Complex have been used to suggest that Li infiltrated over longer distances (60 m) as compared with oxygen (~4 m), which was attribute to a higher diffusivity of Li compared to most other major elements (Ireland and Penniston-Dorland, 2015). Chlorine isotopes have also been used to suggest that the Bushveld Cl is isotopically distinct from the country rocks (Willmore et al., 2002; see Chapter 6).

In contrast are features consistent with a deeper penetration of country fluids into the Bushveld Complex. For example, Schiffries and Skinner (1987) and Schiffries and Rye (1990) noted the presence of high temperature (maximum temperature about 700 °C) saline hydrothermal veining that cut both the intrusion and the underlying contact aureole. Schiffries and Rye (1990) suggested that $\delta^{18}O$ was controlled by isotopic exchange with the Bushveld intrusive rocks, whereas they interpreted the $\delta^{13}C$ to have been inherited from sedimentary carbon of the Transvaal Supergroup and δD by reaction with Transvaal hydrous minerals, results similar to those of Buick et al. (2000). This implies that country fluids will undergo isotopic exchange rather readily for elements that are abundant in the igneous suite (e.g., ^{18}O), but that elements that are in only trace abundance (e.g., Cl, C and H) are less likely to be affected.

Mathez et al. (1994) looked at hydrogen isotope compositions of rocks from the Merensky Reef and related rocks at the Atok mine. They found a small but systematic difference in composition between pegmatoid and pyroxenite of the Merensky reef (δD averaging 65‰) and of the generally more leucocratic rocks elsewhere in the section (δD averaging 79‰) (Figure 13.4). They could not explain the results by a simple mixing of light magmatic water with heavier meteoric-crustal water. Instead, they suggested that the hydrogen isotopic signatures were established during late-stage addition of hydrous vapour to the Merensky 'protoreef' horizon that was degassed from the interstitial liquid below the reef section, and may have included a small portion of water from the country rocks.

Schannor et al. (2018) looked at both O and Sr isotope variations through the UG2 chromitite and found that the initial $^{87}Sr/^{86}Sr$ ratios (Sri) of interstitial plagioclase in the footwall pegmatoidal rocks are significantly higher than in adjacent rocks of the Critical Zone. They suggested that the $\delta^{18}O$ values of plagioclase–pyroxene and plagioclase–chromite mineral pairs are consistent with oxygen isotope exchange from crystallization temperatures down to 700 °C and 800 °C, respectively. They interpreted these result to have been caused by the infiltration of fluid derived (in part) from dehydrating country rocks beneath the Bushveld Complex. The fluid could have been introduced either late or during chromitite crystallization. For the latter, it was suggested that the UG2 chromitite itself could have been the result of incongruent hydration melting.

13.2.2.2 Isotopic Disequilibrium in the Bushveld Complex

More recent work on the Bushveld Complex has shown the presence of significant isotopic heterogeneity, both between coexisting minerals and within single grains. For example, Chutas et al. (2011) did both progressive leaching of plagioclase for Sr isotopes and microdrilling for Sr and Pb isotopes from grains of plagioclase and orthopyroxene from the Critical Zone and the Lower Zone indicates that these minerals are not in isotopic equilibrium (Figure 13.5). Leaching suggests Critical Zone plagioclase either lost Rb or

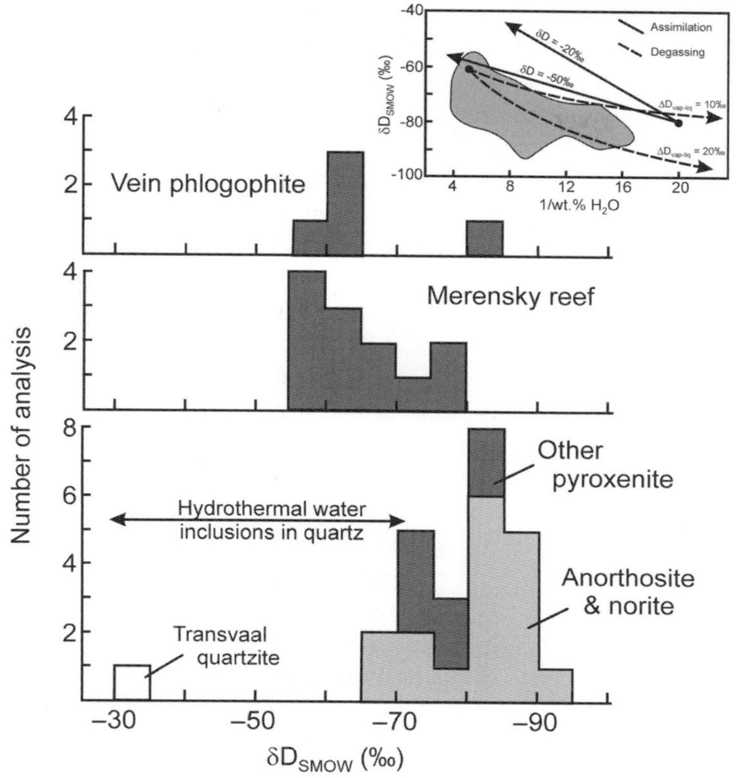

Figure 13.4 Histograms of the bulk δD from the Merensky and Bastard units at Atok mine. Also shown are phlogopite separates from magmatic veins and data from Schiffries and Rye (1990) on waters from the Bushveld hydrothermal system and a Transvaal quartzite. Inset: Whole rock water contents and δD values of rocks from the Merensky and Bastard units (grey field). The solid lines refer to mixing of water of the composition indicated with rock containing 0.05 wt.% water and a δD = -80 ‰. The dashed lines show how the composition of interstitial melt changes during fractionation degassing of vapour for two different vapour–liquid distribution coefficients as indicated. Redrawn after Mathez et al., (1994).

had a more radiogenic $Sr_{initial}$ rim relative to the core, whereas plagioclase from an Upper Zone sample is isotopically homogeneous for $Sr_{initial}$. It also was noted that the isotopic homogeneity of the Upper Zone plagioclase is reflected in the homogeneity of the Upper Zone in $Sr_{initial}$ as a whole, whereas both the Critical Zone plagioclase and the Critical Zone itself are both variable in $Sr_{initial}$. Microdrilling analyses of plagioclase from the Lower and Critical Zones consistently have a higher $Sr_{initial}$ and a less radiogenic modelled $^{238}U/^{204}Pb$ composition than coexisting orthopyroxene. Roelofse and Ashwal (2012) also did micro-drilling and showed that the lower Main Zone exhibits significant Sr-isotopic disequilibrium between coexisting plagioclase, orthopyroxene and clinopyroxene with the latter two mineral typically having a stronger crustal $Sr_{initial}$ signal (Figure 13.3d and 13.3e).

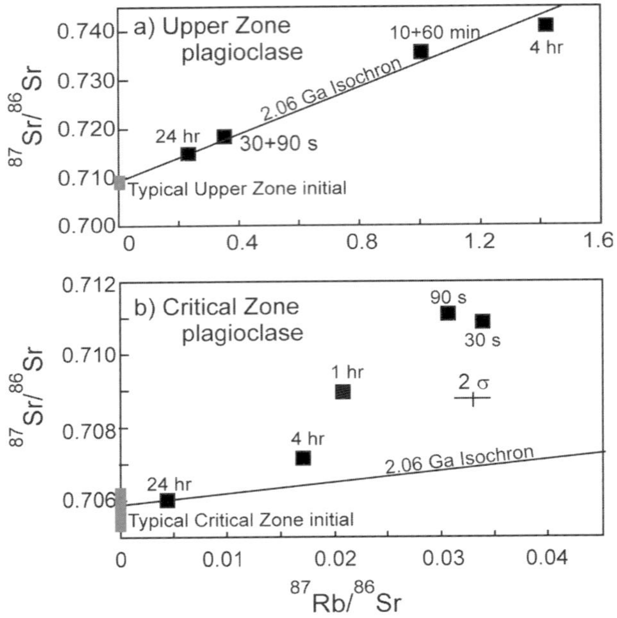

Figure 13.5 Results of progressive leaching experiments on (a) Upper Zone and (b) Critical Zone plagioclase separates. Times denote total leach time. Those with a "+" are samples for which the leachate were combined prior to the analysis owing to low recovery. The grey boxes along the *y*-axis show the typical range of initial Sr values for the two zones. A line with a slope proportional to a 2.06 Ga Bushveld-age isochron is shown for reference. From Chutas et al. (2011), with permission.

A number of studies have focused on short stratigraphic sections that include the ore zone of the Merensky Reef. Mathez and Waight (2003) noted Pb isotopic disequilibrium between plagioclase and sulfide, the later having a country rock signature. Prevec et al. (2005) found the Nd isotopic composition from the Merensky Reef and its immediate footwall suggest that orthopyroxene and plagioclase are not in isotopic equilibrium, with plagioclase having a more radiogenic Nd ($\varepsilon_{Nd} = -1.13$ to -3.37 at 2.06 Ga) relative to pyroxene and whole-rock data ($\varepsilon_{Nd} = -7.46$ and -8.46). Similarly, Seabrook et al. (2005) noted composition and Sr-isotopic variations between the Merensky and Bastard Reefs in mineral separates, noting in particular that the Bastard Reef plagioclase had $Sr_{initial}$ similar to that of the Critical Zone whereas the orthopyroxene was more radiogenic, similar to that of the overlying Main Zone. In contrast, the $Sr_{initial}$ of the Merensky Reef plagioclase and orthopyroxene were more varied. Yang et al. (2013) used *in situ* Sr isotope analyses of plagioclase by laser ablation ICP-MS. They found that both plagioclase core-rim and plagioclase core–core disequilibrium $Sr_{initial}$ isotopic compositions on a ~270 m section from the Upper Critical Zone to the lower part of the Main Zone in the Western Bushveld Complex. An example from Yang et al. is shown in Figure 13.6.

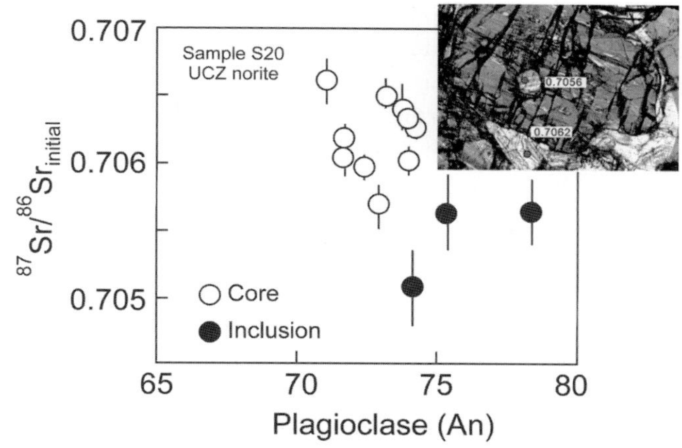

Figure 13.6 Example of isotopic disequilibrium in a norite from the Upper Critical Zone of the Bushveld Complex. Plagioclase inclusions in orthopyroxene have a lower initial Sr-isotropic ratio and have a similar to slightly more anorthitic composition than those outside of the pyroxene. Redrawn after Yang et al. (2013), with permission.

13.2.3 Isotopic Disequilibrium in Other Intrusions

Evidence of isotopic disequilibrium is found in a number of other intrusions as well. For example, Weis and Morse (1993, 1995) analysed mineral separates from the Kiglapait Intrusion and found a marked disequilibrium between the Pb isotope compositions of mafic minerals and plagioclase, and Tepley and Davidson (2003) found both inter-and intracrystalline Sr-isotopic variations between plagioclase and clinopyroxene in the Rum Intrusion. These and other examples suggest that isotopic disequilibrium is a common feature of layered intrusions.

13.3 Explanations for the Isotopic Disequilibrium in Layered Intrusions

While there can be counterexamples, the main observations from the examples of isotopic disequilibrium in the Bushveld Complex and other intrusions can be summarized as follows. For rocks that appear to be normal 'cotectic' precipitates, pyroxene is more likely to show isotopic enrichment in a crustal component than is coexisting plagioclase. However, plagioclase rims are more likely to be enriched in a crustal isotopic component as compared to cores. Bulk rock isotopic compositions generally reflect the average of the modal abundance of these isotopically distinct mineral populations. Because of potential core-rim variations, mineral separates can be as unreliable as are bulk rock isotopic compositions as proxies for parent magma isotopic compositions.

A number of mechanisms have been proposed to explain isotopic disequilibrium and can be summarized as follows (Roelofse et al., 2015): (1) Intrusion of variably contaminated crystal mushes from deeper staging chambers (Roelofse and Ashwal, 2012; Roelofse et al., 2015). (2) Blending of semi-consolidated crystal mushes as a result of slurry flow triggered by subsidence during cooling (Yang et al., 2013). (3) Density-driven mixing of minerals from isotopically distinct magma pulses (Prevec et al., 2005). (4) Contamination of crystals at the roof of the intrusion and mechanical incorporation of such contaminated crystals into the lower crystallization front as a result of gravitational instability at the upper crystallization front (Tepley and Davidson, 2003). (5) Incomplete reaction during infiltration of isotopically distinct fluids or silicate liquids into a partially solidified crystal mush (Mathez and Waight, 2003; Chutas et al., 2011). This may also result from local contamination by xenoliths (McBirney and Creaser, 2003). (6) Late-stage recrystallization-mass transfer metasomatic processes (McBirney and Creaser, 2003). The firsts four mechanisms can be considered together and require the isotopic disequilibrium to have been present in minerals before they were deposited in their current location. In contrast, the last two mechanisms imply that the isotopic disequilibrium was imparted sometime after the crystal mush had formed. Also, there is no reason why it could not be a combination of the above factors.

All of these explanations are based on the assumption that the observed isotopic disequilibrium is the same age as the intrusion itself and not due either to some much younger cryptic alteration event or that proposed xenocrysts are much older than the intrusion. With a few exceptions (e.g., Prevec et al., 2005), there have not been multiphase isotopic measurements to confirm that the isotopic endmember components fall on an intrusion-age isochron. For example, a late crustal contaminant infiltrating along grain boundaries and pockets of interstitial liquid might be expected to affect all late crystallizing minerals such as apatite and biotite even if the cores of the much larger primocryst were largely shielded from this isotopic alteration event. In this case, the interstitial mineral assemblage would lie on a parallel isochron but otherwise have the same age as the primocryst mineral cores. In contrast, if the mineral cores record older xenocrysts, the isochron core ages could be older than that of the intrusion itself.

The suggestion that the minerals represent different crystal populations inherited from blended crystal-phyric magmas does have some support from the core-rim disequilibrium in phenocrysts in volcanic rocks as noted above. This assumes, however, that the isotopic characteristics of the final rock were present in the originally accumulated crystals and did not undergo any significant modification. That is, the preservation of the disequilibrium had to be maintained despite strong evidence of recrystallization driven by processes such as crystal ageing and dissolution-reprecipitation driven by compaction (e.g., Boorman et al., 2004) that would tend to homogenize isotopic differences. Even ignoring this post-accumulation modification, one would expect to see a number of lines of supporting evidence that the assemblage formed from mixed crystal populations.

For example, Chutas et al. (2012), using samples and data from Boorman et al. (2004), noted that the Bushveld crystal size distributions (CSD) suggest a single crystal population. That is, there is no evidence that magmas with many small crystals mixed with magmas with a population of larger crystal sizes. (The exception was for poikilitic rocks in which a smaller plagioclase size fraction is preserved as chadacryts in a larger pyroxene oikocrysts, similar to the situation shown in Figure 13.6. In both cases, this is interpreted to arrested growth of the entrained plagioclase). Chutas et al. thus discounted the possibility that the isotopic differences were inherited from different crystal populations. Furthermore, prismatic minerals such as plagioclase might be expected to become aligned during flow and form strong mineral lineations as well as a lamination in crystal-rich magmas. Those models that suggest that isotopically distinct crystals were deposited from mixed crystal-rich slurries do not report any evidence of this expected flow alignment.

The conventional view is that that the crystallization of interstitial liquid can impart a normally zoned rim to otherwise wide, unzoned to weakly zoned cores. However, plagioclase can show some remarkable complex zoning patterns that are preserved owing to the relatively slow solid-state diffusion. The question is to what degree is this occurring while a plagioclase grain is crystallizing in an open and intermittently contaminated magma (e.g., Font et al., 2008) or during processes occurring within an solidifying mush column (e.g., Hoshide and Obata, 2010). If the former, the blended magma would be intermediate in its isotopic composition and one might expect to see rims enriched in a radiogenic component on minerals from a non-radiogenic magma but the reverse on minerals from the other magma. Instead, the observation that the strongest crustal isotopic signatures are seen on the rims and interstitial minerals would suggest that the contaminant was introduced after the mush had formed.

While there are modal variations present in Bushveld Main Zone, the overall average modes suggest cotectic mineral proportions (Chapter 2, Figure 2.4). Modal variations within a single magma precipitating minerals in cotectic proportions can occur by crystal setting to the process of constitutional zone refining discussed in Chapter 8, but overall the average modal abundance would be as expected from a single, multiply saturated magma. This appears to be true even for the strongly modal variations seen in the Upper Critical Zone (Cawthorn, 2002). This would require that the isotopically distinct pyroxene and plagioclase of the Main Zone would have to mix in rather fortuitous proportions to mimic a cotectic magma. Furthermore, the mineral compositions of the plagioclase and pyroxene are as expected from a typical cotectic magma. That is, one does not observe pyroxene with a high *mg#* similar to that of the Lower Zone coexisting with a low An content plagioclase, for example. This implies that the two magmas, if they existed, were both mono-saturated in a different mineral but that the mineral compositions are fortuitously as expected for a single, multiply saturated magma. Finally, if the isotopically distinct crystals separated from a blended magma, then where are the isotopically homogeneous minerals that would have crystallized from this hybrid magma?

13.4 Evidence for the Late Introduction of a Crustal Isotopic Component

In contrast to those models which suggest that the isotopic disequilibrium was inherited from minerals prior to deposition to form the layered rocks, others have suggested that it was imparted at a later stage as the crystal-liquid mush underwent a prolonged period of crystallization and cooling (e.g., Mathez and Waight, 2003; Chutas et al., 2012; Schannor et al., 2018). In this interpretation, a late interstitial volatile fluid or silicate liquid with a crustal isotopic signature infiltrates and modifies the host minerals. Plagioclase, a stable mineral with slow solid-state diffusion rates, is the least affected except at its margins except in the case of a wholesale recrystallization or remelting event. In contrast, pyroxenes have higher diffusion rates and, perhaps more importantly, undergo a long period of exsolution on cooling that can only enhance isotopic exchange. The can result in a disequilibrium between plagioclase cores and pyroxenes, the latter recording a larger crustal isotopic component.

Although rarely analysed, this would also suggest that the late crystallizing interstitial minerals such as apatite and biotite would also be enriched in a crustal component as compared with plagioclase cores. For example, Schoenberg et al. (1999) note that $Sr_{initial}$ is much more radiogenic in the interstitial minerals of the Upper Critical Zone chromitites (with some >0.720) as compared with typical rocks of the Upper Critical Zone which range from 0.705 to 0.706. They suggested that this was the result of partial melting of roof rock and mixing with the resident magma to both cause an increase in in the $Sr_{initial}$ ratio as well as causing the hybrid magma to precipitate chromite alone. However, this assumes a static mush system and ignores the likelihood that compaction or interstitial liquid advection would move the interstitial liquid away from the chromitites (e.g., Irvine, 1980), even if the chromite initially accumulated as they suggested. It otherwise equally plausible that the interstitial minerals are largely recording the Sr_{intial} of the infiltrant, given that chromite does not contain significant Sr.

13.5 Evidence from Mantle Examples of Isotopic Disequilibrium

The mantle does not have any large open magma chambers or open channelways to carry crystals, thus examples of isotopic disequilibrium must be due to porous flow of isotopically distinct liquids or fluids. For example, Schmidberger et al. (2003) found intergrain Sr variations of up to 1.1 ‰ in clinopyroxene in a garnet peridotite form the Nikos kimberlite, with variations correlating with mineral composition, all of which they attributed to metasomatic reaction with the kimberlite magma host. Tang et al. (2011) found that clinopyroxene has a lower Sr_{intial} and slightly higher ε_{Nd} than coexisting orthopyroxene in mantle peridotites (a situation similar to that seen in many Bushveld Main Zone rocks; Figure 13.3), and suggested that the peridotites were heterogeneously refertilized by multiple metasomatic events.

Re–Os systematics have been used to infer melting and metasomatic events in the mantle and, as for layered intrusions, are strongly influenced by the presence or absence of sulfide. Briefly, the modal abundance of residual mantle sulfide decreases with increasing degree of partial melting and have an unradiogenic $^{187}Os/^{188}Os$ ratio. In contrast, the sulfide content of metasomatized residual peridotite typically increases due to infiltration of silicate liquids or volatile rich fluids and the sulfide has a more radiogenic $^{187}Os/^{188}Os$. This isotopic disequilibrium can be preserved for billions of years (c.f. summary of Bragagni et al., 2017). For example, Burton et al. (1999) noted that the silicates and

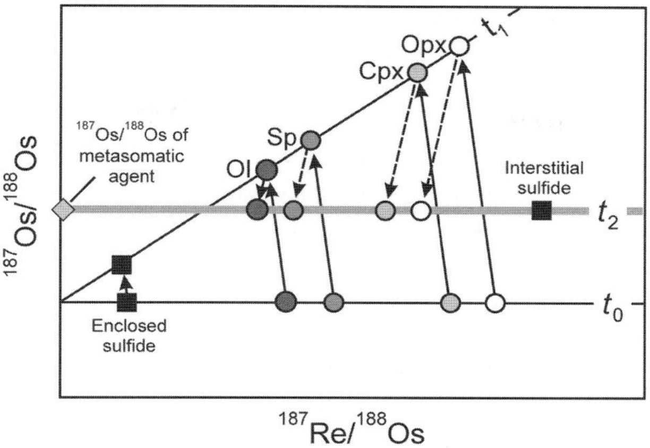

Figure 13.7 Evolution of Re–Os isotopes in minerals of a peridotite at different times. The initial mineral assemblage, which includes a sulfide phase enclosed by another mineral, all have the same initial $^{187}Os/^{188}Os$ at time t_0. All minerals evolve with time by Re decay to the isochron line at t_1. Very soon after, at time t_2 the assemblage interacts with an infiltrating metasomatic melt or fluid with a higher $^{187}Os/^{188}Os$ than the original assemblage. Interaction of the original minerals with the infiltrant resets the Os isotope ratio to that of the infiltrant, but effective re-equilibration could vary. The metasomatic agent may also precipitate an interstitial sulfide with a high $^{187}Re/^{188}Os$ ratio. In contrast, the enclosed sulfide is armoured within its host mineral and preserves its pre-infiltration Os isotope ratio. The result are two populations of sulfide with distinct Re–Os isotopic compositions. Redrawn after Burton et al. (1999), with permission.

interstitial sulfide have similar Os isotope compositions, but that sulfide inclusions in silicate minerals have less radiogenic Os isotope compositions. They suggested that the included sulfide was unable to equilibrate either with other minerals or with an intergranular liquid or fluid, and thus preserve significantly older Re–Os age information than coexisting phases (Figure 13.7). In a similar study, Bragagni et al. (2017) found isotopic disequilibrium in the $^{187}Os/^{188}Os$ isotopic system in different populations of base metal sulfides in peridotite xenoliths from Canada, which they suggest can provide age information on multiple magmatic events that affected the subcontinental lithospheric mantle.

Recent work has attempted to quantify the mechanism by which the both individual isotopic systems can develop isotopic disequilibrium and how different isotopic systems can become decoupled. For example, Le Roux et al. (2009) describe a melt percolation zone in which refractory harzburgites are replaced by a refertilized lherzolite in the Lherz massif in the French Pyrenees that shows a strong decoupling of Hf from Nd and Sr isotopes. Using a one-dimensional infiltration-reaction model, they examined the role of such parameters as diffusivity of the element in minerals, mineral grain size and solid matrix/infiltrating liquid concentrations ratio of the element. They showed that small variations in these parameters can generate significant inter-isotopic decoupling and extensive isotopic variations in mantle rocks that cannot be accounted for by simple mixing involving the harzburgite protolith and the percolating melt. Examples of their results are

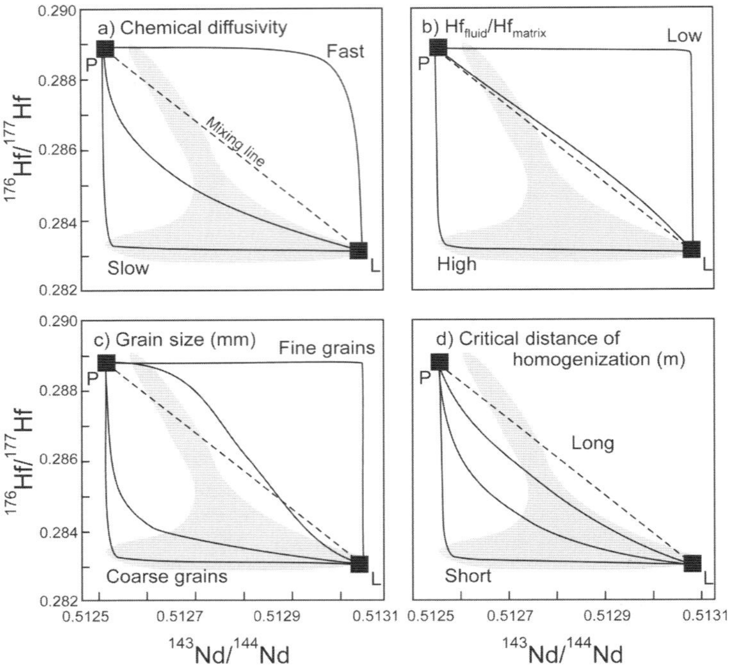

Figure 13.8 ^{176}Hf/^{177}Hf versus ^{143}Nd/^{144}Nd in clinopyroxene separates from Lherz harzburgites and lherzolites (grey field), compared with the mixing curve (dotted line) between a harzburgite protolith (P) and a MORB-type liquid (L), and with the results of the one-dimensional percolation–diffusion modelling (solid lines). The diagrams (a) to (d) illustrate the effects of varying parameters in the model: (a) chemical diffusivity of the elements in minerals; (b) mineral grain size; (c) the Hf_{fluid}/ Hf_{matrix} concentration ratio; (d) the critical distance of isotopic homogenization along the column. Redrawn after Le Roux et al. (2009), with permission.

shown in Figures 13.8 and 13.9. Of particular interest in the application of these results to the Bushveld system, where it is observed that $\varepsilon_{Hf\text{ - }2.06\text{ Ga}}$ shows little variation throughout the complex in contrast to Sr and Nd isotopic systems, is that one can generate strong variations between ^{87}Sr/^{86}Sr and ^{143}Nd/^{144}Nd in the solid assemblage whereas ^{176}Hf/^{177}Hf need not change significantly. These types of quantitative models are so far largely absent in layered intrusion studies.

13.5.1 Layered Intrusions as a Subduction Zone Analogue

Perhaps most importantly for the possible role of country fluids as late contaminants in layered intrusions are the parallels seen in subduction zones. For the latter, it is the movement of fluids from a dehydrating descending lithospheric slab into hotter overlying ultramafic mantle rocks that is the preferred mechanism driving subduction zone melting and arc magmatism. During this process, the slab fluids largely retain their geochemical

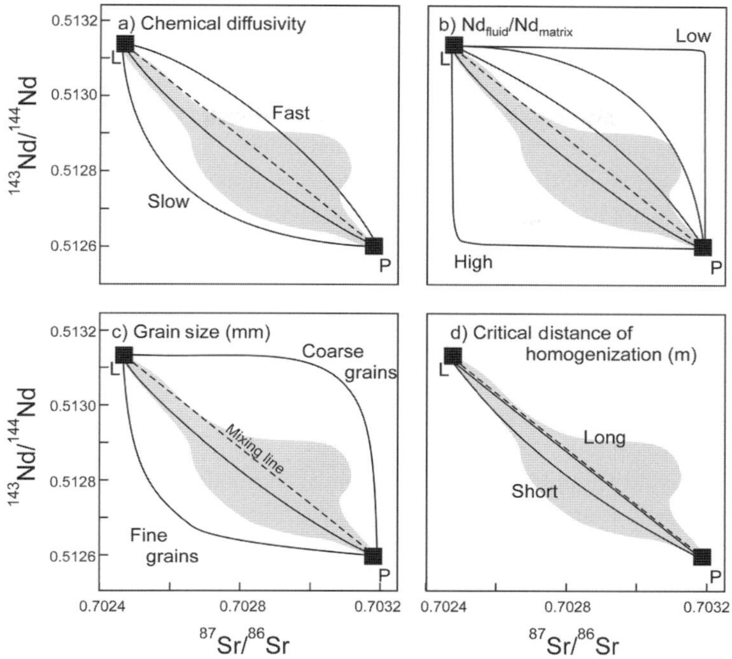

Figure 13.9 $^{143}Nd/^{144}Nd$ versus $^{87}Sr/^{86}Sr$ in clinopyroxene separates from Lherz harzburgites and lherzolites (grey field), compared with the mixing curve (dotted line) between a harzburgite protolith (P) and a MORB-type liquid (L), and with the results of the one-dimensional percolation–diffusion modelling (solid lines). The diagrams (a) to (d) illustrate the effects of varying parameters in the model: (a) diffusivities, D, of the elements in minerals; (b) the mineral grain size; (c) the Nd_{fluid}/Nd_{matrix} ratio; (d) the critical distance of isotopic homogenization along the column. After Le Roux et al. (2009), with permission.

and isotopic characteristics as they migrate into the hotter parts of the mantle wedge to induce hydration melting and arc magmatism (e.g., Mysen and Boettcher, 1975; Hawkesworth et al., 1993; Gaetani and Grove, 1998; Grove et al., 2012). This would suggest that country fluids could migrate into the Bushveld Complex and largely retain their crustal isotopic signatures as they did so. As noted previously in Chapter 3, they Bushveld Complex is among the few thick sill-shaped intrusions that intruded into an unmetamorphosed sedimentary sequence and generated a large volume of metamorphic fluid.

Two mechanisms can limit isotopic equilibration with the country fluids in the lower, cooler sections of a layered intrusion. First, grain-scale interaction of fluids based on isotopic re-equilibration is on the order of 10^3–10^5 years for amphibolite-granulite metamorphic grade rocks (e.g., van Haren et al., 1996; Graham et al., 1998), the longer times about the same as the cooling time for the Bushveld Complex (e.g., Cawthorn and Webb, 2013). Le Roux et al. (2009) summarized slow diffusion rates in mantle minerals that imply that isotopic equilibration may not be reached during partial melting at hot mantle temperature and relatively slow melting rate of an isotopically heterogeneous assemblage. Equilibration would be even less likely in the cooler lithosphere–asthenosphere transition. A rapidly invading country fluid into the lower part of the Bushveld Complex simply may not have time to fully equilibrate with the igneous assemblage until it reached hotter rocks where it would be more reactive for reasons noted below.

Second, as noted above for the stable isotopic signatures of the high temperature veins, some isotopes will not have high bulk rock abundances of the element in question to affect the isotopic character of the infiltrating fluid, particularly if the fluid is otherwise in chemical equilibrium with the host minerals. For example, both Sr and Nd are strongly incompatible in olivine and low Ca pyroxene ($D < 0.01$ and 0.001 for low calcium pyroxene and olivine, respectively) (GERM partition coefficient database, https://earth-ref.org). Hence, the ultramafic rocks that make up the lower portions of the Bushveld and Stillwater complexes do not present a large reservoir of Sr and Nd to exchange with a rapidly moving fluid.

For the Bushveld Complex, the dehydration of the underlying country rock reached its maximum extent at about the time the Upper Critical Zone/Lower Main Zone began to crystallize. The presence of high T veins described above and the diatreme described in Chapter 8 implies that fluid movement in the colder, solidified rocks of the Lower Zone was localized along fractures, faults and pipe-like channels. The rapid injection of country fluid prevented extensive isotopic interaction of this country fluid with the rocks of the Lower and Lower Critical Zones. However, in the Upper Critical Zone these fluids being to interact with a hotter crystal + liquid + igneous vapour assemblage, perhaps even contamination the main magma body by inflow via the diatreme and pothole structures.

13.6 The Mantle Example Applied to Isotopic Disequilibrium in Layered Intrusions

The studies of isotopic disequilibrium in the mantle suggests that isotopic disequilibrium can persist for long periods of time in hot rocks. In part this reflect the fact that the mantle minerals after the infiltration-contamination event are largely in chemical equilibrium with only the isotopic differences to drive re-equilibration. In contrast, layered intrusions that experience the infiltration of a crustal-contaminated fluid during cooling undergo a continuous re-equilibration even down into the subsolidus. This continuous textural and compositional re-equilibration allows a contaminant fluid multiple opportunities to affect the isotopic composition of the original mineral assemblage. This includes the following: (1) Compaction-driven recrystallization during which the rocks develop a foliation as the minerals adjust to the weight of the overlying mush column. (2) Re-equilibration of the primocrysts with the evolving interstitial liquid (the trapped liquid shift effect). (3) Volatile addition that tends to favour one mineral phase over another (constitutional zone refining). (4) Remelting of the original assemblage by volatile fluxing. (5) Subsolidus re-equilibration between mineral phases, e.g., Fe–Mg exchange between chromite and ferro-magnesian silicate minerals. (6) Subsolidus exsolution, as is commonly observed in pyroxenes. (7) Metasomatic replacement where both the mineralogy and the texture of the original assemblage can be substantially altered.

The above discussion suggests several lines of evidence that can distinguish pre-accumulation, isotopically mixed crystal populations from a later metasomatic event. Absent an extensive recrystallization during a metasomatic or remelting event, the infiltration of a fluid or liquid with a crustal signature will most strongly affect the late interstitial minerals (biotite, apatite) and rims on minerals such as plagioclase that have low solid-state diffusion rates. The isotopic signature of these later minerals and plagioclase rims will record a crustal signature that is higher than the cores of any primocryst plagioclase. Owing to the faster diffusion rates, extensive Mg–Fe re-equilibration and exsolution on cooling, pyroxene cores should show a stronger crustal signal than the cores of coexisting plagioclase. This should also be true of sulfides vs. plagioclase as well. However, minerals that are hosted in another mineral phase and isolated from the infiltrant should have an isotopic signature of the original rock or magma.

In contrast, plagioclase that crystallized from a contaminated magma prior to blending with an uncontaminated magma would have the strongest crustal signature in the core, while the rim would reflect an intermediate value that resulted from the mixing event. Mixed populations that result from blending of crystal-phyric magmas should show this in terms of different core plagioclase compositions, bimodal CSD and perhaps different compositional zoning patterns.

In cases where there has been more extensive recrystallization or volatile-induced remelting, the isotopic composition could indicate the contribution of country fluids, which could be much higher than any of the nearby host rocks. For example, Cawthorn et al. (2002) have noted that the $^{187}Os/^{188}Os$ ratio of laurite of the Merensky Reef does not overlap that of the underlying anorthosite (as reported by Hart and Kinloch, 1989), therefore the mineralization could not have come from below as envisioned by the conventional 'Uppers' model. However, the Merensky Reef Os is also more radiogenic than are the overlying rocks (Schoenberg et al., 1999), presenting a similar quandary for the 'Downers' model. Given the evidence for isotopic disequilibrium and flux melting or recrystallization in the Merensky Reef, a plausible fluid transport model would envision early mineralization by a Cl-rich 'igneous' fluid resolved from interstitial silicate liquid, flowed by an influx of later fluids that have a stronger contamination signature. This latter fluid most strongly affected the bulk rock mainly when it resulted in more extensive remelting/recrystallization.

14
Some Objections Considered

14.1 Introduction

There have been a number of objections to the role of fluids as the principle agent of PGE-sulfide mineralization and its role in controlling the development of both PGE-bearing pipes and stratabound PGE reefs. For the most part these objections tend to be stated in as little as one or two sentences and after which the possible role of magmatic fluids is dismissed. In some cases, however, the objections to the hydrothermal transport model have been raised in more detail (e.g., Barnes and Campbell, 1988; Cawthorn et al., 2002; Fonseca et al., 2009; Maier et al., 2013; Godel, 2015; Mungall, 2015). But even in these instances, the authors typically do not otherwise attempt to explain the halogen trends, the fluid inclusion evidence, the formation of PGE-rich discordant pipes and pegmatoids or other evidence.

14.2 Critique of the PGE Pipe Hydromagmatic Formation Models

The hydrothermal metasomatic model for the formation of PGE-bearing pipes of the Bushveld Complex described in Chapter 8 has a number of features in its support. They are only known to occur within the Bushveld Complex and are not seen in the underlying Transvaal sedimentary sequence argues against these bodies having been formed by the influx of an external magma. The evidence for replacement nature of the pipes and IRUP, particularly the presence of remnant chromitite layering in the pipes, argues against the presence of large bodies of liquid and instead support a subsolidus replacement of the host rock by the pipe/IRUP assemblage. A solid-for-solid replacement avoids problems associated with open liquid channels, including the high thermal energy required to melt the host to form the pipes. In contrast, metasomatic replacement is largely thermally neutral as one mineral is replace by another. A replacement mechanism also avoids the difficulty of keeping open liquid channels in a high temperature mush without the mush deforming and closing the channel. An over-pressured system that might keep the channels open should form dilatational sills and dykes and not the approximately constant volume replacements observed.

The Cl-rich nature of the assemblage and the lack of quartz and plagioclase would argue against the infiltration agent being Fe-rich interstitial liquids having percolated downward from the higher, more evolved part of the complex (Scoon and Mitchel, 2009), where the interstitial apatite is distinctly enriched in the fluorapatite component and the assemblage includes quartz. The metal loss seen in the adjacent chromitites in the host layered sequence, the association with Cl- and C-rich minerals, and the pegmatoidal textures are consistent with fluid transport of the PGE. These features are rarely addresses in the various magmatic models. Their apparent uniqueness to the Bushveld Complex, their association with footwall structures and the relatively unique nature of a large, thick sill intruding unmetamorphosed sediments would suggest that degassing of the footwall was important in forming the large pipes. However, there have been a number of objections to the hydrothermal model. These objections, and responses to them, can be summarized as follows.

The Mg-rich nature of the central stock. Scoon and Mitchell (2009) discounted a secondary hydrothermal origin by noting that olivine in the main stock of the pipes is largely constant and independent of host rock, having a higher *mg#* and higher NiO concentration than the host rock pyroxene. They also noted that the stocks contain a Cr spinel that may not be present in the host rock. However, replacement minerals are a function of reaction with a fluid that can increase or decrease the *mg#* depending on the fluid composition; the reaction need not just be a simple loss of silica. Indeed the Fe enrichment in the core of the pipes is likely Fe not accounted for in the main stock.

Because of the extensive volume loss, the Ni now in olivine may be entirely inherited from the host, although the Cl-rich nature of the fluids one cannot discount substantial Ni transported by the fluid. Ni enrichment is observed in a number of hydrothermal ores including black-shale hosted deposits that are usually attributed to a sedimentary-exhalative origin, e.g., the Nick Ni-Zn-PGE-Au prospect, Selwyn Basin, Canada (Hulbert et al., 1992). Experimental work also demonstrates that NiO solubility increases strongly with addition of HCl and NaCl (Tremaine and Leblanc, 1980; Lin and Popp, 1984; Fahlquist and Popp, 1989; Scholten et al., 2018).

As for chromite, pyroxene can contain significant amount of Cr_2O_3, and this can precipitate a Cr spinel during alteration/or incongruent hydration melting (Chapters 9 and 12). Furthermore, experimental demonstrate that Cr can have considerable solubility in aqueous HCl solutions (e.g., Watenphul et al., 2014), and evidence presented by Ari and Akizawa (2014) suggested that chromitite can even be precipitated from hydrothermal solutions seen in clinopyroxenite replacements of gabbros from the Oman ophiolite.

Isotopic evidence. Cawthorn et al. (2000) not that the oxygen isotopes are broadly magmatic and not consistent with and influx of country fluids. However, Schiffries and Skinner (1987) and Schiffries and Rye (1990) noted the presence of high temperature (max T about 700 °C) saline hydrothermal veining that cut both the intrusion and the underlying contact aureole. Schiffries and Rye (1990) note that those that cut the complex have oxygen isotopic signatures similar to the host igneous rocks whereas the C and D isotopes are more like the country rock. They suggested that was $\delta^{18}O$ controlled by isotopic exchange with the Bushveld intrusive rocks whereas the $\delta^{13}C$ δD to have been inherited from Transvaal super group, results similar to that of Buick et al. (2000). This implies that county fluids will undergo isotopic exchange rather readily for elements that are abundant in the igneous suite (e.g., ^{18}O), but that elements that are in only trace abundance (e.g., Cl, C and D) are less likely to be affected. In addition, although country fluids may have initially been responsible for the pipes, the influx of igneous fluids from the host rock would eventually replace this initial fluid.

Cawthorn et al. (2000) also noted the enriched Sr isotopic signature of interstitial plagioclase as compared with the host, which is similar to that of the Upper Zone. However, more recent work has shown that Sr isotopic disequilibrium is a characteristic feature of much of the Bushveld Complex (Chapter 13). For example, e.g., Chutas et al. (2012) have found that the rims of the major modal minerals and the interstitial minerals tend to be enriched compared to the cores, and have attributed this to the infiltration of a late, enriched component. Sr has a relatively low content in Bushveld Ultramafic rocks (a few 10's of ppm), but can have concentrations up to ~0.5 wt.% in country fluids (e.g., Pelch et al., 2015). Thus, unlike oxygen, Sr can preserve its external signature more readily than will oxygen.

Chromitite pods as xenoliths. The chromitite pods found in some of the pipes are displaced to slightly lower levels, suggesting they were xenoliths settled through an invading magma. The blocks are large and dense, their settling velocities would be such that they would sink until they hit some sort of solid floor. Instead, the lower position of the blocks relative to that of the host layers as well and the downturn of the layering in the vicinity of the pipes is consistent with a volume loss during the replacement process.

The lack of textural evidence of olivine replacing pyroxene. This objection assumes that olivine should be rimming pyroxene if the olivine is secondary, in a reverse of the commonly observed orthopyroxene rims on olivine observed during crystallization of a magma. However, loss of one mineral and replacement by another is generally easier if the new mineral can nucleate and grow as a separate grain while the unflavoured grain is free to dissolve. This is what is observed in incongruent melting, for example; olivine in mantle dunite does not have cores of the pyroxene from melting of the original peridotite.

The lack of relic grains of original 'cumulus' plagioclase in the pipes. Cawthorn et al. (2000) suggested there should be locations where the original host norite or individual plagioclase grains have only been partly replaced. However, there are numerous examples where replacement bodies have completely destroyed the original mineralogy and texture. The Stillwater pegmatoid shown in Figure 2.12 and the Skaergaard mafic/felsic segregations shown in Figure 8.12 are but two examples.

The coarse-grained nature of the rocks. Cawthorn et al. (2000) noted the coarse grainsize of the main stock and extensive exsolution features in the pyroxenes and suggested that they are indicative of a high temperature origin. However, a coarsening of texture is an expected consequence of recrystallization in a fluid-rich environment, and minimum re-equilibration temperatures would simply place a lower bound on the temperature that the replacement process (Figure 8.4). There is no reason why the replacement could not have occurred at a supersolidus temperature; indeed, fluid fluxing and partial melting may have accompanied the replacement process, as it appears to do in the mantle (e.g., Keller and Katz, 2016).

14.3 Critique of the PGE Reef Hydromagmatic Formation Models

As for the PGE pipes, the extensive evidence for a role by hydrothermal fluids is straightforward. This evidence includes the coarse textures, abundance of hydrous minerals, the presence of PGE-enriched fluid inclusions in the rocks below the reef, the presence of hydrous melt inclusions in reef minerals and the strong effect that volatiles, especially H_2O, have on affecting the crystallization and flux melting of igneous rocks. Nonetheless, there have been a number of criticisms that some have suggested are fatal to the hydromagmatic model. Some of these objections and possible resolutions are described below.

There are no anomalously-poor PGE concentrations or PGE depletions observed in the rocks below the mineralized horizons. It has been suggested that the rock below the Merensky Reef of the of the Bushveld Complex and the J-M Reef of the Stillwater Complex do not show the expected PGE depletions if the rocks below the ore zones were the source of the ore metals. (Barnes and Maier 2002a,b; Naldrett et al., 2009b; Keays et al., 2012). This objection initially began as an assertion that footwall should only have the PGE budget of a minor trapped liquid component, owing to the presumed incompatible behaviour of the PGE absent sulfide saturation (Barnes and Campbell, 1988). For example, if the magma had 10–20 ppb Pt, and the footwall retained only 10% trapped liquid, then the bulk rock would only contain 1–2 ppb Pt, too low to be an effective source rock for Pt.

However, as noted in Chapter 1, the rocks in the lower part of the Bushveld Complex are unusually enriched in PGE. Barnes et al. (2009) interpret the rocks below the Merensky Reef of the Bushveld Complex to have a 'cumulus' sulfide component based on the observed enrichments in the PGE, although S contents are too low for the magma to have been S saturated. They suggested three possible models to explain the contradiction: (1) migration of sulfide out of the crystal pile, (2) removal of sulfide suspended in a sulfide-saturated magma prior to emplacement and (3) loss of S and Se to a high temperature degassing event. The later was supported by Kanitpanyacharoen and Boudreau (2013) based on textural evidence for S-loss from footwall sulfides. Alternatively, Keays et al. (2012) suggested that the PGE can behave as compatible elements, but do not specify which phase preferentially incorporates the PGE.

Following the observation that the rocks below the platiniferous UG2 and the Merensky Reef is actually a better source rock for the PGE than the presumed parent magma, the objection changed to the apparent lack of depletion of the PGE from the immediate footwall (e.g., Barnes and Maier, 2002a,b; Naldrett et al., 2009; Keays et al., 2012). For example, Maier et al. (2013) suggested that the reefs would require the complete removal of all the PGE from 200 m below the Merensky Reef, although why this thickness was chosen is not made clear. Given that the current PGE and S abundances in the rocks of the Lower and Lower Critical Zones cannot be readily explained by any simple orthomagmatic model (by the trapped liquid abundance, for example) one cannot easily say if there has been some later loss or not. However, Kanitpanyacharoen and Boudreau (2013) noted that adding the ore metals now contained in the UG2 and Merensky Reef back into the footwall would not significantly change the overall average PGE concentration of the footwall; on the typical log PGE concentration plot, the difference between the now observed and the original presumed concentrations are rather negligible (Chapter 1, Figure 1.3).

Ore grade does not correlate with footwall thickness, or locally ore zones sit on or near country rock. A related objection to the footwall being the source of the ore metals involves variable thickness of the underlying mush zone, or ore zones sitting on or near country rock (e.g., the Platreef). For example, Cawthorn et al. (2002) note that the footwall beneath the Rustenburg mine is about twice as thick as that beneath the Western mine, yet it appears that the combined PGE contents of UG2 and Merensky Reefs are not significantly different. If the PGE were scavenged from the footwall, one would expect the ore zones above the thicker sections to have higher grades. A similar point in regard to ore grade and footwall thickness was made by Barnes and Hoatson (1994) who noted that locally the UG2 of the Bushveld Complex sits on top of the footwall metasedimentary rock. The lack of a mush directly below the mineralized zones would imply the mineralization could not have come up from below.

However, as noted in Chapter 5, the highest grades in the Munni Munni (and the Great Dyke as well) occur were the dipping of the layers would produce flow focusing towards the margins of the intrusion and where layering is draped over footwall highs. The lateral movement of fluids may homogenise fluids as the move upward. A similar argument was made in regard to the Platreef of the Bushveld Complex. In the case of the Platreef, mineralization occurs proximal to where the footwall rocks onlap onto the country rock, but disappears in more distal regions where the main zone rocks sit on country rock far from Lower and Critical Zone rocks.

Finally, one needs to consider the compaction behaviour of mush zones of variable thickness. That is, although a thick mush zone may have potentially more metals to move, it will cool slower, allowing compaction to become more effective at removing the interstitial liquid than in a thin compaction zone. The lower amount of interstitial liquid will produce a smaller volume of exsolved fluid, all other factors being equal. The net result is that a thick mush zone may produce the same amount of fluid as does a thinner mush.

Problems with the hydromagmatic model. Cawthorn et al. (2002) also noted that if the PGE were concentrated by a chromatographic mechanism as described in Chapter 10, then metal offsets should be observed in the Merensky Reef. However, as noted by Boudreau and Meurer (1999), the chromatographic front can collapse if it interacts with a discontinuity in the crystal pile. They used the top if the crystal pile as an example of such a discontinuity, where all fluid would redissolve and precipitate metals at the base of the overlying magma body.

Furthermore, the metals do not necessarily need to move upward in a front; the model of Boudreau and Meurer (1999) looked at the case where the original rocks had abundant sulfide controlling PGE partitioning between fluid and the host assemblage and metals are largely moving only at a sulfide dissolution front. It can also be that the fluids are largely in equilibrium with a thick section of mush regardless of the presence of a trace sulfide fraction. This is suggested by the observation from fluid inclusions well below the ore zones that they can have metal ratio similar to the overlying ore zones (Chapter 7).

PGE reefs can extend over long distances. It is suggested that the lateral extent of the ore zones and the observation that the ore deposits may directly overly magmatic unconformities is not consistent with the ore component having been introduced by fluids. Furthermore, it is assumed that the upward migration of fluids would lead to discordant features that are not related to magmatic stratigraphy.

There are several responses to this claim. First, the ore zones are, in detail, discontinuous and discordant. This is particularly true for the Stillwater Complex. Although broadly stratabound, for both the J-M Reef and the Picket Pin deposit the mineralization is locally discontinuous and, for the former, can transgress layering. The Merensky Reef of the Bushveld Complex is also discontinuous, displaced by potholes. Furthermore, ore grade even in normal Merensky Reef shows small-scale variability in the style and grade of mineralization (e.g., Cawthorn et al., 2002).

To the degree that the underlying mush is laterally extensive, then volatile saturation and upwards migration in a fluid-saturation front would be similar, being controlled by the regional thermal profile of the mush zone. Furthermore, as suggested for the formation of Great Dyke-type reefs, it can be the case that it is changes in the original magmatic stratigraphy that act as physio-chemical traps to migrating fluids that lead to precipitation of PGE-sulfide. This can occur either because the fluids react with a changing bulk rock composition or because the fluids redissolve in fluid-undersaturated interstitial liquids.

More importantly, it can be that the magmatic stratigraphy is itself controlled by the influx of fluids. That is, it is the vapour itself may be the source of the crystallization perturbations that define the modal layering, and can actually be responsible for the apparent magmatic unconformity. Thus, for the case of the J-M Reef of the Stillwater Complex, the reappearance of olivine is a direct result of flux melting by the introduced fluids (Boudreau, 1988, 1999).

The insoluble nature of the IPGE. This objection comes from extrapolating a number of older experimental studies done at low to moderate temperatures that show the PGE to have a low solubility in hydrothermal fluids, especially for the IPGE (Ir, Ru and Rh). Related to this, it has been noted that PGE abundances in hydrothermal sulfides associate with PGE-bearing magmatic sulfide typically have very low PGE concentration (e.g., Luguet et al., 2004), again suggesting that the PGE are not very soluble. However these secondary sulfides are associated with alteration assemblages far below typical magmatic temperatures.

Similarly, Fonseca et al. (2009) note the observation of Wood (1987) who suggested that even if the PGE can be transported in Cl-rich brine fluids, Cu, Ni and Fe should be more enriched by hydrothermal fluids than the PGE by 'several factors of tens', which is not observed in the PGE-rich reefs of the Stillwater and Bushveld Complexes. However, more recent work done at magmatic temperatures and summarized in Chapter 4 show that Ir can have a fluid/liquid distribution coefficient $> 10^3$ in Cl-rich fluids. Fluid inclusion in rocks below the J-M Reef have Pd/Cu ratio similar to that of the ore zone (Chapter 7), and sulphide-bearing pegmatoids found well below the J-M Reef have similar bulk rock PGE–base metal concentrations as that of the ore zone (Chapter 8, Figure 8.2). While more experimental work needs to be done for PGE partitioning into fluids at high temperatures, the existing data are pointing to significant solubility for all the PGE for appropriate fluid compositions at magmatic conditions. Once sulfide, if present, is (locally) totally resorbed, the fluid phase is the next best host for all the PGE, hence the metal concentration profile of the fluid should mimic that of the host rock. The exception is the preference of chromite for the IPGE.

The PGE reefs have similar metal patterns as the parent magma. This is a corollary of the previous objection. Thus, it is observed that the mantle-normalized base-metal and PGE patterns of the mineralized horizons show shapes identical to those of the B1 Bushveld magma (Chapter 1, Figure 1.3), suggesting that the same process concentrated all the PGE at the level of the reefs (Fonseca et al., 2009; Maier et al., 2013).

As pointed out in Chapter 1, the PGE concentrations in the UG2 and Merensky Reefs do show a preference for the PPGE. Furthermore, if the ore component is added back into the footwall, the addition is modest but actually makes the PGE profile of the footwall more closely parallel that of the presumed parent magma.

Isotopic changes. This objection is specific to the Bushveld Complex, the osmium isotopic composition of Bushveld rocks shows a sharp change towards more radiogenic compositions at the level of the Merensky Reef (McCandless and Ruiz, 1991), suggesting that the Os cannot have been derived from the underlying cumulates. However, as discussed in Chapter 13, the rocks above the Merensky Reef are also do not have a radiogenic Os isotopic signature and at face value would also suggest that the magma above the reef could not be the source for the OS.

Most of these objections come from studies prior to the discovery of extensive isotopic disequilibrium in the Bushveld Complex. As also described in Chapter 13, interstitial fluids may be contaminated with a crustal component that does not markedly affect the host rock unless there is a recrystallization or remelting event, as suggested to have occurred for normal Merensky Reef.

Problems with bubble transport. In an extensive treatment of bubble formation and migration in crystal-rich systems, Mungall (2015) concludes that bubble will not move more than a few metres from the site of the initial nucleation until the system has cooled below its solidus. The bubbles effectively are trapped by capillary forces in the crystal framework. While conceding that a dynamic system such as a compacting crystal pile may allow for some movement, he effectively discounts this by noting that compaction happens largely at the top of the pile.

This treatment ignores evidence from Stillwater Complex in which the efficiency of compaction is enhanced by local increase in volatile concentrations (Meurer and Boudreau, 1988) and other evidence presented in Chapter 5, and more recent theoretical treatments that point to the ease of fluid migration in compacting systems (e.g., Huber and Parmigiani, 2018).

14.4 Future Work

This book has attempted to highlight the importance of hydromagmatic processes in the crystallization of layered intrusions, but there are a number of broad areas where additional research is indicted. Briefly, these are as follows.

Fluid inclusion studies. A more extensive study of fluid inclusion compositions, and especially their metal concentrations, would help to focus experimental studies as to those fluid compositions that are most likely to transport the PGE. The fluid inclusion evidence noted in Chapter 7 suggests that CO_2-, Cl-rich fluids are particularly favourable for PGE transport. These studies should include the IPGE as well as the PPGE.

Experimental work. More work is needed regarding the solubility of the PGE as a function of fluid composition at magmatic temperatures at mid- to upper-crustal pressures. More broadly, the effects of volatiles on mineral stability and incongruent reaction in crystal-liquid systems is warranted. Most existing experimental work has looked at the effect of volatiles on specific parent liquid compositions but not on crystal-liquid (i.e., 'cumulate') mixtures. Experiments in which fluids migrate through crystal-liquid mixtures may elucidate details of fluid refining and ore element transport mechanisms.

Numerical modelling studies. Numerical modelling of fluid generation, migration and reaction in the mantle and at subduction zones is becoming increasingly sophisticated. Application of these techniques to layered intrusions is a logical extension. This would be particularly useful for the understanding of later migration of fluids and fluid mixing and homogenization.

Detailed petrographic and isotopic studies. Details of fluid and liquid migration requires understanding of small-scale variations of trace minerals and isotopic trends on thin section to outcrop scales. Work such as that of Meurer and Meurer (2006) on apatite distribution and modal variations a small boulder-size sample and that of Yang et al. (2013) on Sr isotopic variations within a thin section allow one to infer silicate liquid migration scales. It is unknown if isotopic disequilibrium also occurs in channelled regions, is random, or pervasive, which has consequences for understanding migrations of fluids with a country rock isotopic component.

It was noted in Chapter 13 that, prior to the advent of extensive isotopic studies, it was thought that the Bushveld Complex was largely the product of injection, mixing and fractional crystallization of a single magma type. With the possibility that much of the observed lithologic and isotopic variation, including isotopic disequilibrium, could be due to syn- and post-emplacement hydromagmatic processes (including a country fluid component), it may be that interpretations return to a single magma model that was strongly modified as it cooled and crystallized.

References

Ahmed, A.H. and Arai, S. (2002). Unexpectedly high-PGE chromitite from the deeper mantle section of the northern Oman ophiolite and its tectonic implications *Contributions to Mineralogy and Petrology*, 143, 263–278.

Aird, H.M. and Boudreau, A.E. (2013). High-temperature carbonate minerals in the Stillwater Complex, Montana, USA. *Contributions to Mineralogy and Petrology*, 166, 1143–1160.

Aird, H.M., Ferguson, K.M., Lehrer, M.L. and Boudreau, A.E. (2017). A study of the trace sulfide mineral assemblages in the Stillwater Complex, Montana. *Mineralium Deposita*, 52(3), 361–382.

Algar, C.K., Boudreau, B.P. and Barry, M.A. (2011). Initial rise of bubbles in cohesive sediments by a process of viscoelastic fracture. *Journal of Geophysical Research*, 116, B04207. DOI:10.1029/2010JB008133.

Almeev, R.R., Holtz, F., Koepke, J., Parat, F. and Botcharnikov, R.E. (2007). The effect of H2O on olivine crystallization in MORB: Experimental calibration at 200 MPa. *American Mineralogist*, 92, 670–674.

Almeev, R.R., Holtz, F., Koepke, J. and Parat, F. (2012). Experimental calibration of the effect of H2O on plagioclase crystallization in basaltic melt at 200 MPa. *American Mineralogist*, 97, 1234–1240.

Anderson, A.T., Jr, (1974). Chlorine, sulfur and water in magmas and oceans. *Geological Society of America Bulletin*, 85, 1485–1492.

Andersen, J.C.Ø. (2006). Postmagmatic sulphur loss in the Skaergaard Intrusion: Implications for the formation of the Platinova Reef. *Lithos*, 92, 198–221.

Andersen, J.C.Ø., Rasmussen, H., Nielsen, T.F.D. and Ronsbo, J.G. (1998). The triple group and the Platinova gold and palladium reefs in the Skaergaard intrusion: Stratigraphic and petrographic relations. *Economic Geology*, 93, 488–509.

Appold, M.S. and Nunn, J.A. (2002). Numerical models of petroleum migration via buoyancy porosity waves in viscously deformable sediments. *Geofluids*, 2, 233–247.

Arai, S. and Akizawa, N. (2014). Precipitation and dissolution of chromite by hydrothermal solutions in the Oman ophiolite, New behavior of Cr and chromite. *American Mineralogist*, 99, 28–34.

Arai, S., Prichard, H.M., Matsumoto, I. and Fisher, P.C. (1999). Platinum-group minerals in podiform chromitite from the Kamuikotan zone, Hokkaido, northern Japan. *Resource Geology*, 49, 39–47.

Aranovich, L.Ya., Zakirov, I.V., Sretenskaya, N.G. and Gerya, T.V. (2010). Ternary system H2O–CO2–NaCl at high T–P parameters: An empirical mixing model. *Geochemistry International*, 48(5), 446–455.

Armienti, P., Francalanci, L. and Landi, P. (2007). Textural effects of steady state behaviour of the Stromboli feeding system. *Journal of Volcanology and Geothermal Research*, 160, 86–98.

Armitage, G.J. (1992). *The petrogenesis of potholes in the UG2 chromitite layer, Crocodile River Mine, Western Bushveld complex.* Ph.D. *dissertation, University of Natal, Pietermaritzburg, South Africa.*

Armstrong, R. and Wilson, A.H. (2000). A SHRIMP U-Pb study of zircons from the layered sequence of the Great Dyke, Zimbabwe, and a granitoid anatectic dyke. *Earth and Planetary Science Letters*, 180, 1–12.

Atkins, F.B. (1969). Pyroxenes of the Bushveld Intrusion, South Africa. *Journal of Petrology*, 10(2), 222–249.

Augé, T. (1988). Platinum-group minerals in the Tiébaghi and Vourinos ophiolitic complexes: Genetic implications. *Canadian Mineralogist*, 26, 177–192.

Azimov, P.Ya. and Bushmin, S.A. (2007). Solubility of minerals of metamorphic and metasomatic rocks in hydrothermal solutions of varying acidity: Thermodynamic modeling at 400–800 °C and 1–5 kbar. *Geochemistry International*, 45(12), 1210–1234.

Bachmann, O. and Bergantz, G.W. (2006). Gas percolation in upper-crustal silicic crystal mushes as a mechanism for upward heat advection and rejuvenation of near-solidus magma bodies. *Journal of Volcanology and Geothermal Research*, 149, 85–102.

Bagdassarov, N., Dorfman, A. and Dingwell, D.B. (2000). Effect of alkalis, phosphorus, and water on the surface tension of haplogranite melt. *American Mineralogist*, 85, 33–40.

Baker, D.R. and Moretti, R. (2011). Modeling the Solubility of Sulfur in Magmas: A 50-Year Old Geochemical Challenge. In Sulfur in Magmas and Melts: Its Importance for Natural and Technical Processes (Behrens, H. and Websater, J.D., eds.). Reviews in Mineralogy and Geochemistry, 73, 167–213.

Ballhaus, C.G. (1988). Potholes of the Merensky Reef at Brakspruit Shaft, Rustenburg Platinum Mines: Primary disturbances in the magmatic stratigraphy. *Economic Geology*, 83, 1140–1158.

Ballhaus, C.G. and Stumpfl, E.F. (1985). Occurrence and petrological significance of graphite in the Upper Critical Zone, Western Bushveld Complex, South Africa. *Earth and Planetary Science Letters*, 74, 58–68.

Ballhaus, C.G. and Stumpfl, E.F. (1986). Sulphide and platinum mineralization in the Merensky Reef: Evidence from hydrous silicates and fluid inclusions. *Contributions to Mineralogy and Petrology*, 94, 193–204.

Ballhaus, C.G. and Ulmer, P. (1995). Platinum-group elements in the Merensky Reef: II. Experimental solubilities of platinum and palladium in Fe1-XS from 950 to 450°C under controlled f(S2) and f(H2). *Geochimica et Cosmochimica Acta*, 59(23), 4881–4888.

Ballhaus, C.G., Ryan, C.G., Mernagh, T.P. and Green, D.H. (1994). The partitioning of Fe, Ni, Cu, Pt, and Au between sulfide, metal, and fluid phases: A pilot study. *Geochimica et Cosmochimica Acta*, 58(2), 811–826.

Barnes, S.J. (1986). The effect of trapped liquid crystallization on cumulus mineral compositions in layered intrusions. *Contributions to Mineralogy and Petrology*, 93 (4), 524–531.

Barnes, S.J. (1993). Partitioning of the platinum group elements and gold between silicate and sulphide magmas in the Munni Munni Complex, Western Australia. *Geochimica et Cosmochimica Acta*, 57, 1277–1290.

Barnes, S.J. and Campbell, I.H. (1988). Role of late magmatic fluids in Merensky-type platinum deposits: A discussion. *Geology*, 16, 488–491.

Barnes, S.J. and Fiorentini, M.L. (2008). Iridium, ruthenium and rhodium in komatiites: Evidence for iridium alloy saturation. *Chemical Geology*, 257, 44–58.

Barnes, S.J. and Hoatson, D.M. (1994). The Munni Munni Complex, Western Australia: Stratigraphy, structure, and petrogenesis. *Journal of Petrology*, 35, 715–751.

Barnes, S.J. and Naldrett, A.J. (1986). Geochemistry of the J-M Reef of the Stillwater Complex, Minneapolis Adit Area II. Silicate Mineral Chemistry and Petrogenesis. *Journal of Petrology*, 27(4), 791–825.

Barnes, S.J., McIntyre, J.R., Nisbet, B.W. and Williams, C.R. (1990). Platinum group element mineralisation in the Munni Munni Complex, Western Australia. *Mineralogy and Petrology*, 42, 141–164.

Barnes, S.J., Keays, R.R. and Hoatson, D.M. (1992). Distribution of sulphides and PGE within the porphyritic websterite zone of the Munni Munni Complex, Western Australia. *Australian Journal of Earth Sciences*, 39(3), 289–302.

Barnes, S.-J. and Fiorentini, M.L. (2008). Iridium, ruthenium and rhodium in komatiites: Evidence for iridium alloy saturation. *Chemical Geology*, 257, 44–58.

Barnes, S.-J. and Gomwe, T.S. (2011). The Pd deposit of the Lac des Iles Complex, Northwestern Ontario. *Reviews in Economic Geology*, 17, 351–370.

Barnes, S-J. and Maier, W.D. (1999). The fractionation of Ni, Cu and the noble metals in silicate and sulfide liquids. In *Dynamic Processes in Magmatic Ore Deposits and Their Application in Mineral Exploration* (Keays, R.R., Lesher, C.M. Lightfoot, P.C. and Farrow, C.E.G., eds). Geological Association of Canada, Short Course Volume 13, 69–106.

Barnes, S.-J. and Maier, W.D. (2002a). Platinum-group element distributions in the Rustenberg Layered Suite of the Bushveld Complex, South Africa. In *The Geology, Geochemistry, Mineralogy and Mineral Beneficiation of Platinum-Group Elements* (Cabri, L.J., ed). Canadian Institute of Mining, Metallurgy and Petroleum Special Volume, 54, 431–458.

Barnes, S.-J. and Maier, W.D. (2002b). Platinum-group elements and microstructures of Normal Merensky Reef from Impala Platinum Mines, Bushveld Complex. *Journal of Petrology*, 43, 103–128.

Barnes, S.-J., Naldrett, A.J. and Gorton, M.P. (1985). The origin of the fractionation of platinum-group elements in terrestrial magmas. *Chemical Geology*, 53, 303–323.

Barnes, S.-J., Savard, D., Bédard, L.P. and Maier, W.D. (2009). Selenium and sulfur concentrations in the Bushveld Complex of South Africa and implications for formation of the platinum-group element deposits. *Mineralium Deposita*, 44, 647–664

Barnes, S.-J., Maier, W.D. and Curl, E.A. (2010). Composition of the marginal rocks and sills of the Rustenburg Layered Suite, Bushveld Complex, South Africa: Implications for the formation of the platinum-group element deposits. *Economic Geology*, 105, 1491–1511.

Barnes, S.-J., Pagé, P., Prichard, H.M., Zientek, M.L. and Fisher, P.C. (2016). Chalcophile and platinum-group element distribution in the Ultramafic series of the Stillwater Complex, MT, USA—Implications for processes enriching chromite layers in Os, Ir, Ru, and Rh. *Mineralium Deposita*, 51(1), 25–47.

Barry, M.A., Boudreau, B.P., Johnson, B.D. and Reed, A.H. (2010). First-order description of the mechanical fracture behavior of fine-grained surficial marine sediments during gas bubble growth. *Journal of Geophysical Research*, 115, F04029, doi:10.1029/2010JF001833.

Batanova, V.G. and Savelieva, G.N (2009). Melt migration in the mantle beneath spreading zones and formation of replacive dunites: A review. *Russian Geology and Geophysics*, 50, 763–778.

Bazarkina, E.F., Pokrovski, G.S. and Hazemann, J.-L. (2014). Structure, stability and geochemical role of palladium chloride complexes in hydrothermal fluids. *Geochimica et Cosmochimica Acta*, 146, 107–131.

Bédard, J.H. (1994). A procedure for calculating the equilibrium distribution of trace elements among the minerals of cumulate rocks, and the concentration of trace elements in the coexisting liquids. *Chemical Geology*, 118, 143–153.

Bédard, J.H., Sparks, R.S.J., Renner, R., Cheadle, M.J. and Hallworth, M.A. (1988). Peridotite sills and metasomatic gabbros in the Eastern Layered Series of the Rhum Complex. *Journal of the Geological Society, London*, 145, 207–224.

Bédard J.H., Marsh, B.D., Hershum, T.G., Naslund, H.R. and Musaka, S.B. (2007). Large scale mechanical redistribution of orthopyroxene and plagioclase in the Basement Sill, Ferrar Dolerites, Antarctica. *Journal of Petrology*, 48, 2289–2326.

Behrens, H. and Webster, J.D., eds. (2011). Sulfur in magmas and melts: Its importance for natural and technical processes. *Mineralogical Society of America, Reviews in Mineralogy*, 30.

Belien, I.B., Cashman, K.V. and Rempel, A.W. (2010). Gas accumulation in particle-rich suspensions and implications for bubble populations in crystal-rich magma. *Earth and Planetary Science Letters*, 297, 133–140.

Bell, A.S., Siman, A. and Guilling, M. (2009). Experimental constraints on Pt, Pd and Au partitioning and fractionation in silicate melt–sulfide–oxide–aqueous fluid systems at 800 C, 150 MPa and variable sulfur fugacity. *Geochimica et Cosmochimica Acta*, 73, 5778–5792.

Berkowitz, B. and Ewing, R.P. (1998). Percolation theory and network modeling applications in soil physics. *Surveys in Geophysics*, 19, 23–72.

Bezman, N.I., Gorbachev, P.N., Shalynin, A.I., Asif, M. and Naldrett, A.J. (2008). Solubility of platinum and palladium in silicate melts under high water pressure as a function of redox conditions. *Petrology*, 16(2), 161–176.

Bezos A., Lorand J.-P., Humler E. and Gros M. (2005). Platinum-group element systematics in Mid-Oceanic Ridge basaltic glasses from the Pacific, Atlantic, and Indian Oceans. *Geochimica et Cosmochimica Acta*, 69, 2613–2627.

Bird, D.K., Arnason, J.G., Brandriss, M.E., Nevle, R.J, Radford, G., Bernstein, S., Gannicott, R.A. and Kelemen, P.B. (1995). A gold-bearing horizon in the Kap Edvard Holm Complex, East Greenland. *Economic Geology*, 90(5), 1288–1300.

Bird, D.K., Brooks, C.K., Gannicott, R.A. and Turner, P.A. (1991). A gold-bearing horizon in the Skaergaard intrusion. *Economic Geology*, 56, 157–166.

Blander, M. and Katz, J.L (1975). Bubble nucleation in liquids. *American Institute of Chemical Engineering Journal*, 21(5), 833–848.

Blaine, F.A. (2010). *The effect of volatiles (H2O, Cl and CO2) on the solubility and partitioning of platinum and iridium in fluid-melt systems*. Ph.D. Disseration, University of Waterloo, Ontario, Canada.

Blaine, F.A., Linnen, R.L., Holtz, F. and Brügmann, G.E. (2005). Platinum solubility in a haplobasaltic melt at 1250°C and 0.2 GPa: The effect of water content and oxygen fugacity. *Geochimica et Cosmochimica Acta*, 69(5), 1265–1273.

Blaine, F.A., Linnen, R.L., Holtz, F. and Brügmann, G.E. (2011). The effect of Cl on Pt solubility in haplobasaltic melt: Implications for micronugget formation and evidence for fluid transport of PGEs. *Geochimica et Cosmochimica Acta*, 75, 7792–7805.

Bodnar, R. J. (1994). Synthetic fluid inclusions; XII, The system H2O-NaCl; experimental determination of the halite liquidus and isochores for a 40 wt. % NaCl solution. *Geochimica et Cosmochimica Acta*, 53, 725–733.

Bodnar, R.J., Burnham, C.W. and Sterner, S.M. (1985). Synthetic fluid inclusions in natural quartz. III. Determination of phase equilibrium properties in the system H2O-NaCl to 1000 C and 1500 bars. *Geochimica et Cosmochimica Acta*, 49(9), 1861–1873.

Boorman, S.L., McGuire, J.B., Boudreau, A.E. and Kruger, F.J. (2003). Fluid overpressure in layered intrusions: Formation of a breccia pipe in the Eastern Bushveld Complex, Republic of South Africa. *Mineralium Deposita*, 38, 356–369.

Boorman, S., Boudreau, A.E. and Kruger, F.J. (2004). The Lower Zone – Critical Zone transition of the Bushveld Complex: A quantitative textural study. *Journal of Petrology*, 45, 1209–1235.

Borisov, A. and Danyushevsky, L. (2011). The effect of silica contents on Pd, Pt and Rh solubilities in silicate melts: An experimental study. *European Journal of Mineralogy*, 23(3), 355–367.

Borisov, S. and Palme, H. (1995). The solubility of iridium in silicate melts: New data from experiments with lr10Pt90 alloys. *Geochimica et Cosmochimica Acta*, 59(3), 481–485.

Borisov, S. and Palme, H. (2000). Solubilities of noble metals in Fe-containing silicate melts as derived from experiments in Fe-free systems. *American Mineralogist*, 85(11–12), 1665–1673.

Boudreau, A.E. (1988). Investigations of the Stillwater Complex; IV, The role of volatiles in the petrogenesis of the J-M Reef, Minneapolis adit section. *Canadian Mineralogist*, 26, 193–208.

Boudreau, A.E. (1992). Volatile fluid overpressure in layered intrusions and the formation of potholes. *Australian Journal of Earth Sciences*, 39, 277–287.

Boudreau, A.E. (1993). Chlorine as an exploration guide for the Platinum-group elements in layered intrusions. *Journal of Geochemical Exploration*, 48, 21–37.

Boudreau, A.E. (1994). Pattern formation during crystallization in two crystal, two component systems. *South African Journal of Geology*, 97, 473–485.

Boudreau, A.E. (1995). Crystal aging and the formation of fine-scale layering. *Mineralogy and Petrology*, 54, 55–69.

Boudreau, A.E. (1999a). Fluid fluxing of cumulates: The J-M Reef and associated rocks of the Stillwater Complex, Montana. *Journal of Petrology*, 40(5), 755–772.

Boudreau, A.E. (1999b). PELE – A version of the MELTS software program for the PC platform. *Computers and Geosciences*, 25, 201–203.

Boudreau, A.E. (2004). PALLADIUM – A program to model the chromatographic separation of the platinum-group elements, base metals and sulfur in a solidifying igneous crystal pile. *Canadian Mineralogist*, 42, 393–403.

Boudreau, A.E. (2008). Modeling the Merensky Reef, Bushveld Complex, South Africa. *Contributions to Mineralogy and Petrology*, 156, 431–437.

Boudreau, A.E. (2009). Transport of the platinum-group elements by igneous fluids in layered intrusions. In *New Developments in Magmatic Ni-Cu and PGE Deposits* (Li, C. and Ripley, E.M., eds.). Beijing: Geological Publishing House, 229–249.

Boudreau, A.E. (2011). The evolution of texture and layering in layered intrusions. *International Geology Review*, 53(3–4), 330–353.

Boudreau, A.E. (2016a). Bubble migration in a compacting crystal-liquid mush. *Contributions to Mineralogy and Petrology*, 171, 32. doi: 10.1007/s00410-016-1237-9.

Boudreau, A.E. (2016b). The Stillwater Complex, Montana – Overview and the significance of volatiles. *Mineralogical Magazine*, 80, 585–637.

Boudreau, A.E. (2016c). *The Bushveld complex as an analog to subduction zone hydrothermal systems*. Penrose Conference on Layered Intrusions, Red Lodge Montana, 6–10 August, 2016 (abstract).

Boudreau, A.E. and Hoatson, D.M. (2004). Halogen variations in the paleoProterozoic layered mafic-ultramafic Intrusions of the East Kimberly, Western Australia: Implications for platinum-group element mineralization. *Economic Geology*, 99, 1015–1026.

Boudreau, A.E. and Kruger, F.J. (1990). Variations in the composition of apatite through the Merensky cyclic unit on the Western Bushveld Complex. *Economic Geology*, 85, 737–745.

Boudreau, A.E. and McBirney, A.R. (1997). The Skaergaard Layered Series. Part III. In Situ Layering. *Journal of Petrology*, 38, 1003–1020.

Boudreau, A.E. and McCallum, I.S. (1985). Features of the Picket Pin Pt-Pd deposit). In *Stillwater Complex, Montana: Geology and Guide* (Czamanske, C.K. and Zientek, M.L., eds.). Montana Bureau of Mines and Geology, Special Publication 92, 346–357.

Boudreau, A.E. and McCallum, I.S. (1986). Investigations of the Stillwater Complex. Part III. The Picket Pin Pt-Pd deposit. *Economic Geology*, 81, 1953–1975.

Boudreau, A.E. and McCallum, I.S. (1989). Investigations of the Stillwater Complex: Part V. Apatites as indicators of evolving fluid composition. *Contributions to Mineralogy and Petrology*, 102, 138–153.

Boudreau, A.E. and McCallum, I.S. (1990). Low temperature alteration of REE-rich chlorapatite from the Stillwater Complex Montana. *American Mineralogist*, 75, 87–693.

Boudreau, A.E. and McCallum, I.S. (1992a). Concentration of platinum-group elements by magmatic fluids in layered intrusions. *Economic Geology*, 87, 1830–1848.

Boudreau, A.E. and McCallum, I.S. (1992b). Infiltration metasomatism in layered intrusions—an example from the Stillwater Complex, Montana. *Journal of Volcanology and Geothermal Research*, 52, 171–183.

Boudreau, A.E. and Meurer, W.P. (1999). Chromatographic separation of the platinum-group elements, gold, base metals and sulfur during degassing of a compacting and solidifying igneous crystal pile. *Contributions to Mineralogy and Petrology*, 134, 174–185.

Boudreau, A.E. and Philpotts, A.J. (2002). Quantitative modeling of compaction in the Holyoke flood basalt flow, Hartford Basin, Connecticut. *Contributions to Mineralogy and Petrology*, 144, 176–184.

Boudreau, A.E. and Simon, A. (2007). Halogen variations and degassing in the Basement Ferrar sill, Antarctica. *Journal of Petrology*, 48, 1369–1386. DOI:10.1093/petrology/egm022.

Boudreau, A.E., Mathez, E.A. and McCallum, I.S. (1986). Halogen geochemistry of the Stillwater and Bushveld Complexes: Evidence for t transport of the platinum-group elements by Cl-rich Fluids. *Journal of Petrology*, 27, 967–986.

Boudreau, A.E., Love, C. and Hoatson, D. (1993). Halogen geochemistry of the Munni Munni Complex and associated intrusions of the Pilbara Block, W. Australia. *Geochimica et Cosmochimica Acta*, 57, 4467–4477.

Boudreau, A.E., Love, C. and Prendergast, M.D. (1995). Halogen geochemistry of the Great Dyke, Zimbabwe. *Contributions to Mineralogy and Petrology*, 122, 289–300.

Boudreau, A.E., Stewart, M.A. and Spivack, A.J. (1997). Stable Cl isotopes and origin of high-Cl magmas of the Stillwater Complex, Montana. *Geology*, 25(9), 791–794.

Boudreau, A.E., Djon, L., Ychalikian, A. and Corkery, J. (2014). The Lac Des Iles Palladium Deposit, Ontario, Canada. Part I. The effect of variable alteration on the Offset Zone. *Mineralium Deposita*, 49, 625–654.

Boudreau, A.E., Butak, K.C., Geraghty, E.P., Holick, P.A. and Koski, M.S. (2019). Mineral Deposits of the Stillwater Complex. In Geology of Montana). Montana Bureau of Mines and Geology (In press).

Boudreau, B.P. (2012). The physics of bubbles in surficial, soft cohesive sediments. *Marine and Petroleum Geology*, 38, 1–18.

Boudreau, B.P., Algar, C., Johnson, B.D., Croudace, I., Reed, A., Furukawa, Y., Dorgan, K.M., Jumar, P.A., Grader, A.S. and Gardiner, B.S. (2005). Bubble growth and rise in soft sediments. *Geology*, 33, 517–520.

Boyce, J.W., Tomlinson, S.M., McCubbin, F.M., Greenwood, J.P. and Treiman, A.H. (2014). The lunar apatite paradox. *Science*, 344, 400–402.

Bragagni, A., Luguet, A., Fonseca, R.O.C., Pearson, D.G., Lorand, J.-P.,Nowell, G.M. and Kjarsgaard, B.A. (2017). The geological record of base metal sulfides in the cratonic mantle: A microscale 187Os/188Os study of peridotite xenoliths from Somerset Island, Rae Craton (Canada). *Geochimica et Cosmochimica Acta*, 216, 264–285.

Brandeis, G. and Jaupart, C. (1986). On the interaction between convection and crystallization in cooling magma chambers. *Earth and Planetary Science Letters*, 77, 345–361.

Braun, K., Meurer, W. and Boudreau, A.E. (1994). Compositions of pegmatoids beneath the J-M Reef of the Stillwater Complex, Montana, U.S.A. *Chemical Geology*, 113, 245–257.

Brearley, M. and Scarfe, C.M. (1986). Dissolution rates of upper mantle minerals in an alkali basalt melt at high pressure—An experimental study and implications for ultramafic xenolith survival. *Journal of Petrology*, 27, 1157–1182.

Brenan, J.M. (1993). Partitioning of fluorine and chlorine between apatite and aqueous fluids at high pressure and temperature: Implications for the F and Cl content of high P-T fluids. *Earth and Planetary Science Letters*, 117, 251–263.

Brenan, J.M. (1994). Kinetics of fluorine, chlorine and hydroxyl exchange in fluorapatite. *Chemical Geology*, 110, 195–210.

Brenan, J.M. and Andrews, D. (2001). High temperature stability of laurite and Ru-Os-Ir alloy and their role in PGE fractionation in mafic magmas. *Canadian Mineralogist*, 39, 341–360.

Brenan, J.M., McDonough, W.F. and Dalpé, C. (2003). Experimental constraints on the partitioning of rhenium and some platinum-group elements between olivine and silicate melt. *Earth and Planetary Science Letters*, 212, 135–150.

Brenan, J.M., McDonough, W.F., and Ash, R. (2005). An experimental study of the solubility and partitioning of iridium, osmium and gold between olivine and silicate melt. *Earth and Planetary Science Letters*, 237, 855–872.

Brenan, J.M., Finnigan, C.S., McDonough, W.F. and Homolova, V., (2012). Experimental constraints on the partitioning of Ru, Rh, Ir, Pt and Pd between chromite and silicate melt: The importance of ferric iron. *Chemical Geology*, 302–303, 16–32.

Broek, D. (1982). *Elementary Engineering Fracture Mechanics*, Fourth Edition. Dordrecht: Kluwer Academic Publishers.

Brown, G. M. and Peckett, A. (1977). Fluorapatites from the Skaergaard intrusion, East Greenland. *Mineralogy Magazine* 41, 227–232.

Büchl, A., Brügmann, G. and Batanov, V.G. (2004). Formation of podiform chromitite deposits: Implications from PGE abundances and Os isotopic compositions of chromites from the Troodos complex, Cyprus. *Chemical Geology*, 208, 217–232.

Buick, I.S., Gibson, R.L., Cartwright, I., Maas, R., Wallmach, T. and Uken, R. (2000) Fluid flow in metacarbonates associated with emplacement of the Bushveld Complex, South Africa. *Journal of Geochemical Exploration*, 69–70, 391–395.

Buntin, T.J., Grandstaff, D.E., Ulmer, G.C. and Gold, D.P. (1985). A pilot study of geochemical and redox relationships between potholes and adjacent normal Merensky Reef of the Bushveld Complex. *Economic Geology*, 80, 975–987.

Burnham, C.W. (1979). Magmas and hydrothermal fluids. In *Geochemistry of Hydrothermal Ore Deposits*, 2nd Edition (Barnes, H.L., ed.). New York: Wiley, 71–136.

Burton, K.W., Schiano, P., Birck, J.-L. and Allègre, C. J. (1999) Osmium isotope disequilibrium between mantle minerals in a spinel-lherzolite. *Earth and Planetary Science Letters*, 172, 311–322.

Cabri, L.J. (1981). The platinum-group minerals. In *Platinum-Group Elements: Mineralogy, Geology, Recovery* (Cabri, L.J., ed.). Canadian Institute of Mining, Metallurgy and Petroleum Special Volume 23, 13–129.

Caltabiano, T., Burton, M., Giammanov, S., Allard, P., Bruno, N., Murè. F. and Romano R. (2004). Volcanic gas emissions from the summit craters and flanks of Mt. Etna, 1987–2000. American Geophysical Union Geophysical Monograph doi:10.1029/143GM08.

Cameron, E.N. (1978). The Lower Zone of the Eastern Bushveld Complex in the Olifants River Trough. *Journal of Petrology*, 19(3), 437–462.

Cameron, E.N. and Desborough, G.A. (1964). Origin of certain magnetite-bearing pegmatites in the eastern part of the Bushveld Complex, South Africa. *Economic Geology*, 59, 197–225.

Campbell, I.H. (1978). Some problems with the cumulus theory. *Lithos*, 11, 311–323.

Campbell, I.H. (1986). A fluid dynamic model for the potholes of the Merensky reef. *Economic Geology*, 81, 1118–1125.

Campbell, I.H. (1987). Distribution of orthocumulate textures in the Jimberlana intrusion. *Journal of Geology*, 95, 35–54.

Campbell, I.H. and Murck, B.W. (1993). Petrology of the G and H Chromitite Zones in the Mountain View Area of the Stillwater Complex, Montana. *Journal of Petrology*, 34 (2), 291–316.

Campbell, I.H. and Naldrett, A.J. (1979). The influence of silicate: Sulfide ratios on the geochemistry of magmatic sulfides. *Economic Geology*, 74(6), 1503–1506.

Campbell, I.H., Roeder, P.L. and Dixon, J.M. (1978). Plagioclase buoyancy in basaltic liquids as determined with a centrifuge furnace. *Contributions to Mineralogy Petrology*, 67, 369–377.

Campbell, I.H., Naldrett, A.J. and Barnes, S.J. (1983). A model for the origin of the platinum-rich sulphide horizons in the Bushveld and Stillwater complexes. *Journal of Petrology*, 24, 33–165.

Candela, P.A. (1986). Toward a thermodynamic model for the halogens in magmatic systems: An application to melt-vapor-apatite equilibria. *Chemical Geology*, 57, 289–301.

Candela, P.A. (1991). Physics of aqueous phase evolution in plutonic environments. *American Mineralogist*, 76, 1081–1091.

Capobianco, C.H. and Drake, M. (1990). Partitioning of ruthenium, rhodium, and palladium between spinel and silicate melt and implications for platinum-group element fractionation trends. *Geochimica et Cosmochimica Acta*, 54, 869–874.

Carr, H.W., Groves, D.I. and Cawthorn, R.G. (1994). The importance of synmagmatic deformation in the formation of Merensky Reef potholes in the Bushveld Complex. *Economic Geology*, 89, 1398–1410.

Cashman, K.V. (1993). Relationship between plagioclase crystallization and cooling rate in basaltic melts. *Contributions to Mineralogy Petrology*, 113, 126–142.

Cashman, K.V. and Ferry, J.M. (1988). Crystal size distribution (CSD) in rocks and the kinetics and dynamics of crystallization. III. Metamorphic crystallization. *Contributions to Mineralogy Petrology*, 99, 401–415.

Cawthorn, R.G. (1976). Some chemical controls in igneous amphibole compositions. *Geochimica et Cosmochimica Acta*, 40, 1319–1328.

Cawthorn, R.G. (1982). An origin for the small scale fluctuations in orthopyroxene composition in the lower and critical zones of the Bushveld Complex, South Africa. *Chemical Geology*, 36, 227–236.

Cawthorn, R.G. (1994). Formation of chlor- and fluor-apatite in layered intrusions. *Mineralogical Magazine*, 58, 299–306.

Cawthorn, R.G. (2002). Delayed accumulation of plagioclase in the Bushveld Complex. *Mineralogical Magazine*, 66(6), 881–893.

Cawthorn, R.G. and Poulton, K.L. (1988). Evidence for fluid in the footwall beneath potholes in the Merensky Reef of the Bushveld Complex. In *Geo-Platinum 87* (Prichard, H.M., Potts, P.J., Bowles, J.F.W. and Cribbs, S.J., eds.), London: Elsevier, 343–356.

Cawthorn, R.G. and Walsh, K.L. (1988). The use of phosphorus contents in yielding estimates of the proportion of trapped liquid in cumulates of the Upper Zone of the Bushveld Complex. *Mineralogical Magazine*, 52, 81–89.

Cawthorn, R.G. and Lee, C. (1998). *Field Excursion Guide to the Bushveld Complex. 8th International Platinum Symposium*. The Geological Society of South Africa and The South African Institute of Mining and Metallurgy.

Cawthorn, R.G. and Boerst, K. (2006). Origin of the pegmatitic pyroxenite in the Merensky Unit, Bushveld Complex, South Africa. *Journal of Petrology*, 47, 1509–1530.

Cawthorn, R.G. and Webb, S.J. (2013). Cooling of the Bushveld Complex, South Africa: Implications for paleomagnetic reversals. *Geology*, 41(6), 687–690.

Cawthorn, R.G., Harris, C. and Kruger, F.J. (2000). Discordant ultramafic pegmatoidal pipes in the Bushveld Complex. *Contributions to Mineralogy and Petrology*, 140, 119–133.

Cawthorn, R.G., Lee, C., Schouwstra, R.P. and Mellowship, P. (2002). Relationship between PGE and PGM in the Bushveld Complex. *The Canadian Mineralogist*, 40, 311–328.

Cawthorn, R.G., Latipov, R., Klemd, R. and Vuthuza, A. (2017). Origin of discordant ultramafic pegmatites in the Bushveld Complex, South Africa. Contributions to Mineralogy and Petrology (In review).

Chauveau, B. and Kaminski, E. (2008). Porous compaction in transient creep regime and implications for melt, petroleum, and CO2 circulation. *Journal of Geophysical Research – Solid Earth*, 113, B09406. doi:10.1029/2007JB005088

Chitiyo, G., Schweitzer, J., de Waal, S., Lambert, P. and Olgilvie, P. (2008). Predictability of pothole characteristics and their spatial distribution at Rustenburg Platinum Mine.

The Journal of The Southern African Institute of Mining and Metallurgy, 108, 733–740.

Chou, I.-M. (1987). Phase relations in the system NaCl-KCl-H2O. III: Solubilities of halite in vapor-saturated liquids above 445°C and redetermination of phase equilibrium properties in the system NaCl-H2O to 1000°C and 1500 bars. *Geochimica et Cosmochimica Acta*, 51(7), 1965–1975.

Chutas, N.I., Bates, E., Prevec, S.A., Coleman, D.S. and Boudreau, A.E. (2012). Sr and Pb isotopic disequilibrium between coexisting plagioclase and orthopyroxene in the Bushveld Complex, South Africa: Microdrilling and progressive leaching evidence for sub-liquidus contamination within a crystal mush. *Contributions to Mineralogy and Petrology*, 163(4), 653–668.

Clennell, M.B., Judd, A. and Hovland, M., (2000). Movement and accumulation of methane in marine sediments: Relation to gas hydrate systems. In *Natural Gas Hydrate in Oceanic and Permafrost Environments*, (Max, M.D., ed.). Dordrecht: Kluwer Academic Publishers, 105–122.

Connolly, J.A.D. (2010). The mechanics of metamorphic fluid expulsion. *Elements*, 6(3), 165–172. doi:10.2113/gselements.6.3.165

Connolly, J.A.D. and Podladchikov, Y.Y. (1998). Compaction – Driven fluid flow in viscoelastic rock. *Geodinamica Acta*, 11(2/3), 55–84.

Connolly, J.A.D. and Podladchikov, Y.Y. (2007). Decompaction weakening and channeling instability in ductile porous media: Implications for asthenospheric melt segregation. *Journal of Geophysical Research – Solid Earth*, 112(B10), 15. doi:10.1029/2005jb004213

Coombs, M.L., Sisson, T.W. and Kimura, J.-I. (2004). Ultra-high chlorine in submarine Kilauea glasses: Evidence for direct assimilation of brine by magma. *Earth and Planetary Science Letters*, 217, 297–313.

Crawford, A.J., Fallon, T.J. and Green, D.H. (1989). Classification, petrogenesis and tectonic setting of boninites. In *Boninites and Related Rocks* (Crawford, A.J., ed.). London: Unwin Hyman, 1–49.

Criscenti, L.J. (1984). *The Origin of Macrorhythmic Units in the Stillwater Complex*. M.S. thesis, University of Washington, Seattle, Washington.

Czamanske, G.K., Zientek, M.L. and Manning, C.E. (1991). Low-K granophyres of the Stillwater Complex, Montana. *American Mineralogist*, 76, 1646–1661.

Damian, S. Smith, D.S., Bassson, I.J. and Reid, D.L. (2003). Normal reef subfacies of the Merensky Reef at Northam Platinum Mine, Zwartklip Facies, Western Bushveld complex, Sourth Africa. *The Canadian Mineralogist*, 42, 243–260.

Davidson, J.P. and Tepley, F.J. (1997). Recharge in volcanic systems: Evidence from isotope profiles of phenocrysts. *Science*, 275, 826–829.

Davies, G. and Tredoux, M. (1985). The platinum-group element and gold contents of the marginal rocks and sills of the Bushveld Complex. *Economic Geology*, 80, 838–848.

Davydov, M.N. (2012). Nucleation and growth of a gas bubble in magma. *Journal of Applied Mechanics and Technical Physics*, 53(3), 324–332.

DePaolo, D.J. and Wasserburg, G.J. (1979). Sm-Nd age of the Stillwater complex and the mantle evolution curve for neodymium. *Geochimica et Cosmochimica Acta*, 43(7), 999–1008.

De Vivo, B., Torok, K., Ayuso, R.A., Lima, A. and Lirer, L. (1995). Fluid inclusion evidence for magmatic silicate/saline/CO2 immiscibility and geochemistry of alkaline xenoliths from Ventotene Island, Italy. *Geochimica et Cosmochimica Acta*, 59(14), 2941–2953.

De Waal, S.A. (1977). Carbon dioxide and water from metamorphic reactions as agents for sulphide and spinel precipitation in mafic magmas. *Geological Society of South Africa Transactions*, 80, 198–196.

Dixon, J. E., Clague, D.A. and Stolper, E.M. (1991). Degassing history of water, sulfur and carbon in submarine lavas from Kilaueau Volcano, Hawaii. *Journal of Geology*, 99 371–394.

Doherty A.L., Webster J.D., Goldoff, B.A. and Piccoli P.M. (2014). Partitioning behavior of chlorine and fluorine in felsic melt–fluid(s)–apatite systems at 50 MPa and 850–950 °C. *Chemical Geology*, 384, 94–109.

Donaldson, M.J. (1974). Petrology of the Munni Munni Complex, Roebourne, Western Australia. *Journal of the Geological Society of Australia*, 21, 1–16.

Dreibus, G., Palme, H., Spettel, B., Zipfel, J. and Wanke, H. (1995). Sulphur and selenium in chondritic meteorites. *Meteoritics*, 30, 439–445.

Drinkwater, J.L., Czamanske, G.K. and Ford, A.B. (1990). Apatite of the Dufek intrusion: Disribution, paragenesis, and chemistry. *Canadian Mineralogist*, 28, 835–854.

Duran, C.J., Barnes, S.-J. and Corkery, J.T. (2015). Chalcophile and platinum-group element distribution in pyrites from the sulfide-rich pods of the Lac des Iles Pd deposits, Western Ontario, Canada: Implications for post-cumulus re-equilibration of the ore and the use of pyrite compositions in exploration. *Journal of Geochemical Exploration*, 158, 223–242.

Duran, C.J., Barnes, S.-J. and Corkery, J.T. (2016). Geology, petrography, geochemistry, and genesis of sulfide-rich pods in the Lac des Iles palladium deposits, western Ontario, Canada. *Mineralium Deposita*, 51(4), 509–532.

Eales, H.V. and Cawthorn, R.G. (1996). The Bushveld Complex. In *Layered Intrusions* (Cawthorn, R.G., ed.). Amsterdam: Elsevier, 181–229.

Eales, H.V., De Klerk, W.J. and Teigler, B. (1990). Evidence for magma mixing processes within the Critical and Lower Zones of the northwestern Bushveld Complex. *Chemical Geology*, 88, 261–278.

Eales, H.V., De Klerk, W.J., Teigler, B. and Maier, W.D. (1994) Nature and origin of orthopyroxenites in the western Bushveld Complex, in the light of compositional data. *South African Journal of Geology*, 97(4), 399–407.

Eales, H.V., Field, M., de Klerk, W.J. and Scoon, R.N. (1988). Regional trends of chemical variation and thermal erosion in the Upper Critical Zone, western Bushveld Complex. *Mineralogical Magazine*, 52: 63–79.

Economou, M.I. (1986). Platinum group elements (PGE) in chromite and sulphide ores within the ultramafic zone of some Greek ophiolite complexes. In *Metallogeny of Basic and Ultrabasic Rocks* (Gallagher, M.J., Ixer, R.A., Neary, C.R. and Prichard, H.M., eds.). London: The Institution of Mining and Metallurgy, 441–453.

Eggler, D.H. and Burnham, C.W. (1973). Crystallization and fractionation trends in the system andesite-H2O-CO2-O2 at pressures to 10 kilobars. *Geological Society of America Bulletin* 84, 2517–2532.

Elliot, W.C., Grandstaff, D.E., Ulmer, G.C. and Gold, D.P. (1982). An intrinsic oxygen fugacity study of platinum-carbon associations in layered intrusions. *Economic Geology*, 77, 1493–1510.

Eriksson, P.G., Hattingh, P.J. and Altermann, W. (1995). An overview of the geology of the Transvaal Sequence and Bushveld Complex. *Mineralium Deposita*, 30, 98–111.

Eriksson, P.G., Schweitzer, K., Bosch, P.J.A., Schereiber, U.M., Van Deventer, J. and Hatton, C.J. (1993). The Transvaal Sequence: an overview. *Journal of African Earth Sciences*, 16(1/2), 25–51.

Ertel, W., O'Neill, H.St.C., Sylvester, P.J. and Dingwell, D.B. (1999). Solubilities of Pt and Rh in a haplobasaltic silicate melt at 1300°C. *Geochimica et Cosmochimica Acta*, 63(16), 2439–2449.

Fahlquist, L.S. and Popp, R.K. (1989). The effect of NaCl on bunsenite solubility and Ni complexing in supercritical aqueous fluids. *Geochimica et Cosmochimica Acta*, 53, 989–995.

Fehlhaber, K. and Bird, D. K. (1991). Oxygen-isotope and mineral alteration in gabbros of the Lower Layered Series, Kap Edvard Holm Complex, East Greenland. *Geology*, 19, 819–822.

Feig, S.T, Koepke, J. and Snow, J.E. (2010). Effect of oxygen fugacity and water on phase equilibria of a hydrous tholeiitic basalt. *Contributions to Mineralogy and Petrology*, 160, 551–568

Ferguson, J. and McCarthy, T.S. (1970). Origin of an ultramafic pegmatoid in the eastern part of the Bushveld Complex. *Geological Society of South Africa Special Pub*, 1, 74–79.

Ferris, J., Johnson, A. and Strey, B. (1998). Form and extent of the Dufek intrusion, Antarctica, from newly compiled aeromagnetic data. *Science Letters*, 154, 185–202.

Fiege, A., Behrens, H., Holtz, F. and Adams, F. (2014). Kinetic vs. thermodynamic control of degassing of H2O-S ± Cl-bearing andesitic melts. *Geochimica et Cosmochimica Acta*, 125, 241–264

Finnigan, C.S., Brenan, J.M., Mungall, J.E. and McDonough, W.F. (2008). Experiments and models bearing on the role of chromite as a collector of platinum group minerals by local reduction. *Journal of Petrology*, 49, 1647–1665.

Fiorentini, M.L, Stone, W.E., Beresford, S.W. and Barley, M.E. (2004). Platinum-group element alloy inclusions in chromites from Archaean mafic-ultramafic units: Evidence from the Abitibi and the Agnew-Wiluna Greenstone Belts. *Mineralogy and Petrology*, 82, 341–355.

Fleet, M.E. and Wu, T.-W. (1993). Volatile transport of platinum-group elements in sulfide-chloride assemblages at 1000 °C. *Geochimica et Cosmochimica Acta*, 57, 3519–3531.

Fleet, M.E. and Wu, T.-W. (1995). Volatile transport of precious metals at 1000 °C: Speciation, fractionation, and effect of base metal sulfide. *Geochimica et Cosmochimica Acta*, 59, 487–495.

Fleet, M.E., Crocket, J.H. and Stone, W.E. (1996). Partitioning of platinum-group elements (Os, Ir, Ru, Pt, Pd) and gold between sulfide liquid and basalt melt. *Geochimica et Cosmochimica Acta*, 60, 2397–2412.

Fonseca, R.O.C., Campbell, I.H., O'Neill, H.S.C. and Allen, C.M. (2009). Solubility of Pt in sulfide mattes: Implications for the genesis of PGE-rich horizons in layered intrusions. *Geochimica et Cosmochimica Acta*, 73(19), 5764–5777.

Font, L., Davidson, J.P., Pearson, D.G., Nowell, G.M., Jerram, D.A. and Ottley, C.J. (2008). Sr and Pb isotope micro-analysis of Plagioclase Crystals from Skye Lavas: An insight into open-system processes in a Flood Basalt Province. *Journal of Petrology*, 49(8), 1449–1471.

Ford, C.E., Biggar, G.M., Humphries, D.J., Wilson, G., Dixon, D. and O'Hara, M.J. (1972). Role of water in the evolution of the lunar crust; an experimental study of sample 14310; an indication of lunar calc-alkaline volcanism. Proceedings of the Third Lunar Science Conference. Geochimica et Cosmochimica Acta, Suppl. 3, 207–229.

Frost, B.R. and Touret, J.L.R. (1989). Magmatic CO2 and saline melts from the Sybille monzosyenite, Laramie anorthosite complex, Wyoming. *Contributions to Mineralogy and Petrology*, 103, 178–186.

Fowler, A.C., Rust, A.C. and Vynnycky, M. (2015). The formation of vesicular cylinders in pahoehoe lava flows. *Geophysical and Astrophysical Fluid Dynamics*, 109, 39–61.

Gaetani, G.A. and Grove, T.L. (1998). The influence of water on melting of mantle peridotite. *Contributions to Mineralogy and Petrology*, 131, 323–346.

Gaetani, G.A., Grove, T.L. and Bryan, E.B. (1994). Experimental phase relations of basaltic andesite from Hole 839B under hydrous and anhydrous conditions). In *Proceedings of the Ocean Drilling Program, Scientific Results, 135* (Hawkins, J., et al., eds). College Station, Ocean Drilling Program, 557–564.

Gafeira, J., Dolan, M.F.J. and Monteys, X. (2018). Geomorphic characterization of pockmarks by using a GIS-based semi-automatic toolbox. *Geosciences*, 8. doi:10.3390/geosciences8050154

Gain, S.B. (1985). The geologic setting of the platiniferous UG2 chromitite layer on the Farm Maandagshoek, eastern Bushveld Complex. *Economic Geology*, 80, 925–943.

Gammons, C.H., Bloom, M.S. and Yu, Y. (1992). Experimental investigation of the hydrothermal geochemistry of platinum and palladium: I. Solubility of platinum and palladium sulfide minerals in NaCl/H2SO4 solutions at 300°C. *Geochimica et Cosmochimica Acta*, 56, 3881–3894.

Gardner, J.E., M. Hilton, M. and Carroll, M.R. (1999). Experimental constraints on degassing of magma: Isothermal bubble growth during continuous decompression from high pressure. *Earth and Planetary Science Letters*, 168, 201–218.

Garuti, G., Bea, F., Zaccarini, F. and Montero, P. (2001). Age, geochemistry and petrogenesis of the ultramafic pipes in the Ivrea Zone, NW Italy. *Journal of Petrology*, 42, 433–457.

Garuti, G., Zaccarini, F., Cabella, R., Fershtater, G., (1997). Occurrence of unknown Ru–Os–Ir–Fe oxides in the chromitites of the Nurali ultramafic complex, Southern Urals, Russia. *Canadian Mineralogist*, 35, 1431–1439.

Garuti, G., Zaccarini, F., Moiloshaq, V. and Alimov, V. (1999). Platinum-group minerals as indicators of sulfur fugacity in ophiolitic upper mantle; an example from chromitites of the Ray-Iz ultramafic complex, Polar Urals, Russia. *Canadian Mineralogist*, 37, 1099–1115.

Gee, J.S., Cheadle, M.J., Meurer, W.P. and Grimes, C.B. (2016). Magnetic Constraints on the Thermal History of the Dufek Intrusion. Geological Society of America Penrose Conference on Layered Mafic Intrusions and Associated Economic Deposits, Red Lodge, Montana, August 8–12, 2016 (Abstract).

Gerlach, T.M. and Graber, E.J. (1985). Volatile budget of Kilauea volcano. *Nature*, 313, 273–277.

Ghiorso, M.S. and Sack, R.O. (1995). Chemical mass transfer in magmatic processes. IV. A revised and internally consistent thermodynamic model for the interpretation and extrapolation of liquid-solid equilibria in magmatic systems at elevated temperatures and pressures. *Contributions to Mineralogy and Petrology*, 119, 197–212.

Ghiorso, M.S., Hirschman, M. and Sack, R.O. (1994). MELTS: Software for thermodynamic modeling of magmatic systems. *EOS*, 75, 571.

Gijbels, R.H., Millard, H.T. Jr., Desborough, G.A. and Bartel, A.J. (1974). Osmium, ruthenium, iridium and uranium in silicates and chromite from the eastern Bushveld Complex, South Africa. *Geochimica et Cosmochimica Acta*, 38, 319–337.

Gleason, J.D., Gutzmer, J., Kesler, S.E. and Zwingmann, H. (2011). 2.05-Ga Isotopic Ages for Transvaal Mississippi Valley–Type Deposits: Evidence for Large-Scale Hydrothermal Circulation around the Bushveld Igneous Complex, South Africa. *Journal of Geology*, 119, 69–80.

Glebovitsky, V.A., Semenov, V.S., Belyatsky, B.V., Koptev-Dvornikov, E.V., Pchelintseva, N.F., Kireev, B.S. and Koltsov, A.B. (2001). The structure of the Lukkulaisvaara Intrusion, Oulanka Group, Northern Karelia: Petrological Implications. *Canadian Mineralogist*, 39, 607–637.

Godard, M., Bodinier, J.-L. and Vasseur, G. (1995). Effects of mineralogical reactions on trace element redistributions in mantle rocks during percolation processes: A chromatographic approach. *Earth and Planetary Science Letters*, 133, 449–461.

Godel B. (2015). Platinum-Group Element Deposits. In *Layered Intrusions: Recent Advances in the Understanding of the Ore Forming Processes* (Charlier B., Namur O., Latypov R. and Tegner C., eds.). Dordrecht: Springer Geology, 379–432.

Godel, B., Barnes, S.-J. and Maier, W.D. (2006). 3-D distribution of sulphide minerals in the Merensky Reef (Bushveld Complex, South Africa) and the J-M Reef (Stillwater Complex, USA) and their relationship to microstrucures using X-ray computed tomography. *Journal of Petrology*, 47, 1853–1872.

González-Jiménez, J.M., Griffin, W.L., Gervilla, F., Proenza, J.A., O'Reilly, S.Y. and Pearson, N.J. (2013). Chromitites in ophiolites: How, where, when, why? Part I. A review and new ideas on the origin and significance of platinum-group minerals. *Lithos*, 189, 127–139.

Graham, C.M., Valley, J.W., Eiler, J.M. and Wada, H. (1998). Timescales and mechanisms of fluid infiltration in a marble: An ion microprobe study. *Contributions to Mineralogy and Petrology*, 132, 371-389.

Grant, N.K. and Moiling, P.A. (1981). A strontium isotope and trace element profile through the Partridge River tractolite, Duluth Complex, Minnesota. *Contributions to Mineralogy and Petrology*, 77, 296–305.

Green, D.H., Schmidt, M.W. and Hibberson, W.O. (2004). Island-arc Ankaramites: Primitive Melts from Fluxed Refractory Lherzolitic Mantle. *Journal of Petrology*, 45, 392–403.

Grove T.L., Till, C.B. and Krawczynski, M.J. (2012). The role of H2O in subduction zone magmatism. *Annual Review of Earth and Planetary Sciences*, 40, 413–439.

Gurioli, L., Harris, A.J.L., Houghton, B.F., Polacci, M. and Ripepe, M. (2008). Textural and geophysical characterization of explosive basaltic activity at Villarrica volcano. *Journal of Geophysical Research*, 113, B08206. doi:10.1029/2007JB005328.

Halkoaho T.A. and Alapieti, T.T. (1993). Variation in F and Cl-contents of apatite in the Penikat layered intrusion, Northern Finland. In Symposium on layering in igneous complexes, September, 1993, Johannesburg, South Africa, Abstracts Volume, 18–19.

Hamilton, D.L. and Anderson, G.M. (1967). Effects of water and oxygen pressure on the crystallization of basaltic magmas, In *Basalts* (Hess, H.H. and Poldervaart, A., eds.). New York: Wiley Interscience, 445–482.

Hamilton, J. (1977). Sr isotope and trace element studies of the Great Dyke and Bushveld mafic phase and their relation to early Proterozoic magma genesis in southern Africa. *Journal of Petrology*, 18, 24–52.

Hamlyn, P.R. and Keays, R.R. (1986). Sulfur saturation and second-stage melts: Application to the Bushveld platinum metal deposits. *Economic Geology*, 81, 1431–1445.

Hanley, J.J. and Gladney, E.R. (2011). The presence of carbonic-dominant volatiles during the crystallization of sulfide-bearing mafic pegmatites in the North Roby Zone, Lac des Iles Complex, Ontario. *Economic Geology*, 106, 33–54.

Hanley J., Pettke T., Mungall J. and Spooner E.T.C. (2005a) Fluid and melt inclusion evidence for platinum-group element transport by high salinity fluids and halide melts below the J-M Reef, Stillwater Complex, Montana, U.S.A., 10th Platinum Symposium (abstract)

Hanley, J.J., Mungall, J.E., Pettke, T., Spooner, E.T.C. and Bray, C.J. (2005b). Ore metal redistribution by hydrocarbon–brine and hydrocarbon–halide melt phases, North Range footwall of the Sudbury Igneous Complex, Ontario, Canada. *Mineralium Deposita*, 40, 237–256.

Hanley, J.J., Pettke T., Mungall, J.E. and Spooner, E.T.C. (2005c). The solubility of platinum and gold in NaCl brines at 1.5 kbar, 600 to 800°C: A laser ablation ICP-MS pilot study of synthetic fluid inclusions. *Geochimica et Cosmochimica Acta*, 69 (10), 2593–2611.

Hanley, J.J., Mungall, J.E., Pettke, T., Spooner, E.T.C. and Bray, C.J. (2008). Fluid and halide melt inclusions of magmatic origin in the Ultramafic and Lower Banded series, Stillwater Complex, Montana, USA. *Journal of Petrology*, 49(6), 1133–1160.

Hanley, J.J., Ames, D., Barnes, J., Sharp, Z. and Guillong, M. (2011). Interaction of magmatic fluids and silicate melt residues with saline groundwater in the footwall of the Sudbury Igneous Complex, Ontario, Canada: New evidence from bulk rock geochemistry, fluid inclusions and stable isotopes. *Chemical Geology*, 281, 1–25.

Hardy, S.C. and Voorhees, P.W. (1988). Ostwald ripening in a system with a high volume fraction of coarsening phase. *Metallurgical and Materials Transactions A*, 19A, 2713–2721.

Harper, M.P. (2004). *Platinum Group Element Mineralization in "Ballrooms" of the J-M Reef of the Stillwater Complex, Montana*. M.S. thesis, Brigham Young University, Provo.

Harmer, R.E. and Sharpe, M.R. (1985). Field relations and Sr isotope systematics of the marginal rocks of the eastern Bushveld Complex. *Economic Geology*, 80, 813–837.

Harmer, R.E., Auret, J.M. and Eglington, B.M. (1995). Lead isotope variations within the Bushveld complex, Southern Africa: A reconnaissance study. *Journal of African Earth Sciences*, 21(4), 595–606.

Harris, C. and Chaumba, J.B. (2001). Crustal contamination and the fluid-rock interaction during the formation of the Platreef, Northern limb of the Bushveld Complex, South Africa. *Journal of Petrology*, 42, 1321–1347.

Harris, C., Pronost, J.J.M., Ashwal, L.D. and Cawthorn, R.G. (2005). Oxygen and Hydrogen Isotope Stratigraphy of the Rustenburg Layered Suite, Bushveld Complex: Constraints on Crustal Contamination. *Journal of Petrology*, 46(3), 579–601.

Harris, N., McMillan, A., Holness, M., Uken, R., Watkeys, M., Rogers, N. and Fallick, A. (2003). Melt generation and fluid flow in the thermal aureole of the Bushveld Complex. *Journal of Petrology*, 44(6), 1031–1054.

Hart, S.R. and Ravizza, G.E. (1996). Os partitioning between phases in lherzolite and basalt. *Geophysical Monographs*, 95, 123–134.

Hart, S.R. and Kinloch, E.D. (1989). Osmium isotope systematics in Witwatersrand and Bushveld ore deposits. *Economic Geology*, 84, 1651–1655.

Harvey, J., Warren, J.M. and Shirey, S.B. (2016). Mantle sulfides and their role in Re–Os and Pb isotope geochronology. *Reviews in Mineralogy and Geochemistry*, 81, 579–649.

Haskin, L.A. and Salpas, P.A. (1992). Genesis of compositional characteristics of Stillwater AN-I and AN-II thick anorthosite units. *Geochimica et Cosmochimica Acta*, 56, 1187–1212.

Hatch, F.H., Wells, A.K. and Wells, M.K. (1974). *Petrology of the Igneous Rocks*. London: Thomas Murphy and Company.

Hauri, E.H., Gaetani, G.A. and Green, T.H. (2006). Partitioning of water during melting of the Earth's upper mantle at H2O-undersaturated conditions. *Earth and Planetary Science Letters*, 248, 715–734.

Haughton, D.R., Roeder, P.L. and Skinner, B.J. (1974). Solubility of sulfur in mafic magmas. *Economic Geology*, 69, 451–467.

Hawkesworth, C.j., Gallagher, K., Hergt, J.M. and McDermott, F. (1993). Manlte and slab contributions in arc magmas. *Annual Review of Earth and Planetary Sciences*, 21, 175–204.

Helz, R.T. (1976). Phase relations of basalts in their melting ranges at PH2O = 5 kb. Part II. Melt compositions. *Journal of Petrology*, 17(2), 139–193.

Helz, R.T. (1985). Compositions of fine-grained mafic rocks from sills and dikes associated with the Stillwater complex. In *Stillwater complex, Geology and Guide* (Czamanske, G.K. and Zientek, M.L., eds.). Montana Bureau of Mines and Geology Special Publication 92, pp.97–117.

Henderson, P. (1970). The significance of mesostasis of basic layered igneous rocks. *Journal of Petrology*, 11(3), 463–473.

Hermann, J., Spandler, C. Hack, A. and Korsakov, A.V. (2006). Aqueous fluids and hydrous melts in high-pressure and ultra-high pressure rocks: Implications for element transfer in subduction zones. *Lithos*, 92, 399–417.

Hess, H.H. (1960). Stillwater igneous complex, Montana – A quantitative mineralogical study. *Geological Society of America, Memoir* 80.

Hiemstra, S. (1979). The role of collectors in the formation of the platinum deposits in the Bushveld Complex. *Canadian Mineralogist*, 17, 469–482.

Higgins, M.D. (1991). The origin of laminated and massive anorthosite, Sept Iles Layered Intrusion, Quebec, Canada. *Contributions to Mineralogy and Petrology*, 106, 340–354.

Higgins, M.D. (1998). Origin of anorthosite by textural coarsening: Quantitative measurements of a natural sequence of textural development. *Journal of Petrology*, 39, 1307–1325.

Higgins, M.D. (2002). The role of textural coarsening in the development of the Kiglapait layered mafic intrusion, Labrador, Canada: A crystal size distribution study. *Contributions to Mineralogy and Petrology*, 144, 314–330.

Higgins, M.D. (2006). *Quantitative Textural Measurements in Igneous and Metamorphic Petrology: Cambridge*, Cambridge University Press, 275 p.

Higgins, M.D. (2011). Textural coarsening in igneous rocks: *International Geology Review*, 53(3–4), 354–376.

Hoatson, D.M. (1991). The petrology and platinum-group element geochemistry of the Munni Munni and Mount Sholl layered mafic-ultramafic intrusions of the West Pilbara Block, Western Australia. Unpublished PhD Dissertation, University of Melbourne.

Hoatson, D.M. and Keays, R.R., (1989). Formation of platiniferous sulfide horizons by crystal fractionation and magma mixing in the Munni Munni layered intrusion, West Pilbara Block, Western Australia. *Economic Geology*, 84, 1775–1804.

Hoatson, D.M., Sun, S.-S. and Keays, R.R. (1992). Petrogenesis of the west Pilbara layered mafic-ultramafic intrusions. In *Petrology and Platinum-Group-Element Geochemistry of Archaean Layered Mafic-Ultramafic Intrusions, West Pilbara Block, Western Australia* (Hoatson, D.M. et al., eds.). Australian Geological Survey Organization Bulletin, 242, 150–66.

Holland, H.D. (1972). Granites, solutions, and base metal deposits. *Economic Geology*, 67, 281–301.

Holland, T. and Blundy, J. (1994). Non-ideal interactions in calcic amphiboles and their bearing on amphibole-plagioclase thermometry. *Contributions to Mineralogy and Petrology*, 116, 433–447.

Holloway, J.R. and Burham, C.W. (1972). Melting relations of basalt with equilibrium water pressure less than total pressure. *Journal of Petrology*, 13, 1–29.

Holness, M.B., Hallworth, M.A., Woods, A. and Sides, R.E. (2007). Infiltration Metasomatism of Cumulates by Intrusive Magma Replenishment: The Wavy Horizon, Isle of Rum, Scotland. *Journal of Petrology*, 48, 563–587.

Holtz, F., Pichavant, M., Barbey, P. and Johannes, W. (1992). Effects of H2O on liquidus phase relations in the haplogranite system at 2 and 5 kbar. *American Mineralogist*, 77, 1223–1241.

Holzheid A. and Grove, T.L. (2002) Sulfur saturation limits in silicate melts and their implications for core formation scenarios for terrestrial planets. *American Mineralogist*, 87, 227–237.

Hoshide, T. and Obata, M. (2010). Zoning and resorption of plagioclase in a layered gabbro, as a petrographic indicator of magmatic differentiation. *Earth and Environmental Science Transactions of the Royal Society of Edinburgh*, 100, 235–249.

Hovis, G.L. and Harlov, D.E. (2010). Solution calorimetric investigation of fluor-chlorapatite crystalline solutions. *American Mineralogist*, 95, 946–952.

Hovland, M. and Judd, A.G. (1988). *Seabed Pockmarks and Seepages: Impact on Geology, Biology and the Marine Environment*. London: Graham and Trotman (Kluwer).

Hsu, L.C., Lechler, P.J. and Nelson, J.H. (1991). Hydrothermal solubility of palladium in chloride solutions from 300° to 700°C: Preliminary results. *Economic Geology*, 86, 422–427.

Huber, C., Bachmann, O., Vigneresse, J.L., Dufek, J. and Parmigiani, A. (2012). A physical model for metal extraction and transport in shallow magmatic systems. *Geochemistry, Geophysics, Geosystems*, 13, Q08003. doi:10.1029/2012GC004042.

Huber, C. and Parmigiani, A. (2018). A physical model for three-phase compaction in Silicic Magma Reservoirs. *Journal of Geophysical research: Solid Earth*, 123. doi. org/10.1002/2017JB015224

Hulbert, L.J., Carne, R.C., Gregoire, C.D. and Paktunc, D. (1992). Sedimentary nickel, zinc and platinum-group-element mineralization in Devonian black shales at the Nick Property, Yukon, Canada: A new deposit type. *Exploration and Mining Geology*, 1, 39–62.

Hulbert, L.J. and Von Gruenewaldt, G. (1985) Textural and compositional features of chromite in the Lower and Critical zones of the Bushveld complex south of Potgietersrus. *Economic Geology*, 80, 872–895.

Hunter, R.H., (1987). Textural equilibrium in layered igneous rocks. In Origins of Igneous Layering (Parsons, I., ed.). *NATO ASI Series C*, 196, 473–503.

Huntington, H. D. (1979). Kiglapait mineralogy I: apatite, biotite, and volatiles. *Journal of Petrology*, 20, 625–652.

Hurwitz, S. and Navon, O. (1994). Bubble nucleation in rhyolitic melts: Experiments at high pressure, temperature, and water content. *Earth and Planetary Science Letters*, 122, 267–280.

Ireland, R.H.P. and Penniston-Dorland, S.C. (2015). Chemical interactions between a sedimentary diapir and surrounding mafic magma: Evidence from the Phepane Dome and Bushveld Complex, South Africa. *American Mineralogist*, 100, 1985–2000.

Irvine, T.N. (1965). Chromian spinel as a petrogenetic indicator. Part I: Theory. *Canadian Journal of Earth Sciences*, 2(6), 648–672.

Irvine, T.N. (1967). Chromian spinel as a petrogenetic indicator. Part II: Petrologic implications. *Canadian Journal of Earth Sciences*, 4, 71–103.

Irvine, T.N. (1974). Petrology of the Duke Island Ultramafic Complex, Southeastern Alaska. *Geological Society of America Memoir* 128, 240 p.

Irvine, T.N. (1975). Crystallization sequences in the Muskox intrusion and other layered intrusions. II. Origin of chromite layers and similar deposits of other magmatic ores. *Geochimica et Cosmochimica Acta*, 39, 991–1020.

Irvine, T.N. (1977). Origin of chromite layers in the Muskox intrusion and other stratiform intrusions: A new interpretation. *Geology*, 5, 273–277.

Irvine, T.N. (1980). Magmatic infiltration metasomatism, double diffusive fractional crystallization and adcumulus growth in the Muskox Intrusion and other layered intrusions. In *Physics of Magmatic Processes* (Hargraves, R.B., ed.). Princeton: Princeton University Press, 325–384.

Irvine, T.N., Keith, D.W. and Todd, S.G. (1983). The J-M platinum-palladium reef of the Stillwater Complex, Montana: II. Origin by double-diffusive convective magma mixing and implications for the Bushveld Complex. *Economic Geology*, 78, 1287–1334.

Jackson, E.D. (1961). Primary Textures and Mineral Associations in the Ultramafic Zone of the Stillwater Complex: Montana. United States Geological Survey Professional Paper 358.

Jackson, E.D. (1963) Stratigraphic and lateral variation of chromite composition in the Stillwater Complex). Mineralogical Society of America Special Paper, 1, 46–54.

Jain, A. and Juanes, R. (2009). Preferential mode of gas invasion in sediments: Grain-scale mechanistic model of coupled multiphase fluid flow and sediment mechanics. *Journal of Geophysical Research – Solid Earth*, 114, B08101. http://dx.doi.org/10.1029/2008JB006002.

Jaupart, C. and Brandeis, G. (1986). The stagnant bottom layer of convecting magma chambers. *Earth and Planetary Sciences Letters*, 80, 183–199.

Joachim, B., Pawley, A., Lyon, I.C., Marquardt (neé Hartmann) K., Henkel, T., Clay P.L., Ruzié, L., Burgess, R. and Ballentine, C.J. (2015). Experimental partitioning of F and Cl between olivine, orthopyroxene and silicate melt at Earth's mantle conditions. *Chemical Geology*, 416, 65–78.

Johnson, J.W., Oelkers, E.H. and Helgeson H.C. (1992). SUPCRT92: A software package for calculating the standard molal thermodynamic properties of minerals, gases, aqueous species and reactions from 1 to 5000 bars and 0° to 1000 °C. *Computers and Geosciences*, 18, 899–947.

Johnson, B.D., Boudreau, B.P., Gardiner, B.S. and Maass, R. (2002). Mechanical response of sediments to bubble growth. *Marine Geology*, 187, 347–363.

Jones, M. (1994). Mechanical principles of sediment deformation. In *The Geological Deformation of Sediments* (Maltman, A., ed.). London: Chapman and Hall, 37–71.

Kamenetsky, V.S. (2006). Melt inclusion record of magmatic immiscibility in crustal and mantle magmas. In *Melt Inclusions in Plutonic Rocks* (Webster, J.D., ed.). Mineralogical Association of Canada Short Course Series, 36, 81–98.

Kanitpanyacharoen, W. and Boudreau, A. (2013). Sulfide-associated mineral assemblages in the Bushveld Complex, South Africa: Platinum-group element enrichment by vapor refining by chloride–carbonate fluids. *Mineralium Deposita*, 48, 193–210.

Keays R.R. (1982). Palladium and iridium in komatiites and associated rocks: Application to petrogenetic problems. In *Komatiites* (Arndt, N.T. and Nisbet, E.G., eds.). Berlin: Springer, 435–457.

Keays, R.R. (1989). Formation of platiniferous sulfide horizons by crystal fractionation and magma mixing in the Munni Munni layered intrusion, West Pilbara Block, Western Australia. *Economic Geology*, 84, 1775–1804.
Keays, R.R. (1995). The role of komatiitic and picritic magmatism and S-saturation in the formation of ore deposits. *Lithos*, 43(1), 1–18.
Keays, R.R. and Lightfoot, P.C. (2010). Crustal sulfur is required to form magmatic Ni–Cu sulfide deposits: Evidence from chalcophile element signatures of Siberian and Deccan Trap basalts. *Mineralium Deposita*, 45(3), 241–257.
Keays, R.R. and Tegner, C. (2016). Magma Chamber Processes in the Formation of the Low-sulphide Magmatic Au–PGE Mineralization of the Platinova Reef in the Skaergaard Intrusion, East Greenland. *Journal of Petrology*, 56, 2319–2340.
Keays, R.R., Lightfoot, P.C. and Hamlyn, P.R. (2012). Sulfide saturation history of the Stillwater Complex, Montana: Chemostratigraphic variation in platinum group elements. *Mineralium Deposita*, 47, 151–173.
Kelemen, P.B., Dick, H.J.B. and Quick, J.E. (1992). Formation of harzburgite by pervasive melt/rock reaction in the upper mantle. *Nature*, 358, 635–641.
Kelemen, P., Hirth, G., Shimizu, N., Spiegelman, M. and Dick, H. (1997). A review of melt migration processes in the adiabatically upwelling mantle beneath oceanic spreading ridges. *Philosophical Transactions of the Royal Society of London, Series A*, 355, 283–318.
Kelemen, P.B., Shimizu, N. and Salters, V.J.M. (1995). Extraction of mid-ocean-ridge basalt from the upwelling mantle by focused flow of melt in dunite channels. *Nature*, 375, 747–753.
Keller, T. and Katz, R.F. (2016). The role of volatiles in reactive melt transport in the asthenosphere. *Journal of Petrology*, 57(6), 1073–1108.
Keller, T., Katz, R.F. and Hirschmann, M.M. (2017). Volatiles beneath mid-ocean ridges: Deep melting, channelised transport, focusing, and metasomatism. *Earth and Planetary Science Letters*, 464, 55–68.
Khitarov, N.I., Lebedev, Ye.B., Dorfman, A.M. and Bagdasarov, N.Sh. (1979). Effects of temperature, pressure, and volatiles on the surface tension of molten basalt. *Geochemistry International*, 16(5), 78–86.
Kinloch, E.D. (1982). Regional trends in the platinum-group mineralogy of the Critical zone of the Bushveld Complex, South Africa. *Economic Geology*, 77, 1328–1347.
Kinloch, E.D. and Peyerl, W. (1990). Platinum-group minerals in various rock types of the Merensky Reef: Genetic implications. *Economic Geology*, 85, 537–555.
Konnerup-Madsen, J. and Rosew-Hansen, J. (1982). Volatiles associated with alkaline igneous rift activity: Fluid inclusions in the Ilimaussaq intrusion and Gardar granite complexes (South Greenland). *Chemical Geology*, 37, 79–93.
Korzhinskii, D.S. (1959). *Physicochemical Basis of the Analysis of the Paragenesis of Minerals*. London, Chapman and Hall.
Korzhinskii, D.S. (1968). The theory of metasomatic zoning. *Mineralium Deposita*, 3, 222–231.
Kruger, F.J. (1990). The stratigraphy of the Bushveld Complex: A reappraisal and the relocation of the Main Zone boundaries. *South African Journal of Geology*, 93, 376–381.
Kruger, F.J. (1994). The Sr-isotopic stratigraphy of the western Bushveld Complex. *South African Journal of Geology*, 97, 393–398.
Kruger, F.J. and Marsh, J.S. (1982). Significance of 87Sr/86Sr ratios in the Merensky cyclic unit of the Bushveld Complex. *Nature*, 298, 53–55.

Kusebauch, C., John, T., Whitehouse, M.J., Klemme, S. and Putnis, A. (2015). Distribution of halogens between fluid and apatite during fluid-mediated replacement processes. *Geochimica et Cosmochimica Acta*, 170, 225–246.

Kushiro, I. (1975). On the nature of silicate melt and its significance in magma genesis: Regularities in the shift of the liquidus boundaries involving olivine, pyroxene, and silica minerals. *American Journal of Science*, 275, 411–431.

Kushiro, I., Yoder, H.S. and Nishikawa, M. (1968). Effect of water on the melting of enstatite. *Geological Society of America Bulletin*, 79, 1685–1692.

Labotka, T.C. and Kath R.L. (2001). Petrogenesis of the contact-metamorphic rocks beneath the Stillwater Complex, Montana. *Geological Society of America Bulletin*, 113, 1312–1323.

Lackey, J.S. and Valley, J.W. (2004). Complex patterns of fluid flow during wollastonite formation in calcareous sandstones at Laurel Mountain, Mt. Morrison Pendant, California. *Geological Society of America Bulletin*, 116(1/2), 76–93.

Lake, E.T. (2013). Crystallization and saturation front propagation in silicic magma chambers. *Earth and Planetary Science Letters*, 383, 182–193.

Lambert, D.D. and Simmons, E.C. (1988). Magma evolution in the Stillwater Complex, Montana: II. Rare earth element evidence for the formation of the J-M Reef. *Economic Geology*, 83, 1109–1126.

Langmuire, C.H. (1989). Geochemical consequences of in situ crystallization. *Nature*, 340, 199–205.

Larsen, R.B., Brooks, C.K. and Bird, D. K. (1992). Methane-bearing, aqueous, saline solutions in the Skaergaard intrusion, east Greenland. *Contributions to Mineralogy and Petrology*, 112, 428–437.

Latypov, R., O'Driscoll, B. and Lavrenchuk, A. (2013). Towards a model for the in situ origin of PGE reefs in layered intrusions: Insights from chromitite seams of the Rum Eastern Layered Intrusion, Scotland. *Contributions to Mineralogy and Petrology*, 166, 309–327.

Latypov, R. and 16 others (2015). A fundamental dispute: A discussion of "On some fundamentals of igneous petrology" by Bruce D. Marsh, Contributions to Mineralogy and Petrology (2013) 166: 665–690. Contributions to Mineralogy and Petrology, 169, DOI 10.1007/s00410-015-1108-9

Latypov, R., Costin, G., Chistyakova, S., Hunt, E.J., Mukherjee, R. and Naldrett, A. (2018). Platinum-bearing chromite layers are caused by pressure reduction during magma ascent. Nature Communications. DOI: 10.1038/s41467-017-02773-w

Laurenz, V., Fonseca, R.O.C., Ballhaus, C. and Sylvester, P.J. (2010). The solubility of palladium in picritic melts: 1. The effect of iron. *Geochimica et Cosmochimica Acta*, 74, 2989–2998.

Laurenz, V., Fonseca, R.O.C., Ballhaus, C., Jochum, K.P., Heuser, A. and Sylvester, P.J. (2013). The solubility of palladium and ruthenium in picritic melts: 2. The effect of sulfur. *Geochimica et Cosmochimica Acta*, 108, 172–183.

Le Guen, Y., Renard, F., Hellmann, R., Brosse, E., Collombet, M., Tisserand, D. and Gratier, J.P. (2007). Enhanced deformation of limestone and sandstone in the presence of high PCO2 fluids. *Journal of Geophysical Research – Solid Earth*, 112(B5), B05421.

Le Roux, V., Bodinier, J.-L. Alard, O., O;Reilly, S.Y. and Griffin, W.L. (2009). Isotopic decoupling during porous melt flow: A case-study in the Lherz peridotite. *Earth and Planetary Science Letters*, 279, 76–85.

Lee, C.A. and Butcher, A.R. (1990). Cyclicity in the Sr isotope stratigraphy through the Merensky and Bastard Reef units, Atok section, eastern Bushveld Complex. *Economic Geology*, 85, 877–883.

Lee, C.-T.A. and Morton, D.M. (2015). High silica granites: Terminal porosity and crystal settling in shallow mama chambers. *Earth and Planetary Science Letters*. 409, 23–31.

Leeman, W.P. and Dasch, E.J. (1978). Strontium, lead, and oxygen isotopic investigation of the Skaergaard Intrusion, East Greenland. *Earth and Planetary Science Letters*, 41, 47–59.

Lenormand, R., Touboul, E. and Zarcone, C. (1988). Numerical models and experiments on immiscible displacements in porous media. *Journal of Fluid Mechanics*, 189, 165–187.

Li, C. and Boudreau, A.E. (2017). The origin of high Cu/S sulfides by shallow level degassing in the Skaergaard Intrusion, East Greenland. *Geology*, 45(12), 1075–1078.

Li, C. and Ripley, E.M. (2005). Empirical equations to predict the sulfur content of mafic magmas at sulfide saturation and applications to magmatic sulfide deposits. *Mineralium Deposita*, 40, 218–230.

Li, C. and Ripley, E.M. (2009). Sulfur contents at sulfide-liquid or anhydrite saturation in silicate melts: Empirical equations and example applications. *Economic Geology*, 104, 405–412.

Li, C., Ripley, E.M., Oberthür, T, Miller, J.D. and Joslin, G.D. (2008). Textural, mineralogical and stable isotope studies of hydrothermal alteration in the main sulfide zone of the Great Dyke, Zimbabwe and the precious metals zone of the Sonju Lake Intrusion, Minnesota, USA. *Mineralium Deposita*, 43, 97–110.

Li, C., Ripley, E.M., Sarkar, A., Shin, D. and Maier, W.D. (2005). Origin of phlogopite-orthopyroxene inclusions in chromites from the Merensky Reef of the Bushveld Complex, South Africa. *Contributions to Mineralogy Petrology*. 150, 119–130.

Li, H. and Hermann, J. (2015). Apatite as an indicator of fluid salinity: An experimental study of chlorine and fluorine partitioning in subducted sediments. *Geochimica et Cosmochimica Acta*, 166, 267–297.

Liebscher, A., Barnes, J.D. and Sharp, Z.D. (2006). Chlorine isotope vapor-liquid fractionation during experimental fluid-phase separation at 400 °C/23 MPa to 450 °C/42 MPa. *Chemical Geology* 234, 340–345.

Lin, S. and Popp, R.K. (1984). Solubility and complexing of Ni in the system NiO-H2O-HCl. *Geochimica et Cosmochimica Acta*, 48, 2713–2722.

Lindsley, D.H. (1983). Pyroxene thermometry. *American Mineralogist*, 68, 477–493.

Lipin, B.R. (1993) Pressure increases, the formation of chromite seams, and the development of the Ultramafic series in the Stillwater Complex, Montana. *Journal of Petrology*, 34, 955–976.

Lipman, P.W., Banks, N.G. and Rhodes, J.M. (1985). Degassing-induced crystallization of basaltic magma and effects on lava rheology. *Nature*, 317, 604–607

Liu, Y., Samaha, N.-T. and Baker, D.R. (2007). Sulfur concentration at sulfide saturation (SCSS) in magmatic silicate melts. *Geochimica et Cosmochimica Acta*, 71, 1783–1799.

Locmelis, M., Pearson, N.J., Barnes, S.J. and Fiorentini, M.L. (2011). Ruthenium in komatiitic chromite. *Geochimica et Cosmochimica Acta*, 75 (13), 3645–3661.

Locmelis, M., Fiorentini, M.L., Barnes, S.J. and Pearson, N.J. (2013). Ruthenium variation in chromite from komatiites and komatiitic basalts – A potential mineralogical indicator for nickel sulfide mineralization. *Economic Geology*, 108, 355–364.

Locmelis, M., Fiorentini, M.L., Rushmer, T., Arevalo Jr., R., Adamc, J. and Denyszyn, S.W. (2016). Sulfur and metal fertilization of the lower continental crust. *Lithos*, 244, 74–93.

Loferski, P.J. and Arculus, R.J. (1993). Multiphase inclusions in plagioclase from anorthosites in the Stillwater Complex, Montana: Implications for the origin of the anorthosites. *Contributions to Mineralogy and Petrology*, 114, 63–78.

Lofersky, P.J., Lipin, B.R. and Cooper, R.W. (1990). Petrology of chromite-bearing rocks from the lowermost cyclic units in the Stillwater Complex, Montana. *U.S. Geological Survey Bulletin*, 1674E, E1–E25.

Longhi, J. (1982). Effects of fractional crystallization and cumulus processes on mineral composition trends of some lunar and terrestrial rock series. Proceedings of the 13th Lunar Planet Science Conference, *Journal of Geophysical Research* 87, A54–A64.

Lorand, J.-P. and Alard, O. (2001). Platinum-group element abundances in the upper mantle: New constraints from in situ and whole-rock analyses of massif central xenoliths (France). *Geochimica et Cosmochimica Acta*, 65, 2789–2805.

Lorenz, V. (1975). Formation of phreatomagmatic maar-diatreme volcanoes and its relevance to kimberlite diatremes. *Physical Chemistry of the Earth*, 9, 17–27.

Løseth, H., Wensaas, L., Arntsen, B., Hanken, N.-M., Basire, C. and Graue, K. (2011). 1000 m long gas blow-out pipes. *Marine and Petroleum Geology*, 28, 1047–1060.

Løvoll, G., Méheust, Y., Toussaint, R., Schmittbuhl, J. and Måløy, K.J. (2004). Growth activity during fingering in a porous Hele-Shaw cell. *Physical Review E*, 70, 02630–1. doi:10.1103/PhysRevE.70.026301

Lowenstein, J.B. (1995). Application of silicate-melt inclusions to the study of magmatic volatiles. In *Magmas, Fluids, and Ore Deposits* (Thompson, J.F.H., ed.). Mineralogical Association of Canada, Short Course Series 23, 71–99.

Lubetkin, S.D. (2003). Why is it much easier to nucleate gas bubbles than theory predicts? *Langmuir*, 19, 2575–2587.

Luguet, A., Lorand, J.-P., Alard, O. and Cottin, J.-Y. (2004). A multi-technique study of platinum-group elements systematic in some ligurian ophiolitic peridotites, Italy. *Chemical Geology*, 208, 175–194.

Luguet, A., Shirey, S.B., Lorand, J.-P., Horan, M.F. and Carlson, R.W. (2007). Residual platinum-group minerals from highly depleted harzburgites of the Lherz massif (France) and their role in HSE fractionation of the mantle. *Geochimica et Cosmochimica Acta*, 71, 3082–3097.

Maier, D.W., Arndt, N.T. and Curl, E.A. (2000). Progressive crustal contamination of the Bushveld Complex: Evidence from Nd isotopic analyses of the cumulate rocks. *Contributions to Mineralogy and Petrology*, 140(3), 316–327.

Maier, W.D. and Barnes, S.-J. (1998). Concentrations of rare earth elements in silicate rocks of the Lower, Critical and Main Zones of the Bushveld Complex. *Chemical Geology*, 150, 85–103.

Maier, W.D. and Barnes, S.-J. (2004). Pt/Pd and Pt/Ir Ratios in mantle-derived magmas: A possible role for mantle metasomatism. *South African Journal of Geology*, 107, 333–340.

Maier, W.D. and Barnes, S.-J. (2008). Platinum-group elements in the UG1 and UG2 chromitites, and the Bastard reef, at Impala platinum mine, western Bushveld Complex, South Africa: Evidence for late magmatic cumulate instability and reef constitution. *South African Journal of Geology*, 111, 159–176.

Maier, W.D., Barnes, S.-J. and Groves, D.I. (2013). The Bushveld Complex, South Africa: Formation of platinum-palladium, chrome and vanadium-rich layers via hydrodynamic sorting of a mobilized cumulate slurry in a large, relatively slowly cooling, subsiding magma chamber. *Mineralium Deposita*, 48, 1–56.

Maier, W.D., Karykowski, B.T. and Yang, S.-H. (2016). Formation of transgressive anorthosite seams in the Bushveld Complex via tectonically induced mobilisation of plagioclase-rich crystal mushes. *Geoscience Frontiers*, 7, 875–889.

Maier, W.D., Prichard, H.M., Barnes, S.J. and Fisher, P.C. (1999). Compositional variation of laurite at Union Section in the western Bushveld Complex. *South African Journal of Geology*, 102, 286–292.

Makovicky, E. and Karup-Møller, S. (2000). Phase relations in the metal-rich portions of the phase system Pt-Ir-Fe-S at 1000°C and 1100°C. *Mineralogical Magazine*, 64(6), 1047–1056.

Mangan, M. and Sisson, T. (2004). Delayed, disequilibrium degassing in rhyolite magma: Decompression experiments and implications for explosive volcanism. *Earth and Planetary Science Letters*, 183, 441–455.

Mangan, M. and Sisson, T. (2005). Evolution of melt-vapor surface tension in silicic volcanic systems: Experiments with hydrous melts. *Journal of Geophysical Research*, 110, B01202. doi:10.1029/2004JB003215.

Mangan, M., Mastin, L. and Sisson, T. (2004). Gas evolution in eruptive conduits: Combining insights from high temperature and pressure decompression experiments with steady-state flow modeling. *Journal of Volcanology and Geothermal Research*, 129, 23–36.

Magenheima, A., Spivack, A.J., Michael, P.J. and Gieskes, J.M. (1995). Chlorine stable isotope composition of the oceanic crust: Implications for Earth's distribution of chlorine. *Earth and Planetary Science Letters*, 131(3–4), 427–432.

Manning C. E. (1994). The solubility of quartz in the lower crust and upper mantle. *Geochimica et Cosmochimica Acta*, 58(22), 4831–4839.

Marsh, B.D. (2004). A magmatic mush column Rosetta stone: The McMurdo Dry Valleys of Antarctica. *EOS*, 85, 497–502.

Marsh, B.D. (2013). On some fundamentals of igneous petrology. *Contributions to Mineralogy and Petrology*, 166(3), 665–690.

Mather, T.A., Witt, M.L.I., Pyle, D.M., Quayle, B.M., Aiuppa, A., Bagnato, E., Martin, R.S., Sims, K.W.W., Edmonds, M., Sutton, A.J. and Ilyinskaya, E. (2012). Halogens and trace metal emissions from the ongoing 2008 summit eruption of Kīlauea volcano, Hawai`i. *Geochemica et Cosmochimica Acta*, 83, 292–323.

Mathez, E.A. (1995). Magmatic metasomatism and formation of the Merensky reef, Bushveld Complex. *Contributions to Mineralogy and Petrology*, 119, 277–286.

Mathez, E.A. and Marcantonio, F. (1995). Sr isotopes and magma mixing in the Bushveld Complex. *EOS Transactions, American Geophysical Union*, 76, F641.

Mathez, E.A. and Waight, T.E. (2003). Lead isotopic disequilibrium between sulfide and plagioclase in the Bushveld Complex and the chemical evolution of large layered intrusions. *Geochimica et Cosmochimica Acta*, 67, 1875–1888.

Mathez, E.A. and Webster, J. D. (2005). Partitioning behavior of chlorine and fluorine in the system apatite-silicate melt-fluid. *Geochimica et Cosmochimica Acta*, 69, 1275–1286.

Mathez, E.A. and Kent A.J.R. (2007). Variable initial Pb isotopic compositions of rocks associated with the UG2 chromitite, eastern Bushveld Complex. *Geochimica et Cosmochimica Acta*, 71, 5514–5527.

Mathez, E.A. and Kinzler, R.J. (2017). Metasomatic chromitite seams in the Bushveld and Rum Layered Intrusions. *Elements*, 13(6), 397–402.

Mathez, E.A., Dietrich, V.J., Holloway, J.R. and Boudreau, A.E. (1989). Carbon distribution in the Stillwater Complex and evolution of vapour during crystallization of Stillwater and Bushveld magmas. *Journal of Petrology*, 30(1), 153–173.

Mathez, E.A., Agrinier, P. and Hutchinson, R. (1994). Hydrogen isotopic composition of the Merensky reef and related rocks, Atok section, Bushveld Complex. *Economic Geology*, 89, 791–802.

Mazzini, A., Svensen, H.H., Forsberg, C.F., Linge, H., Lauritzen, S.-E., Haflidason, H., Hammer, Ø., Planke, S. and Tjelta, T.I. (2017). A climatic trigger for the giant Troll pockmark field in the northern North Sea. *Earth and Planetary Science Letters*, 464, 24–34.

McBirney, A.R. (1975). Differentiation of the Skaergaard Intrusion. *Nature*, 253, 691–694.

McBirney, A.R. (1987). Constitutional zone refining of layered intrusions. In Origins of Igneous Layering (Parsons, I., ed.). *NATO ASI Series C*, 196, 437–452.

McBirney, A.R., (1996). The Skaergaard Intrusion. In *Developments in Petrology 15 – Layered Intrusions* (Cawthorn, R.G., ed.). Amsterdam: Elsevier Science, 147–180.

McBirney, A.R. (2009). Factors governing the textural development of Skaergaard gabbros: A review. *Lithos*, 111, 1–5.

McBirney, A.R. and Creaser, R.A. (2003). The Skaergaard Layered Series, Part VII: Sr and Nd isotopes. *Journal of Petrology*, 44(4), 757–771.

McBirney, A.R. and Hunter, R.H. (1995). The cumulate paradigm reconsidered. *The Journal of Geology*, 103, 114–122.

McBirney, A.R. and Noyes, R.M. (1979). Crystallization and layering of the Skaergaard Intrusion. *Journal of Petrology*, 20, 487–564.

McBirney, A.R. and Sonnenthal, E.L. (1990). Metasomatic replacement in the Skaergaard Intrusion, East Greenland: Preliminary observations. *Chemical Geology*, 88, 245–260.

McCallum, I.S. (1996). The Stillwater Complex. In *Developments in Petrology 15 – Layered Intrusions* (Cawthorn, R.G., ed.). Amserdam: Elsevier Science, 441–484.

McCallum, I.S., Raedeke, L.D. and Mathez, E.A. (1977). Stratigraphy and petrology of the Banded zone of the Stillwater Complex, Montana. *EOS*, 58, 1245.

McCallum, I.S., Raedeke, L.D. and Mathez, E.A. (1980). Investigations in the Stillwater Complex: Part I. Stratigraphy and structure of the Banded zone. In *The Jackson Volume, vol. 280-A of American Journal of Science* (Irving, A. and Dungan, M., eds.). New Haven, Connecticut, Kline Geology Laboratory, Yale University, 59–87.

McCallum I.S., Thurber D.W., O'Brien H.E. and Nelson B.K. (1999). Lead isotopes in sulfides from the Stillwater Complex, Montana: Evidence for subsolidus remobilization. *Contributions to Mineralogy and Petrology*, 137, 206–219.

McCallum, M.E. (1985). Experimental evidence for fluidization processes in breccia pipe formation. *Economic Geology*, 80, 1523–1543.

McCandless, T.E. and Ruiz, J. (1991). Osmium isotopes and crustal sources for platinum-group element mineralization in the Bushveld Complex, South Africa. *Geology*, 19, 1225–1228.

McCandless, T.E., Ruiz, J., Adair, B.I. and Freydier, C. (1999). Re–Os isotope and Pd/Ru variations in chromitites from the Critical zone, Bushveld Complex, South Africa. *Geochimica et Cosmochimica Acta*, 63, 911–923.

McCubbin, F.M., Vander Kaaden, K.E., Tartese, R., Boyce, J.W., Mikhail, S., Whitson, E.S., Bell, A.S., Anand, M., Franchi, I.A., Wang, J. and Hauri E.H. (2015). Experimental investigation of F, Cl and OH partitioning between apatite and Fe-rich basaltic melt at 1.0–1.2 GPa and 950–1000 °C. *American Mineralogist*, 100(8–9), 1790–1802.

McDonald, I., Holwell, D.A. and Armitage, P.E.B. (2005) Geochemistry and Mineralogy of the Platreef and "Critical Zone" of the Northern Lobe of the Bushveld Complex, South Africa: Implications for Bushveld Stratigraphy and the Development of PGE Mineralization. *Mineralium Deposita*, 40, 526–549.

McDonough, W.F., Sun, S.-S. (1995). The composition of the Earth. *Chemical Geology*, 120, 223–253.

Mcllveen, C.L. (1996). *Anomalous platinum-group element occurrence below the JM Reef, Stillwater Complex, Montana*. M.S. Thesis, University of Montana.

McInnes, B.I.A., McBride, J.S., Evans, N.J., Lambert, D.D. and Andrew, A.S. (1999). Osmium isotope constraints on ore metal recycling in subduction zones. *Science*, 286, 512–516.

McKenzie, D. (1984). The generation and compaction of partially molten rock. *Journal of Petrology*, 25, 713–765.

McKenzie, D. (2011). Compaction and crystallization in magma chambers: Towards a model of the Skaergaard Intrusion. *Journal of Petrology*, 52, 905–930.

Médard, E. and Grove, T.L. (2007). Water in basaltic melts: Effect on the liquidus temperature, olivine-melt thermometry and mantle melting. In *Workshop on Water in Planetary Basalts*, 26–27. LPI Contribution 373, Lunar and Planetary Institute, Houston.

Melcher, F., Grum, W., Simon, G., Thalhammer, T.V. and Stumpfl, E.F. (1997). Petrogenesis of the Ophiolitic Giant Chromite Deposits of Kempirsai, Kazakhstan: A Study of Solid and Fluid Inclusions in Chromite. *Journal of Petrology*, 38, 1419–1458.

Mei, Y., Etschmann, B., Liu, W., Sherman, D.M., Barnes, S.J., Fiorentini, M.L., Seward, T.M., Tesemale, D. and Brugger, J. (2015). Palladium complexation in chloride- and bisulfide-rich fluids: Insights from ab initio molecular dynamics simulations and X-ray absorption spectroscopy. *Geochimica et Cosmochimica Acta*, 161, 128–145.

Merkle, R.K.W. (1992). Platinum-group minerals in the middle group of chromitite layers at Marikana, western Bushveld Complex: Indications for collection mechanisms and post-magmatic modification. *Canadian Journal of Earth Sciences*, 29, 209–221.

Métrich N. and Wallace, P. (2008). Volatile abundances in basaltic magmas and their degassing paths tracked by melt inclusions. In *Minerals, Inclusions and Volcanic Processes* (Putirka, K. and Tepley, F., eds,). Mineralogical Society of America, Reviews in Mineralogy and Geochemistry, 69, 363–402

Métrich, N. and Rutherford, M. J. (1992). Experimental study of chlorine behavior in hydrous silicic melts. *Geochimica et Cosmochimica Acta*, 57, 607–616.

Métrich, N., Bertagnini, A., Landi, P. and Rosi, M. (2001). Crystallization driven by decompression and water loss at Stromboli volcano (Aeolian Islands, Italy). *Journal of Petrology*, 42(8), 1471–1490.

Meurer, W.P. and Boudreau A.E. (1996). Compaction of density-stratified cumulates: Effect on trapped-liquid distribution. *Journal of Geology*, 104, 115–120.

Meurer, W.P. and Boudreau, A.E. (1998a). Compaction of Igneous Cumulates Part I: Geochemical Consequences for Cumulates and Liquid Fractionation Trends. *Journal of Geology*, 106, 281–292.

Meurer, W.P. and Boudreau, A.E. (1998b). Compaction of igneous cumulates. Part II – Compaction and the development of igneous foliation. *Journal of Geology*, 106, 293–304.

Meurer, W.P., Klaber, S.A. and Boudreau, A.E. (1997). Discordant bodies from Olivine-Bearing zones III and IV of the Stillwater complex, Montana – Evidence for post-cumulus fluid migration in layered intrusions. *Contributions to Mineralogy and Petrology*, 130, 81–92.

Meurer W.P. and Meurer, M.E.S. (2006). Using apatite to dispel the "trapped liquid" concept and to understand the loss of interstitial liquid by compaction in mafic cumulates: An example from the Stillwater Complex, Montana. *Contributions to Mineralogy Petrology*, 151, 187–201.

Meurer, W.P., Willmore, C.C. and Boudreau, A. E. (1998). Metal redistribution during fluid exsolution and migration in the Middle Banded series of the Stillwater complex, Montana. *Lithos*, 47, 143–156.

Mondal, S.K. and Mathez, E.A. (2007). Origin of the UG2 chromitite layer, Bushveld Complex. *Journal of Petrology*, 48(3), 495–510.

Morse, S.A. (1980). *Basalts and Phase Diagrams.* New York: Springer, 493.

Morse, S.A. (1983). Strontium isotope fractionation in the Kiglapait Intrusion. *Science*, 220, 193–195.

Mourtada-Bonnefoi, C.C. and Laporte, D. (1999). Kinetics of bubble nucleation in a rhyolitic melt: An experimental study of the effect of ascent rate. *Earth and Planetary Science Letters*, 218, 521–537.

Mukherjee, R., Latypov, R. and Balakrishna, A. (2017). An intrusive origin of some UG-1 chromitite layers in the Bushveld Igneous Complex, South Africa: Insights from field relationships. *Ore Geology Reviews*, 90, 94–109.

Mungall, J.E. (2002). Kinetic controls on the partitioning of trace elements between silicate and sulfide liquids. *Journal of Petrology*, 43(5), 749–768.

Mungall, J.E. (2015). Physical controls of nucleation, growth and migration of vapor bubbles in partially molten cumulates. In *Layered Intrusions* (Charlier, B., Namur, O., Latypov, R. and Tegner, C., eds.). Dordrecht: Springer Geology, 331–378.

Munoz, J.L. and Swenson, A. (1981). Chloride-hydroxyl exchange in biotite and estimate of relative HCl/HF activities in hydrothermal fluids. *Economic Geology*, 76, 2212–2221.

Murase, T. and McBirney, A.R. (1973). Properties of some common igneous rocks and their melts at high temperatures. *Geological Society of America Bulletin*, 84, 3563–3592.

Mutanen, T., Törnroos, R. and Johanson, B. (1988). The Significance of Cumulus Chlor-apatite and High-temperature Dashkesanite to the Genesis of PGE Mineralization in the Koitelainen and Keivitsa-Satovaara Complexes, Northern Finland. In *Geo-Platinum 87* (Prichard, H.M., Potts, P.J., Bowles J.F.W. and Cribb S.J., eds.). Dordrecht: Springer,

Mysen, B.O. and Boettcher, A.L. (1975). Melting of a hydrous mantle: I. Phase relations of natural peridotite at high pressures and temperatures with controlled activities of water, hydrogen, and carbon dioxide. *Journal of Petrology*, 16, 520–548.

Nakagawa, M. and Franco, H.E.A. (1997). An assessment of placer Os–Ir–Ru alloys and sulfides as an indicator of sulfur fugacity in a primitive stage of ophiolite mantle. *Journal of Asian Earth Sciences*, 15, 311–315.

Naldrett, A.J. and Wilson, A.H. (1990). Horizontal and vertical variations in noble metals in the Great Dyke of Zimbabwe: A model for the origin of PGE mineralization by fractional segregation. *Chemical Geology*, 88, 279–300.

Naldrett, A.J., Kinnaird, J., Wilson, A. and Chunnett, G. (2008). The concentration of PGE in the Earth's crust with special reference to the Bushveld Complex. *Earth Science Frontiers*, 15(5), 264–297.

Naldrett, A.J., Wilson, A., Kinnaird, J. and Chunnett, G. (2009). PGE tenor and metal ratios within and below the Merensky Reef, Bushveld Complex: Implications for its genesis. *Journal of Petrology*, 50, 625–659.

Naldrett, A.J., Wilson, A., Kinnaird, J., Yudovskaya, M. and Chunnett, G. (2012). The origin of chromitites and related PGE mineralization in the Bushveld Complex: New mineralogical and petrological constraints. *Mineralium Deposita*, 47, 209–232.

Nash, W.P. (1976). Fluorine chlorine and OH-bearing minerals in the Skaergaard Intrusion. *American Journal of Science*, 276, 546–57.

Naslund, H.R. (1986). Disequilibrium partial melting and rheomorphic layer formation in the contact aureole of the Basistoppen sill, East Greenland. *Contributions to Mineralogy and Petrology*, 93, 359–367.

Navon, O. and Lyakhovsky, V. (1998). Vesiculation processes in silicic magmas. *Geological Society Special Publication*, 145, 27–50.

Newton, R.C., Aranovich, L.Ya., Hansen, E.C. and Vandenheuvel, B.A. (1998). Hypersaline fluids in Precambrian deep-crustal metamorphism. *Precambrain Research*, 91(1–2), 41–63.

Newton, R.C. and Manning C.E. (2000). Quartz solubility in H2O-NaCl and H2O-CO2 solutions at deep crust-upper mantle pressures and temperatures: 2–15 kbar and 500–900°C. *Geochimica et Cosmochimica Acta*, 64(17), 2993–3005.

Newton, R.C. and Manning C.E. (2002). Experimental determination of calcite solubility in H2O-NaCl solutions at deep crust/upper mantle pressures and temperatures: Implications for metasomatic processes in shear zones. *American Mineralogist*, 87, 1401–1409

Newton, R.S., Cunningham, R.C. and Schubert, C.E. (1980). Mud volcanoes and pockmarks: Seafloor engineering hazards or geological curiosities? Proceedings 12th Offshore Technical Conference, Houston Texas, 3729, 425–429.

Nicholson, D.M. and Mathez, E.A. (1991). Petrogenesis of the Merensky Reef in the Rustenburg section of the Bushveld Complex. *Contributions to Mineralogy and Petrology*, 107, 293–309.

Nielsen, T.F.D. (2004). The shape and volume of the Skaergaard Intrusion, Greenland: Implications for mass balance and bulk composition. *Journal of Petrology*, 45, 507–530.

Nielsen, T.F.D., Andersen, J.C.Ø., Holness, M.B., Keidin, J.K., Rudashevsky, N.S., Rudashevsky, V.N., Salmonsen, L.P., Tegner, C. and Veksler, I.V. (2015). The Skaergaard PGE and Gold Deposit: The Result of in situ Fractionation, Sulphide Saturation, and Magma Chamber-scale Precious Metal Redistribution by Immiscible Fe-rich Melt, *Journal of Petrology*, 56, 1643–1676.

O'Driscoll, B., Emeleus, C.H., Donaldson, C.H. and Daly, J.S. (2010). Cr-spinel seam petrogenesis in the Rum Layered Suite, NW Scotland: Cumulate assimilation and in situ crystallization in a deforming crystal mush. *Journal of Petrology*, 51(6), 1171–1201.

O'Driscoll, B. and González-Jiménez, J.M. (2016). Petrogenesis of the Platinum-Group Minerals. *Reviews in Mineralogy and Geochemistry*, 81, 489–578.

O'Neill, H.St.C. and Mavrogenes, J.A. (2002). The sulfide capacity and the sulfur content at sulfide saturation of silicate melts at 1400 °C and 1 bar. *Journal of Petrology*, 43, 1049–1087.

Oberthür, T. (2011). Platinum-group element mineralization of the Main Sulfide Zone, Great Dyke, Zimbabwe. *Reviews in Economic Geology*, 17, 329–349.

Oberthür, T., Cabri, L.J., Weiser, T.W., McMahon, G. and Müller, P. (1997). Pt, Pd and other trace elements in sulfides of the Main Sulfide Zone, Great Dyke, Zimbabwe – A reconnaissance study. *Canadian Mineralogist*, 35, 597–609.

Oberthür, T., Weiser, T.W., Müller, P., Lodziak, J. and Cabri, L.J. (1998). New observations on the distribution of platinum group elements (PGE) and minerals (PGM) in the

MSZ at Hartley Mine, Great Dyke, Zimbabwe. In *8th Intl Pt Symposium Abstracts*, South African Institute of Mining and Metallurgy, Johannesburg, 293–296.

Oberthür, T., Davis, D.W., Blenkinsop, T.G. and Höhndorf, A. (2002). Precise U–Pb mineral ages, Rb–Sr and Sm–Nd systematics for the Great Dyke, Zimbabwe – Constraints on late Archean events in the Zimbabwe Craton and Limpopo Belt. *Precambrian Research*, 113, 293–305.

Oberthür, T., Weiser, T.W., Gast, L. and Kojonen, K. (2003). Geochemistry and mineralogy of platinum-group elements at Hartley Platinum Mine, Zimbabwe. *Mineralium Deposita*, 38, 327–343.

O'Neill, H.St.C., Dingwell, D.B., Borisov, A., Spettle, B. and Palme, H. (1995). Experimental petrochemistry of some high siderophile elements at high temperatures and some implications for core formation and the mantle's early history. *Chemical Geology*, 120, 255–273.

Oppenheimer, J., Rust, A.C., Cashman, K.V. and Sandnes, B. (2015). Gas migration regimes and outgassing in particle-rich suspensions. *Frontiers in Physics*, 3: 1–15.

Orville, P.M. (1972). Plagioclase cation exchange equilibria with aqueous chloride solution: Results at 700 °C and 2000 bars in the presence of quartz. *American Journal of Science*, 272, 234–272.

Page, N.J, Shimek, R. and Huffman, C. (1972). *Grain-size variations within an olivine cumulate, Stillwater Complex, Montana*. United State Geological Survey Professional Paper, 800-C, C29-C37.

Page. N.J. and Zientek, M.L. (1987). *Composition of primary postcumulus amphibole and phlogopite within an olivine cumulate in the Stillwater Complex, Montana*. United States Geological Survey Bulletin, 1674-A.

Pagé, P. and Barnes, S.-J. (2013). Improved in-situ determination of PGE concentration of chromite by LA-ICP-MS: Towards a better understanding. Ore deposits associated with mafic and ultramafic rocks, 12th SGA Biennial Meeting 2013. *Proceedings*, 3, 1050–1053.

Pagé, P. and Barnes, S.-J., (2016). The influence of chromite on osmium, iridium, ruthenium and rhodium distribution during early magmatic processes. *Chemical Geology*, 420, 51–68.

Pagé, P., Barnes, S.J. and Zientek, M.L. (2011). Formation and evolution of the chromitites of the Stillwater Complex: A trace element study. In *Let's Talk Ore Deposits: Proceedings of the 11th SGA Biennial Meeting, Antofagasta, Chile* (Barra, F., ed.). Society for Geology Applied to Mineral Deposits, 678–680.

Pagé, P., Barnes, S.-J., Bedard, J.H. and Zientek, M.L. (2012). In situ determination of Os, Ir and Ru in chromites formed from komatiite, tholeiite and boninite magmas: Implications for chromite control of Os, Ir and Ru during partial melting and crystal fractionation. *Chemical Geology*, 302–303, 3–15.

Pagé, P., Barnes, S.-J., Méric, J. and Houlé, M.G. (2015). Geochemical composition of chromite from Alexo komatiite in the western Abitibi greenstone belt: Implications for mineral exploration, In *Targeted Geoscience Initiative 4: Canadian Nickel-Copper-Platinum Group Elements-Chromium Ore Systems — Fertility, Pathfinders, New and Revised Models*, (Ames, D.E. and Houlé, M.G., eds.). Geological Survey of Canada, Open File 7856, 187–195.

Palacz, Z.A. and Tait, S.R. (1985). Isotopic and geochemical investigation of unit 10 from the Eastern Layered Series of the Rhum Intrusion, northwest Scotland. *Geological Magazine*, 122, 485–490.

Palme, H. and O'Neill, H.St.C. (2003). Cosmochemical Estimates of Mantle Composition. In *Treatise on Geochemistry*, Volume 2 (Carlson, R.W., ed). Oxford: Elsevier-Pergamon, 1–38.

Pan, P. and Wood, S.A. (1994). Solubility of Pt and Pd sulfides and Au metal in aqueous bisulfide solutions. *Mineralium Deposita*, 29(5), 373–390.

Park, J.-W., Campbell, I.H. and Eggins, S.M. (2012). Enrichment of Rh, Ru, Ir and Os in Cr spinels from oxidized magmas: Evidence from the Ambae volcano, Vanuatu. *Geochimica et Cosmochimica Acta*, 78, 28–50.

Parmigiani, A., Huber, C., Bachmann, O. and Chopard, B. (2011). Pore-scale mass and reactant transport in multiphase porous media flow. *Journal of Fluid Mechanics*, 686, 40–76.

Patten, C., Barnes, S.-J. and Mathez, E.A. (2012). Textural variations in MORB sulfide droplets due to differences in crystallization history. *Canadian Mineralogist*, 50, 675–692.

Peach, C.L, Mathez, E.A. and Keays, R.R. (1990). Sulfide melt-silicate melt distribution coefficients for noble metals and other chacophile elements as deduced from MORB: Implications for partial melting. *Geochimica et Cosmochimica Acta*, 54, 3379–3389.

Peach C.L., Mathez E.A., Keays R.R. and Reeves S.J. (1994). Experimentally determined sulfide melt-silicate melt partition coefficients for iridium and palladium. *Chemical Geology*, 117, 361–377.

Pebane, M. and Latypov, R. (2017). The significance of magmatic erosion for bifurcation of UG1 chromitite layers in the Bushveld Complex. *Ore Geology Reviews*, 90, 65–93.

Peck, D.C. and Keays, R.R. (1990). Geology, geochemistry and origin of platinum-group element-chromitite occurrences in the Heazlewood River Complex, Tasmania. *Economic Geology*, 85, 765–793.

Peck, D.C, Keays. R.R. and Ford, R.J. (1992). Direct crystallization of refractory platinum-group elements alloys from boninitic magmas: Evidence form western Tasmania. *Australian Journal of Earth Sciences*, 39, 373–387.

Pedersen, A.K., Watt, M., Watt, W.S. and Larsen, L.M. (1997). Structure and stratigraphy of early Tertiary basalts of the Blosseville Kyst, East Greenland. *Journal of the Geological Society, London*, 154(3), 565–570.

Pelch, M.A., Appold, M.S., Emsbo, P. and Bodnar, R.J. (2015). Constraints from fluid inclusion compositions on the origin of Mississippi valley-type mineralization in the Illinois-Kentucky District. *Economic Geology*, 110, 787–808.

Peslier, A.H. (2010). A review of water contents of nominally anhydrous natural minerals in the mantles of Earth, Mars and the Moon. *Journal of Volcanology and Geothermal Research*, 197, 239–258.

Petford, N. and Mirhadizadeh, S. (2017). Image-based modelling of lateral magma flow: The Basement Sill, Antarctica. Royal Society Open Science, 4, 161083). http://dx.doi.org/10.1098/rsos.161083.

Peyerl, W. (1982). The influence of the Driekop dunite pipe on the platinum-group mineralogy of the UG-2 chromtite in its vicinity. *Economic Geology*, 77, 1432–1438.

Philpotts, A.R., Brustman, C.M., Shi, J., Carlson, W.D. and Denison, C. (1999). Plagioclase-chain networks in slowly cooled basaltic magma. *American Mineralogist*, 84, 1819–1829.

Philpotts, A.R., Shi, J. and Brustman, C. (1998). Role of plagioclase crystal chains in the differentiation of partly crystallized basaltic magma. *Nature*, 395, 343–346.

Philpotts, A.R., Carroll, M. and Hill, J.M. (1996). Crystal-mush compaction and the origin of pegmatitic segregation sheets in a thick flood-basalt flow in the Mesozoic Hartford Basin, Connecticut. *Journal of Petrology*, 37, 811–836.

Piccoli P.M. and Candela, P.A. (1994). Apatite in felsic rocks; a model for the estimation of initial halogen concentrations in the Bishop Tuff (Long Valley) and Tuolumne Intrusive Suite (Sierra Nevada Batholith) magmas. *American Journal of Science*, 294, 92–135.

Piccoli, P.M. and Candela, P.A. (2002). Apatite in igneous systems. *Reviews in Mineralogy and Geochemistry*, 48, 255–292.

Podmore, F. and Wilson, A.H. (1987). A reappraisal of the structure, geology and emplacement of the Great Dyke, Zimbabwe. In *Mafic Dyke Swarms* (Halls, H.C. and Fahrig, W.F., eds.). Geological Association of Canada Special Paper, 34, 317–330.

Polacci, M., Corsaro, R.A. and Andronico, D. (2006). Coupled textural and compositional characterization of basaltic scoria: Insights into the transition from Strombolian to fire fountain activity at Mount Etna, Italy. *Geology*, 34(3), 201–204.

Prendergast, M.D. (1988). The geology and economic potential of the PGE-rich Main Sulphide Zone of the Great Dyke, Zimbabwe. In *Geo-Platinum'87* (Prichard, H.M., Potts, P.J., Bowles, J.F.W. and Cribb, S.J., eds). Barking, Essex, Elsevier Applied Science, 281–302.

Prendergast, M.D. and Keays, R.R. (1989). Controls of platinum-group element mineralisation and the origin of the PGE-rich Main Sulphide Zone in the Wedza Subchamber of the Great Dyke, Zimbabwe: Implications for the genesis of, and exploration for, stratiform PGE mineralisation in layered intrusions. In *Magmatic Sulphides – The Zimbabwe Volume* (Prendergast, M.D. and Jones, M.J., eds). London, The Institution of Mining and Metallurgy, 43–69.

Prendergast, M.D. and Wilson, A.H. (1989). The Great Dyke of Zimbabwe - II: Mineralisation and mineral deposits. In *Magmatic Sulphides – The Zimbabwe Volume* (Prendergast, M.D. and Jones, M.J., eds). London: The Institution of Mining and Metallurgy, 21–42.

Prevec, S.A., Ashwal, L.D. and Makaza, M.S. (2005). Mineral disequilibrium in the Merensky Reef, western Bushveld Complex, South Africa: New Sm–Nd isotopic evidence. *Contributions to Mineralogy and Petrology*, 149(3), 306–315.

Prichard, H.M., Potts, P.J. and Neary, C.R. (1981). Platinum group element minerals in the Unst chromite, Shetland Isles. *Transactions of the Institute of Mining and Metallurgy, Section B*, 90, 186–188.

Prichard, H.M., Sa, J.H.S. and Fisher, P.C. (2001). Platinum-group mineral assemblages and chromite composition in the altered and deformed Bacuri complex, Amapa, northeastern Brazil. *Canadian Mineralogist*, 39, 377–396.

Proenza, J.A., Zaccarini, F., Lewis, J.F., Longo, F. and Garuti, G. (2007). Chromian spinel composition and the platinum-group minerals of the PGE-rich Loma Peguera chromitites, Loma Caribe peridotite, Dominican Republic. *The Canadian Mineralogist*, 45, 631–648.

Proussevitch, A.A. and Sahagian, D.L. (1998). Dynamics and energetics of bubble growth in magmas: Analytical formulation and numerical modeling. *Journal of Geophysical Research*, 103, 18223–18251.

Puchtel, I.S. and Humayun, M. (2001). Platinum group element fractionation in a komatiitic basalt lava lake. *Geochimica et Cosmochimica Acta*, 65, 2979–2993.

Raedeke, L.D. (1982). *Petrogenesis of the Stillwater Complex*. Ph.D. Dissertation, University of Washington.

Raedeke, L.D. and McCallum, I.S. (1980). A comparison of fractionation trends in the lunar crust and the Stillwater Complex. In *Proceedings of the conference on the lunar highlands crust* (Merrill, R.B. and Papike, J.J. eds.). Geochimica et Cosmochimica Acta, supplement 12, 133–153.

Raedeke, L.D. and McCallum, I.S. (1984). Investigations in the Stillwater Complex: Part II. Petrology and petrogenesis of the ultramafic series. *Journal of Petrology*, 25, 395–420.

Raedeke, L.D. and Vian, R.W. (1986). A three dimensional view of mineralization in the Stillwater J-M Reef. *Economic Geology*, 81, 1187–1195.

Reid, D.L. and Basson, I.J. (2002). Iron-rich ultramafic pegmatite replacement bodies within the Upper Critical Zone, Rustenburg Layered Suite, Northam Platinum Mine, South Africa. *Mineralogical Magazine*, 66(6), 895–914.

Reisberg, L., Tredoux, M., Harris, C., Coftier, A. and Chaumba, J. (2011). Re and Os distribution and Os isotope composition of the Platreef at the Sandsloot–Mogolakwena mine, Bushveld complex, South Africa. *Chemical Geology*, 24(3–4), 352–363.

Ribe, N.M. (1985). The generation and composition of partial melts in the earth's mantle. *Earth and Planetary Science Letters*, 73, 361–376.

Righter, K., Campbell, A.J., Humayun, M. and Hervig, R.L. (2004). Partitioning of Ru, Rh, Pd, Ir and Au between Cr-bearing spinel, olivine, pyroxene and silicate melts. *Geochimica et Cosmochimica Acta*, 68, 867–880.

Righter, K., Humayun, M. and Danielson, L. (2008). Partitioning of palladium at high pressures and temperatures during core formation. *Nature Geoscience*, 1, doi:10.1038/ngeo180.

Ripley, E.M. (2005). Re/Os isotopic and fluid inclusion studies of fluid-rock interaction in the contact aureole of the Duluth Complex, Minnesota. *Geochimica et Cosmochimica Acta*, 69(10) Supplement, 332.

Rivalta, E. and Dahm, T. (2006). Acceleration of buoyancy-driven fractures and magmatic dikes beneath the free surface. *Geophysical Journal International*, 166, 1424–1439.

Roelofse, F. and Ashwal, L.D. (2012). The Lower Main Zone in the northern limb of the Bushveld complex – A >1.3 km thick sequence of intruded and variably contaminated crystal mushes. *Journal of Petrology*, 53(7), 1449–1476.

Roelofse, F., Ashwal, L.D. and Romer, R.L (2015). Multiple, isotopically heterogeneous plagioclase populations in the Bushveld Complex suggest mush intrusion. *Chemie der Erde*, 75, 357–364.

Rudashevsky, N.S., Avdontsev, S.N. and Dneprovskaya, M.B. (1992). Evolution of PGE mineralization in hortonolitic Ddunites of the Mooihoek and Onverwacht Pipes, Bushveld Complex. *Mineralogy and Petrology*, 47, 37–54.

Rüpke, L.H., Morgan, J.P., Hort, M. and Connolly, J.A.D. (2004). Serpentine and the subduction zone water cycle. *Earth and Planetary Science Letters*, 223, 17–34.

Rust, A.C. and Cashman, K.V. (2004). Permeability of vesicular silicic magma: Inertial and hysteresis effects. *Earth and Planetary Science Letters*, 228, 93–107.

Ryder, G. and Spettel, B. (1985). The parental magma for some rocks from the Norite 1 subzone of the Stillwater Complex: A lunar analog study. Proceedings of the 15th Lunar and Planetary Science Conference, Part 2, *Journal of Geophysical research* 90 (supplement), C545–C559.

Ryerson, F.J. and Watson, E.B. (1987). Rutile saturation in magmas, implications for Ti-Nb-Ta depletion in island-arc basalts. *Earth Planetary Science Letters*, 86, 225–239.

Sahagian, D.L., Proussevitch, A.A. and Carlson, W.D. (2002). Analysis of vesicular basalts and lava emplacement processes for application as a paleobarometer/Ppaleoaltimeter. *Journal of Geology*, 110: 671–685.

Salpas, P.A., Haskin, L.A. and McCallum, I.S. (1983). Stillwater Anorthosites: A lunar analog? Proceedings of the 14th Lunar and Planetary Science Conference, Part 1. *Journal of Geophysical Research*, 88(suppl), B27-B39.

Sassani, D.C. and Shock, E.L. (1998). Solubility and transport of platinum-group elements in supercritical fluids: Summary and estimates of thermodynamic properties for ruthenium, rhodium, palladium, and platinum solids, aqueous ions, and complexes to 1000 °C and 5 kbar. *Geochimica et Cosmochimica Acta*, 62, 2643–2671.

Sato, H., Holtz, F., Behrens, H., Botcharnikov, R. and Nakada, S. (2005). Experimental petrology of the 1991–1995 Unzen dacite, Part II: Cl/OH partitioning between hornblende and melt and its implications for the origin of oscillatory zoning of hornblende Phenocrysts. *Journal of Petrology*, 46, 339–354.

Schannor, M., Veksler, I.V., Hecht, L., Harris, C. and Romer, R.L. (2018). Small-scale Sr and O isotope variations through the UG2 in the eastern Bushveld Complex: The role of crustal fluids. *Chemical Geology*, 485, 90–99.

Schiffries, C.M. (1982). The petrogenesis of a platiniferous dunite pipe in the Bushveld complex: Chloride complexing and cation exchange metasomatism. *Economic Geology*, 77, 1439–1453.

Schiffries, C.M. (1990). Liquid absent aqueous fluid inclusions and phase equilibria in the system CaCl2-NaCl-H2O. *Geochimica et Cosmochimica Acta*, 54, 611–619.

Schiffries, C.M. and Rye, D.M. (1989). Stable isotopic systematics of the Bushveld Complex: I. Constraints of magmatic processes in layered intrusions. *American Journal of Science*, 289, 841–873.

Schiffries, C.M. and Rye, D.M. (1990). Stable isotopic systematics of the Bushveld Complex: II. Constraints on hydrothermal processes in layered intrusions. *American Journal of Sciences*, 290, 209–245.

Schiffries, C.M. and Skinner B.J. (1987). The Bushveld hydrothermal system: Field and petrologic evidence. *American Journal of Sciences*, 287, 566–595.

Schisa, P., Boudreau, A.E., Djon L., Ychalikian, A. and Corkery, J. (2015). The Lac des Iles palladium deposit, Ontario, Canada. Part II. Halogen variations in apatite. *Mineralium Deposita*, 50, 339–355.

Schmidberger, S.S., Simonetti, A. and Francis, D. (2003). Small-scale Sr isotope investigation of clinopyroxenes from peridotite xenoliths by laser ablation MC-ICP-MS – Implications for mantle metasomatism. *Chemical Geology*, 199, 317–329.

Schoenberg, R., Kruger, F.J., Nägler, T.F., Meisel, T. and Kramers, J.D. (1999). PGE enrichment in chromitite layers and the Merensky Reef of the western Bushveld Complex; a Re–Os and Rb–Sr isotope study. *Earth and Planetary Science Letters*, 172, 49–64.

Scholten, L., Watenphulb, A., Beermanna, O., Testemale, D., Ames, D. and Schmidt, C. (2018). Nickel and platinum in high-temperature H2O + HCl fluids: Implications for hydrothermal mobilization. *Geochimica et Cosmochimica Acta*, 224, 187–199.

Scoates, J.S. (2000). The plagioclase-magma density paradox re-examined and the crystallization of Proterozoic anorthosites. *Journal of Petrology*, 41, 627–649.

Scoates, J.S. and Friedman, R.M. (2008). Precise age of the platiniferous Merensky reef, Bushveld Complex, South Africa, by the U–Pb zircon chemical abrasion ID–TIMS technique. *Economic Geology*, 103, 465–471.

Scoon, R.N. (1987). Metasomatism of cumulus magnesian olivine by iron-rich postcumulus liquids in the upper Critical Zone of the Bushveld Complex. *Mineralogical Magazine*, 51, 389–396.

Scoon, R.N. and Mitchell, A.A. (1994). Discordant Iron-rich ultramafic pegmatites and their relationship to iron-rich intercumulus and residual liquids. *Journal of Petrology*, 35, 881–917.

Scoon, R.N. and Mitchell, A.A. (2004a). The platiniferous dunite pipes in the eastern limb of the Bushveld Complex: Review and comparison with unmineralized discordant ultramafic bodies. *South African Journal of Geology*, 107, 505–520.

Scoon, R.N. and Mitchell, A.A., (2004b). Petrogenesis of discordant magnesian dunites from the central sector of the Eastern Bushveld Complex with emphasis on the Winnaarshoek pipe and disruption of the Merensky Reef. *Economic Geology*, 99, 517–541.

Scoon, R.N. and Mitchell, A.A. (2009). A multi-stage orthomagmatic and partial melting hypothesis for the Driekop platiniferous dunite pipe, eastern limb of the Bushveld Complex, South Africa. *South African Journal of Geology*, 112, 187–196.

Scoon, R.N. and Mitchell, A.A. (2010). The principle geologic features of the Onverwacht platiniferous dunite pipe, eastern limb of the Bushveld complex. *South African Journal of Geology*, 113, 155–168.

Scoon, R.N. and Mitchell, A.A. (2011). The principle geologic features of the Mooihoek platiniferous dunite pipe, eastern limb of the Bushveld complex, and similaritites with replaced Merensky Reef at the Amandelbult Mine, South Africa. *South African Journal of Geology*, 114, 15–40.

Scoon, R.N. and Teigler, B. (1994). Platinum-group element mineralization in the critical zone of the Western Bushveld Complex: 1. Sulfide poor chromitites below the UG2. *Economic Geology*, 89, 1094–1121.

Scoon, R.N. and Teigler, B. (1994). A new LG-6 chromite reserve at Eerste Geluk in the boundary zone between the central and southern sectors of the eastern Bushveld Complex. *Economic Geology*, 90, 969–982.

Scott, D.R., and Stevenson, D.J. (1984) Magma solitons: Geophysical Research Letters, 11, 1161–1164.

Seabrook, C.L., Cawthorn, R.G. and Kruger, F.J. (2005). The Merensky Reef, Bushveld Complex: Mixing of minerals not mixing of magmas. *Economic Geology*, 100, 1191–1206.

Segerstrom, K. and Carlson, R.R., 1982. Geologic map of the banded upper zones of the Stillwater Complex and adjacent rocks, Stillwater, Sweet Grass, and Park Counties, Montana. United States Geological Survey Map I-1383.

Self, S., Blake, S., Sharma, K., Widdowson, M. and Sephton, S. (2008). Sulfur and chlorine in Late Cretaceous Deccan magmas and eruptive gas release. *Science*, 391 (5870), 1654–1657,

Sessa, G., Moroni, M., Tumiati, S., Caruso, S. and Fiorentini, M.L. (2017). Ni-Fe-Cu-PGE ore deposition driven by metasomatic fluids and melt-rock reactions in the deep crust: The ultramafic pipe of Valmaggia, Ivrea-Verbano, Italy. *Ore Geology Reviews*, 90, 307–321.

Sharpe, M.R. (1985). Strontium isotope evidence for preserved density stratification in the Main Zone of the Bushveld Complex, South Africa. *Nature*, 316, 119–126.

Sharpe, M.R. and Irvine, T.N. (1983). Melting relations of two Bushveld chilled margin rocks and implications for the origin for the origin of chromitite. *Carnegie Institute of Washington Yearbook*, 82, 295–300.

Shinohara, H. (2008). Excessive degassing from volcanoes and its role on eruptive and intrusive activity. *Reviews of Geophysics*, 46(4), RG4005. doi:10.1029/2007RG000244.

Shinohara, H., Ilyama, J.T. and Matsuo, S. (1989). Partition of chlorine compounds between silicate melt and hydrothermal solutions: I. Partition of NaCl-KCl. *Geochimica et Cosmochimica Acta*, 53, 2617–2630.

Shirley, D.N. (1986). Compaction of igneous cumulates. *Journal of Geology*, 94: 795–809.

Shirley, D.N. (1987). Differentiation and compaction in the Palisades Sill, New Jersey. *Journal of Petrology*, 28, 835–865.

Sisson, T.W. and Grove, T.L. (1993). Experimental investigations of the role of water in calc-alkaline differentiation and subduction zone magmatism. *Contributions to Mineralogy and Petrology*, 113, 143–166.

Skelton, A.D.L. (1996). The timing and direction of metamorphic fluid flow in Vermont. *Contributions to Mineralogy and Petrology*, 125, 75–84.

Skelton, A.D.L., Graham, C.M. and Bickle, M.J. (1995). Lithological and Structural Controls on Regional 3-D Fluid Flow Patterns during Greenschist Facies Metamorphism of the Dalradian of the SW Scottish Highlands. *Journal of Petrology*, 36, 563–586.

Smith, D.S. and Basson, I.J. (2006). Shape and distribution analysis of Merensky Reef potholing, Northam Platinum Mine, western Bushveld Complex: Implications for pothole formation and growth. *Mineralium Deposita*, 41(3), 281–295.

Smythe, D.J., Wood, B.J. and Kiseeva, E.S. (2017). The S content of silicate melts at sulfide saturation: New experiments and a model incorporating the effects of sulfide composition. *American Mineralogist*, 102, 795–803.

Söhnge, P.G. (1963). Genetic problems of pipe deposits in South Africa. *Proceedings of the Geological Society of South Africa*, 66, 59–72.

Somarin, A.K., Kissin, S.A., Heerema, D.D. and Bihari, D.J. (2009). Hydrothermal alteration, fluid inclusion and stable isotope studies of the North Roby zone, Lac des Iles PGE mine, Ontario, Canada. *Resource Geology*, 59, 107–120.

Sonnenthal, E.L. (1992). Geochemistry of dendritic anorthosites and associated pegmatites in the Skaergaard Intrusion, East Greenland: Evidence for metasomatism by a chlorine-rich fluid. *Journal of Volcanology and Geothermal Research*, 52, 209–230.

Sorensen, H.S. and Wilson, J.R. (1995). A strontium and neodymium isotopic investigation of the Fongen–Hyllingen Layered Intrusion, Norway. *Journal of Petrology*, 36, 161–187.

Spandler, C., Mavrogenes, J. and Arculus, R. (2005). Origin of chromitites in layered intrusions: Evidence from chromite-hosted melt inclusions from the Stillwater Complex. *Geology*, 33(11), 893–896.

Stelling, J., Botcharnikov, R.E., Beerman, O. and Nowak, M. (2008). Solubility of H2O- and chlorine-bearing fluids in basaltic melt of Mount Etna at T = 1050–1250 °C and P = 200 MPa. *Chemical Geology*, 256(3–4), 102–110.

Stewart, B.W. and DePaolo, D.J. (1990). Isotopic studies of processes in mafic magma chambers: II. The Skaergaard Intrusion, East Greenland. *Contributions to Mineralogy and Petrology*, 104, 125–141.

Stormer, J.C. Jr., Pierson, M.L. and Tacker, R.C. (1993). Variation of F and Cl X-ray intensity due to anisotropic diffusion in apatite during electron microprobe analysis. *American Mineralogist*, 78, 641–648.

Stumpfl, E.F. (1961). Some new platinoid-rich minerals, identified with the electron microanalyser. *Mineralogical Magazine*, 32, 833–847.

Stumpfl, E.F. (1962). Some aspects of the genesis of platinum deposits. *Economic Geology*, 57, 619–625.

Stumpfl, E.F. and Rucklidge, J.C. (1982). The platiniferous dunite pipes of the eastern Bushveld Complex. *Economic Geology*, 77, 1419–1431.

Stumpfl, E.F. and Tarkian, M. (1976). Platinum genesis: New mineralogical evidence. *Economic Geology*, 71, 1451–1460.

Sourirajan, S. and Kennedy, G.C. (1962). The system H2O-NaCl at elevated temperatures and pressures. *American Journal of Science*, 260, 115–141.

Sun, S.-S., Nesbitt, R.W. and McCulloch, M.T. (1989). Geochemistry and petrogenesis of Archaean and early Proterozoic siliceous high magnesian basalts. In *Boninites and related rocks* (Crawford, A., ed.). London: Allen and Unwin, 148–173.

Tacker, R.C. and Stormer, J.C. (1989). A thermodynamic model for apatite solid solutions applicable to high temperature geologic problems. *American Mineralogist*, 74, 877–888.

Tacker, R.C. and Stormer, J.C. (1993). Thermodynamics of mixing of liquids in the system Ca3(PO4)2 – CaF2 – CaCl2 – Ca(OH)2. *Geochimica et Cosmochimica Acta*, 57, 4663–4676.

Tagirov, B.R., Baranova, N.N., Zotov, A.V., Akinfiev, N.N., Polonyanko, N.A., Shikina, N.D., Koroleva, L. A., Shvarov, Y.V. and Basrakov, E.V. (2013). The speciation and transport of palladium in hydrothermal fluids: Experimental modeling and thermodynamic constraints. *Geochimica et Cosmochimica Acta*, 117, 348–373.

Talkington, R.W. and Lipin, B.R. (1986). Platinum-group minerals in chromite seams of the Stillwater Complex, Montana. *Economic Geology*, 81, 1179–1186.

Talkington, R.W. and Watkinson, D.H. (1984). Trends in the distribution of the precious metals in the Lac-Des-Iles Complex, Northwestern Ontario. *Canadian Mineralogist*, 22, 125–136

Talkington, R.W., Watkinson, D.H., Whittaker, P.J. and Jones, P.C. (1984). Platinum-Group Minerals and Other Solid Inclusions in Chromite of Ophiolitic Complexes: Occurrence and Petrological Significance. *Tschermaks Mineralogische und Petrographische Mitteilungen*, 32, 285–301.

Tang, Y.-J., Zhang, H.-F., Nakamura, E. and Ying, J.-F. (2011). Multistage melt/fluid-peridotite interactions in the refertilized lithospheric mantle beneath the North China Craton: Constraints from the Li–Sr–Nd isotopic disequilibrium between minerals of peridotite xenoliths. *Contributions to Mineralogy and Petrology*, 151, 845–861.

Taylor, H.P. Jr. and Forester, R. W. (1979). An oxygen and hydrogen isotope study of the Skaergaard Intrusion and its country rocks: A description of a 55-MY-old fossil hydrothermal system. *Journal of Petrology*, 20, 355–419.

Teague, A.J., Kohn, S.C., Klimm, K. and Botcharnikov, R.E. (2008). Sulphur solubility in Mount Hood andesites and CO2 fluids: Implications for volcanic degassing. *EOS Transactions of the American Geophysical Union*, 89(53), Fall Meeting Suppl. Abstract V21B-2086.

Tegner, C., Thy, P., Holness, M.B., Jakobsen, J.K. and Lesher, C.E. (2009). Differentiation and compaction in the Skaergaard Intrusion. *Journal of Petrology*, 50(5), 813–840.

Teigler, B. and Eales, H.V. (1993). Correlation between chromite composition and PGE mineralization in the Critical Zone of the western Bushveld Complex. *Mineralium Deposita*, 28, 291–302.

Teigler, B. and Eales, H.V. (1996). The Lower and Critical Zones of the western limb of the Bushveld Complex as intersected by the Nooitgedacht boreholes. *Geological Survey of South Africa Bulletin*, 111, 126.

Tepley, F.J., III and Davidson, J.P. (2003). Mineral-scale Sr-isotope constraints on magma evolution and chamber dynamics in the Rum layered intrusion, Scotland. *Contributions to Mineralogy and Petrology*, 145, 621–648.

Tepley, F.J. and Davidson, J.P. (2003). Mineral-scale Sr-isotope constraints on magma evolution and chamber dynamics in the Rum layered intrusion, Scotland. *Contributions to Mineralogy and Petrology*, 145(5), 628–641.

Toramaru, A. (1995). Numerical study of nucleation and growth of bubbles in viscous magmas. *Journal of Geophysical Research*, 100, 1913–1931.

Tredoux, M., Lindsay, N.M., Davies, G. and Mcdonald, I. (1995). The fractionation of platinum-group elements in magmatic systems, with the suggestion of a novel causal mechanism. *South African Journal of Geology*, 98, 157–167.

Tremaine P.R. and Leblanc J.C. (1980). The solubility of nickel oxide and hydrolysis of Ni2+ in water to 573 K. *Journal of Chemical Thermodynamics*, 12, 521–538.

Tsuji, T., Jiang, F. and Christenson, K.Y (2016). Characterization of immiscible fluid displacement processes with various capillary numbers and viscosity ratios in 3D natural sandstone. *Advances in Water Resources*, 95, 3–15.

Tsujimori, T. and Ernst, W.G. (2014). Lawsonite blueschists and laalwsonite eclogites as proxies for palaeo-subduction zone processes: A review. *Journal of Metamorphic Geology*, 32, 437–454.

van Haren, J.L.M., Ague, J.J. and Rye, D.M. (1996). Oxygen isotope record of fluid infiltration and mass transfer during regional metamorphism of pelitic schist, Connecticut, USA. *Geochimica et Cosmochimica Acta*, 60(18), 3487-3504.

Van Orman, J.A., Keshav, S. and Fei, Y. (2008). High-pressure solid/liquid partitioning of Os, Re and Pt in the Fe-S system. *Earth and Planetary Science Letters*, 274, 250–257.

Vidal, O. and Durin, L. (1999). Aluminium mass transfer and diffusion in water at 400–550°C, 2 kbar in the K2O–Al2O3–SiO2–H2O system driven by a thermal gradient or by a variation of temperature with time. *Mineralogical Magazine*, 63(5), 633–647.

Vigneresse, J.-L. (2015). Textures and melt-crystal-gas interactions in granites. *Geoscience Frontiers*, 6(5), 635–663.

Viljoen, M.J. (1999). The nature and origin of the Merensky Reef of the western Bushveld Complex based on geological facies and geophysical data. *South African Journal of Geology*, 102, 221–239

Viljoen, M.J. and Scoon, R.N. (1985). The distribution and main geologic features of discordant bodies of iron-rich ultramafic pegmatite in the Bushveld Complex. *Economic Geology*, 80, 1109–1128.

Volborth, A. and Housley, R.M. (1984). A preliminary description of complex graphite, sulfide, arsenide and platinum group element mineralization in a pegmatoid pyroxenite of the Stillwater Complex, Montana, U.S.A. *Tschermaks Mineralogische und Petrographische Mitteilungen*, 33, 213–230.

Volfinger, M., Roberts, J.-L., Vielzeuf, D. and Neiva, A.M.R. (1985). Structural controls of the chlorine content of OH-bearing silicates (micas and amphiboles). *Geochimica et Cosmochimica Acta*, 49, 37–48.

von Gruenewaldt, G., Hatton, C.J., Merkle, R.K.W. and Gain, S.B. (1986). Platinum-group element-chromite associations in the Bushveld Complex. *Economic Geology*, 81, 1067–1079

von Gruenewaldt, G., Hulbert, L.J. and Naldrett, A.J. (1989). Contrasting platinum-group element concentration patterns in cumulates of the Bushveld Complex. *Mineralium Deposita*, 24, 219–229.

Voordouw, R., Gutzmer, J. and Beukes, N.J. (2009). Intrusive origin for Upper Group (UG1, UG2) stratiform chromitite seams in the Dwars River area, Bushveld Complex, South Africa. *Mineralogy and Petrology*, 97, 75–94.

Wager, L.R. (1960). The major element variation of the Layered Series of the Skaergaard Intrusion and a re-estimation of the average composition of the Hidden Layered Series and of the successive residual magmas. *Journal of Petrology*, 1, 364–398.

Wager, L.R. and Brown, G.M. (1968). *Layered Igneous Rocks*. Edinburgh: Oliver &Boyd

Wager, L.R., Brown, G.M. and Wadsworth, W.J. (1960). Types of igneous cumulates. *Journal of Petrology*, 1, 73–85.

Wager, L.R. and Deer, W.A. (1939). Geological investigations in east Greenland, Part III. Petrology of the Skaergaard intrusion, Kangerdlugssuaq, east Greenland. *Meddelelser om Gronland*, 105(4), 1–352.

Wager, L.R., Vincent, E.A. and Smales, A.A. (1957). Sulphides in the Skaergaard Intrusion, East Greenland. *Economic Geology*, 52, 855–903.

Wagner, P.A. (1929). *The Platinum Deposits and Mines of South Africa: Edinburgh*, Oliver and Boyd, 326 p.

Wall, C.J. and Scoates, J.S. (2016). High-precision U-Pb Zircon-Baddeleyite dating of the J-M Reef platinum group element deposit in the Stillwater Complex, Montana (USA). *Economic Geology*, 111(3), 771–782.

Wall, C.J. and Scoates, J.S., Weis, D., Friedman, R.M., Amini, M. and Meurer, W.P. (2018). The Stillwater Complex: Integrating zircon geochronological and geochemical constraints on the age, emplacement history and crystallization of a large, open-system layered intrusion. *Journal of Petrology*, 59(1), 153–190.

Wallace, P.J. and Anderson, A.T. Jr. (1998). Effects of eruption and lava drainback on the H2O contents of basaltic magmas at Kilauea Volcano. *Bulletin of Volcanology*, 59, 327–344.

Wallace, P.J., Plank, T., Edmonds, M. and Hauri, E.H. (2015). Volatiles in Magmas. In *Encyclopaedia of Volcanoes* – 2nd ed. (Sigurdsson, H., Houghton, B., McNutt, S., Rymer, H. and Stix, J. eds). Amsterdam: Academic Press, 163–184.

Walraven, F. (1988). Notes on the age and genetic relationships of the Makhutso granite, Bushveld Complex, South Africa. *Chemical Geology*, 72, 17–28.

Watenphul, A., Schmidt, C. and Jahn, S. (2014). Cr(III) solubility in aqueous fluids at high pressures and temperatures. *Geochimica et Cosmochimica Acta*, 126, 212–227.

Waters, C. and Boudreau, A.E. (1996). A re-evaluation of crystal-size distributions in chromite cumulates. *American Mineralogist*, 81, 1452–1459.

Webster, J.D. and Botcharnikov, R.E. (2011). Distribution of Sulfur between melt and fluid in S-O-H-C-Cl-bearing magmatic systems at shallow crustal pressures and temperatures. *Reviews in Mineralogy and Geochemistry*, 73, 247–283.

Webster, J.D., Kinzler, R.J. and Mathez, E.A. (1999). Chloride and water solubility in basalt and andesite liquids and implications for magmatic degassing. *Geochimica et Cosmochimica Acta*, 63, 729–738.

Webster, J.D. (2004). The exsolution of magmatic hydrosaline melts. *Chemical Geology*, 21, 33–48.

Webster, J.D. and DeVivo, B. (2002). Experimental and modeled solubilities of chlorine in aluminosilicate melts, consequences of magma evolution, and implications for exsolution of hydrous chloride melt at Mt. Somma-Vesuvius, Italy. *American Mineralogist*, 87, 1046–1061.

Webster, J.D. and Piccoli, P.M. (2015). Magmatic apatite: A powerful, yet deceptive mineral. *Elements*, 11, 177–182

Webster, J.D., Kinzler, R.J. and Mathez, E.A. (1999). Chloride and water solubility in basalt and andesite melts and implications for magmatic degassing. *Geochimica et Cosmochimica Acta*, 63, 729–738.

Webster, J.D., Thomas, R., Förster H.-J., Seltmann, R. and Tappen, C. (2004). Geochemical evolution of halogen-enriched, granite magmas and mineralizing fluids of the Zinnwald Tin-Tungsten Mining District, Erzgebirge, Germany. *Mineralium Deposita*, 39, 452–472.

Webster, J.D., Tappen, D. and Mandeville, C.W. (2009). Partitioning behavior of chlorine and fluorine in the system apatite-melt-fluid: II. Felsic silicate systems at 200 MPa. *Geochimica et Cosmochimica Acta*, 73, 559–581.

Webster, J.D., Vetere, V., Botcharnikov, R.E., Goldoff, B. McBirney, A. and Doherty, A.L. (2015). Experimental and modeled chlorine solubilities in aluminosilicate melts at 1 to 7000 bars and 700 to 1250 °C: Applications to magmas of Augustine Volcano, Alaska. *American Mineralogist*, 100, 522–535.

Weiss, D. and Morse, S.A. (1995). Mafic–felsic isotopic disparity in the Kiglapait Intrusion: textures vs. diffusion? *Terra Abstracts*, 7, 18.

Weiss, D. and Morse, S.A. (1993). Disparate U-Pb systematics of mafics and plagioclase in the Kiglapait Intrusion, Labrador. *EOS Transactions, American Geophysical Union*, 74, 623.

Wernette, B. and Boudreau, A.E. (2018). The Skaergaard Intrusion: Using Petrography to Characterize an Exsolved Magmatic Volatile Phase. Goldschmidt Conference, Boston, abstract.

Wheeler, S.J. (1988). A conceptual model for soils containing large gas bubbles. *Géotechnique*, 38, 389–397.

Wheeler, S.J. (1990). Movement of large gas bubble in unsaturated fine-grained sediment. *Marine Geotechnology*, 9, 113–129.

Wheeler, S.J. and Gardner, T.N. (1989). The elastic moduli of soils containing large gas bubbles. *Géotechnique*, 39, 333–342.

White, J.A. (1994). The Potgietersrus prospect–geology and exploration history. 15th Council of Mining and Metallurgical Institutions Congress, *South African Institute of Mining and Metallurgy*, 3, 173–181.

Willmore, C.C., Boudreau, A.E. and Kruger, F.J. (2000). The halogen geochemistry of the Bushveld Complex, Republic of South Africa: Implications for chalcophile element distribution in the Lower and Critical zones. *Journal of Petrology*, 41, 1517–1539

Willmore, C.C., Boudreau, A.E., Spivack, A. and Kruger, F.J. (2002). Halogens of the Bushveld Complex, South Africa: δ37Cl and Cl/F evidence for hydration melting of the source region in a back-arc setting. *Chemical Geology*, 182, 503–511.

Wilson A.H. (1982). The geology of the Great Dyke, Zimbabwe: The ultramafic rocks. *Journal of Petrology*, 23, 240–292.

Wilson, A.H. (1992). The geology of the Great Dyke, Zimbabwe: Crystallization, layering, and cumulate formation in the PI pyroxenite of Cyclic Unit 1 of the Darwendale subchamber. *Journal of Petrology*, 33, 611–663.

Wilson, A.H. (1996). The Great Dyke of Zimbabwe. In *Layered Intrusions* (Cawthorn, R.G., ed.). Amserdam: Elsevier, 365–402.

Wilson, A.H. (2001). Compositional and lithological controls on the PGE-bearing sulphide zones in the Selukwe subchamber, Great Dyke: A combined equilibrium-Rayleigh fractionation model. *Journal of Petrology*, 42, 1845–1867.

Wilson, A.H. (2012). A Chill Sequence to the Bushveld Complex: Insight into the first stage of emplacement and Implications for the parental magmas. *Journal of Petrology*, 53(6), 1123–1168.

Wilson, A.H. and Prendergast, M.D. (1989). The Great Dyke of Zimbabwe-I: Tectonic setting, stratigraphy, petrology, structure, emplacement and crystallisation. In *Magmatic sulphides – The Zimbabwe volume* (Prendergast, M.D. and Jones, M.J., eds.). London, Institute of Mining and Metallurgy, 1–20.

Wilson, A.H. and Prendergast, M.D. (2001). Platinum-group element mineralization in the Great Dyke, Zimbabwe, and its relationship to magma evolution and magma chamber structure. *South African Journal of Geology*, 104, 319–342.

Wilson, A.H., Naldrett, A.J. and Tredoux, M. (1989). Distribution and controls of platinum-group element and base-metal mineralization in the Darwendale subchamber of the Great Dyke, Zimbabwe. *Geology*, 17, 649–652.

Wilson, A.H., Murahwi, C.Z. and Coghill, B. (2000). Stratigraphy, geochemistry and platinum group element mineralisation of the central zone of the Selukwe Subchamber, Great Dyke, Zimbabwe. *Journal of African Earth Sciences*, 30, 833–853.

Wilson, C.R., Spiegelman, M., van Keken, P.E. and Hacker, B.R. (2014). Fluid flow in subduction zones: The role of solid rheology and compaction pressure. *Earth and Planetary Science Letters*, 401, 261–274.

Wilson, J.R., Robins, B., Nielsen, F.M., Duchesne, J.C. and Auwers, J.V. (1996). The Bjerkreim–Sokndal Layered Intrusion, southwest Norway. In *Layered Intrusions* (Cawthorn, R.G., ed.). Amsterdam: Elsevier, 231–255.

Woods, W. and Norris, S. (2010). On the role of caprock and fracture zones in dispersing gas plumes in the subsurface. *Water Resources Research*, 46: W08522. doi:10.1029/2008WR007568.

Worst, B.G. (1960). The Great Dyke of Southern Rhodesia. *Southern Rhodesia Geological Survey Bulletin*, 47, 234 pp.

Yang, S.-H., Maier, W.D., Lahaye, Y. and O'Brien, H. (2013). Strontium isotope disequilibrium of plagioclase in the Upper Critical Zone of the Bushveld Complex: Evidence for mixing of crystal slurries. *Contributions to Mineralogy and Petrology*, 166(4), 959–974.

Yardley, B.W.D. (2005). Metal Concentrations in Crustal Fluids and Their Relationship to Ore Formation. *Economic Geology*, 100, 613–632.

Zhang, C., Oostrom, M. Wietsma, T.W., Grate, J.W. and Warner, M.G. (2011). Influence of viscous and capillary forces on immiscible fluid displacement: Pore-scale experimental study in a water-wet micromodel demonstrating viscous and capillary fingering. *Energy and Fuels*, 25, 3493–3505.

Zhu, C. and Sverjensky, D.A. (1992). F-Cl-OH partitioning between biotite and apatite. *Geochimica et Cosmochimica Acta*, 56(9), 3435–3468.

Zientek, M.L., Cooper, R.W., Corson, S.R. and Geraghty, E.P. (2002). Platinum-group element mineralization in the Stillwater Complex, Montana. In *The Geology, Geochemistry, Mineralogy and Mineral Beneficiation of Platinum-Group Elements* (Cabri, L.J., ed.). Canadian Institute of Mining and Metallurgy Special Volume 54, 459–481.

Index

Note: Page numbers in italics indicate figures.